D1258413

Optical Fibre Communications

The Wiley Series in Solid State Devices and Circuits

Edited by

M. J. Howes and D. V. Morgan

Department of Electrical and Electronic Engineering, University of Leeds

Microwave Devices

Variable Impedance Devices

Charge-coupled Devices and Systems

Optical Fibre Communications

Optical Fibre Communications

Devices, Circuits, and Systems

Edited by

M. J. Howes
D. V. Morgan
Department of Electrical and Electronic Engineering, University of Leeds

A Wiley–Interscience Publication

JOHN WILEY & SONS
Chichester · New York · Brisbane · Toronto

British Library Cataloguing in Publication Data:

Optical fibre communications. – (The Wiley series in solid state devices and circuits).
1. Data transmission systems
2. Fibre optics
I. Howes, Michael John II. Morgan, David Vernon
621.38′028 TK5105 79-40512

ISBN 0 471 27611 1

Typeset by Preface Ltd., Salisbury, Wilts., and printed by Page Bros. (Norwich) Ltd., Norwich

Contributors

R. DAVIS — *Allen Clark Research Centre, The Plessey Company, Towcester, Northants*

I. GARRETT — *Post Office Research Centre, Martlesham Heath, Ipswich IP5 7RE*

R. C. GOODFELLOW — *Allen Clark Research Centre, The Plessey Company, Towcester, Northants*

C. K. KAO — *ITT Electro Optical Products Group, 7635 Plantation Road, Roanoke, Virginia 24019, USA*

J. E. MIDWINTER — *Head of Optical Communications Systems Division, Post Office Research Centre, Martlesham Heath, Ipswich IP5 7RE*

D. B. OSTROWSKY — *Université de Nice, UERMSP, Laboratoire D'Electro-Optique, Parc Valrose, 06034 Nice-Cedex, France*

T. PEARSALL — *Thompson CSF, Laboratoire Central de Recherches, Orsay, France*

P. RUSSER — *Optical Communications Group, AEG Telefunken, Research Institute, ULM, Germany*

Series Preface

The Oxford Dictionary defines the word revolution as 'a fundamental reconstruction'; these words fittingly describe the state of affairs in the electronic industry following the advent of solid state devices. This 'revolution', which has taken place during the past 25 years, was initiated by the discovery of the bipolar junction transistor in 1948. Since this first discovery there has been a worldwide effort in the search for new solid state devices and, although there have been many notable successes in this search, none have had the commercial impact which the transistor has had. Possibly no other device will have such an impact; but the commercial side of the electronics industry stands poised, awaiting the discovery of new devices as significant perhaps as the transistor.

Research and development in the field of solid state devices has concerned itself with two important problems. On the one hand we have device physics, where the aim is to understand in terms of basic *physical concepts* the mode of operation of the various devices. In this way one seeks to optimize the technology in order to achieve the best performance from each device. The second aspect of this work is to consider the important contribution of the circuit to the operation of a device. This problem has been called *device circuit interaction*. It is a great pity that in the past these two major aspects of the one problem have been tackled by separate groups of scientists with little exchange of ideas. In recent years, however, this situation has been somewhat remedied, the improvement being due directly to the very rigorous system specifications demanded by industry. Such demands constantly require greater performance from devices, which can only be brought about by coordinated team work.

The objective of this new series of books is to bring together the two aspects of this problem: device physics and device circuit interactions. We hope to achieve, by coordinated co-authorship of leading experts in the respective fields, a varied and balanced review of past and current work. The books in the series will cover many aspects of device research and will deal with both the commercially successful and the more speculative devices. Each volume will be an in-depth account of one or more devices centred on some common theme. The level of the text is designed to be suitable for the graduate student or research worker wishing to enter the field of research concerned. Basic physical concepts in semiconductors and elementary ideas in passive and active circuit theory will be assumed as a starting point.

M. J. HOWES
D. V. MORGAN

University of Leeds
December 1978

Preface

The application of glass fibres as a medium for transmission lines in optical communication systems has been under intensive study during the past ten years. The practical realization of high-quality components such as optical source devices, planar waveguide components, photodiodes and low-loss optical fibres has enabled this area of engineering to expand rapidly. It is now at the point where practical optical-fibre systems are being constructed and evaluated. The field is therefore past the 'basic research' stage and is commanding attention from the engineering profession. In this, the fourth volume in the series on 'Solid State Devices and Circuits' we cover all aspects of Optical Fibre Communications. Chapter 1 introduces the topic, whilst Chapters 2 to 5 deal with the individual elements of a complete system. Chapter 2 is an appraisal of source devices and includes LEDs and lasers. Chapter 3 covers photodiodes. In Chapter 4 (Optical Wavelength Components) and Chapter 5 (Optical Fibres and Cables) the problems of transmission and the manipulation of optical signals are conditioned in detail. The final chapter deals with Optical Communications Systems and the authors concerned direct their attention to the current state of the art, in full communication systems.

The editors wish to thank most warmly the authors contributing to this volume. They are particularly indebted to Professor A. E. Ash FRS (University of London) and Professor J. O. Scanlan (University of Eire, Dublin) for their valuable advice and encouragement in preparing this volume. It is again a pleasure to thank our wives Jean (Morgan) and Dianne (Howes) for their valuable assistance in checking the text.

M. J. Howes
D. V. Morgan

University of Leeds
December 1978

Contents

Optical Fibre Communications
Edited by M. J. Howes and D. V. Morgan

CHAPTER 1

Introduction to optical communications

P. RUSSER

1.1 HISTORY AND METHODS OF OPTICAL COMMUNICATIONS

The transmission of information by means of light has a much longer history than electrical communications. At the end of the sixth century B.C. Aeschylus mentioned passing the news on of Troys downfall by fire signals via a long chain of relay stations from Asia Minor to Argos. In the second century B.C. Polybius described an arrangement by which the whole Greek alphabet could be transmitted by fire signals using a two-digit, five-level code.[1] To our knowledge this was the first optical communications link which allowed the transmission of messages not previously agreed upon. At the end of the eighteenth century A.D. the optical telegraph by Claude Chappe allowed the transmission of a signal over the 423 km distance from Paris to Strasbourg within a time of six minutes.[2] Chappe's telegraph used movable signal elements which were observed from the subsequent relay station by telescopes. In the middle of the nineteenth century, optical telegraphy was replaced by electrical telegraphy, which at this time allowed a faster signal transmission and required fewer skilled personnel.

However, although optical communication exhibited low practical importance in the next decades, its development proceeded. In 1880 Graham Bell reported the transmission of speech over a beam of light.[3] To modulate a beam of sunlight, Graham Bell used a diaphragm-mirror, against the back of which the speakers voice was directed. The beam was received upon a parabolic reflector, in the focus of which was placed a sensitive selenium photoresistor, connected with a battery and a telephone. Besides a number of other means for modulating a light beam, Bell had already proposed modulators based upon the Faraday effect and Kerr effect. In the first half of the twentieth century optical communications has been used only on a small-scale in mobile low-bandwidth and short-distance communication links.[4,5]

The situation changed rapidly with the invention of the laser. In 1958 Schawlow and Townes proposed the extension of the maser principle into the optical region;[6] in 1960 Maiman reported the generation of coherent light pulses by a ruby laser;[7]

and one year later Javan and coworkers realized a continuously operating helium–neon laser with a linewidth of only a few tens of kilohertz.[8] The availability of coherent light sources greatly stimulated the research into optical communications for the following reasons: the carrier frequencies between 10^{13} und 10^{15} Hz yield a high available bandwidth for signal transmission. To make use of this high available bandwidth it is necessary that the carrier bandwidth should not exceed the signal bandwidth by orders of magnitude. For example, if optical frequency multiplexing is used, the carrier wavelength spacing is determined by the carrier bandwidth if the carrier bandwidth is considerably larger than the signal bandwidth. Also in the case of time-division multiplexing over long transmission lines, a higher carrier bandwidth will result in a higher pulse spreading due to line dispersion. Since the relative bandwidth of incoherent light sources is usually of the order of a few per cent, the carrier bandwidth is higher by orders of magnitude than the signal bandwidth. The attempt to realize incoherent light sources with narrow spectral width is not the right way of solving these difficulties. On the one hand, the transmitted power would be greatly reduced by narrow-bandwidth optical filtering, and on the other hand, due to the fundamental differences in the statistical properties of coherent and incoherent light, a low optical carrier bandwidth also requires coherence in order to obtain a high signal-to-noise ratio.[9,10] We shall come back to this topic in Section 1.2.

A further advantage of laser light is its spatial coherence, which yields a lower beam divergence. This is advantageous not only for free-space communications but also for the coupling efficiency into optical beam waveguides. The reason is the following: an incoherent radiation source emits light in many radiation modes, the number of which is in the order of the ratio of the light-emitting area to the square of the light wavelength. As a result of this, the radiation source emits light in the hemisphere according to Lambert's cosine law. The reduction of the beam divergence is always connected with an inversely proportional increase of the beam diameter. The coherent laser light source, on the other hand exhibits not only a low beam diameter but also a low beam divergence, especially if the laser oscillates in its fundamental transverse mode only. A further decrease of the beam divergence by optical means is also possible here. If a monomode optical beam waveguide is used in order to achieve maximum bandwidth, besides narrow spectral source width, transverse single-mode emission is also required in order to obtain a good coupling efficiency into the optical beam waveguide.

As is the case at radio frequencies, optical communications using wave propagation in free space as well as in light waveguides is possible. For atmospheric free-space optical transmission, several windows of high transparency exist in the visible-wavelength range and in some wavelength regions between 1 and 12 μm.[11–14] Unfortunately, the transmission loss is increased considerably by fog or precipitation. In addition, clear-weather changes in refractive index, caused by temperature gradients and turbulences in the air, degrade the transmission properties of atmospheric communication channels.[13,14] For all these regions, atmospheric optical transmission systems need a high transmitter power and a close repeater spacing. An interesting field of application of free-space optical communications are satellite links in outer space.[15] In satellite communications, the low beam divergence achievable with small antenna areas is advantageous.

Optical waveguides have been developed in order to avoid disturbances by the atmosphere. Whereas light waves in free space propagate in a straight line, waveguides in all practical cases are forced to follow a curved path as determined by the terrain contours and diverse physical obstacles. Miller has shown theoretically that for any electromagnetic waveguide having transverse planes in which the field is essentially equiphase, the minimum bending radius R_{min} and the maximum abrupt change δ_{max} for a wavelength λ, a beam diameter $2a$, and for $\lambda < a$ are determined by[16]

$$R_{min} = 2(a^3/\lambda^2) \tag{1.1}$$

$$\delta_{max} = \tfrac{1}{2}(\lambda/a) \tag{1.2}$$

For a He–Ne-laser light beam with 0.6328 μm wavelength and a 1 mm beam diameter, $\delta_{max} = 0.036°$ and $R_{min} = 600$ m. It is clear that such a high minimum bending radius and low maximum abrupt angular change would impose severe restrictions on the path taken by an optical link and would also require high precision components and installation.

Lightwave guidance can be performed discontinuously or continuously. Discontinuous waveguides consist of a sequence of uniformly spaced irises or lenses (Figure 1.1).[17,18] An experimental light waveguide comprising 10 lenses within 1 km length and enclosing the light beam within a 10.2 mm aluminium pipe has been investigated by Goubeau and Christian.[18] In spite of the doubly-shielded light path, temperature-dependent inhomogeneities of the air caused beam deflections, making a remote lens adjustment necessary. A modified kind of lens waveguide is the tubular gas lens waveguide.[19] The gas lens is formed by blowing a cool gas through a warmer tube. The gas near the axis has a lower temperature and consequently a higher density and a higher refractive index than that near the walls. Therefore the tube acts as focusing lens. The use of gas lenses instead of glass lenses avoids reflection losses.

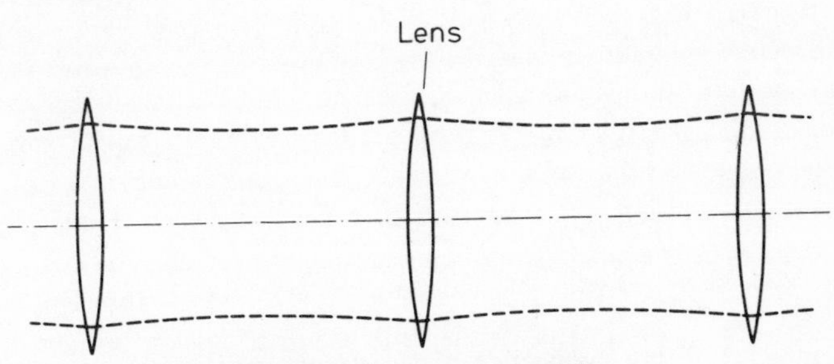

Figure 1.1 Lens waveguide

Early proposals for continuous light waveguides were hollow metallic and dielectric waveguides.[20,21] Since the attenuation of each mode in such waveguides is inversely proportional to the cube of the inner diameter, the inner diameter must not be too small. For a dielectric waveguide made of glass with a refractive index of 1.5, an inner diameter of 2 mm and a wavelength of 1 μm, Marcatili and Schmeltzer have calculated an attenuation of 1.85 dB/km for the EH_{11} mode, which doubled even for a bending radius of 10 km.[21] Although the hollow metallic waveguide turned out to be less sensitive to bending, the minimum radius is also in the order of tens of metres.

All the above concepts for optical waveguides suffered from the disadvantages of high precision requirements and consequently high manufacturing costs as well as high installation costs due to their large minimum curvatures. Their application only seemed to be appropriate for communication links with extremely high bandwidths. The breakthrough of guided optical communications came about with the development of the optical-fibre waveguides. The ability of a dielectric rod with a higher dielectric constant than its surrounding to guide electromagnetic waves had already been known for a long time. In 1910 Hondros and Debye solved Maxwell's equations for wave propagation in a circular dielectric rod.[22] First experimental investigations on radio-frequency electromagnetic wave propagation in a dielectric waveguide were published by Schriever in 1920.[23] Early applications of glass fibres in the optical region were to transport light or optical images across short distances.[24] Van Heel first suggested coating fibres with a layer of lower refractive index to ensure total reflection at the core-cladding interface and in this way to isolate optically individual fibres within a bundle from their neighbours.[25] The first suggestions to use fibres as a transmission medium for optical communications were made in 1966 by Kao and Hockham,[26] Werts,[27] and Boerner.[28] These authors proposed the use of glass fibres with a core of higher refractive index and a cladding

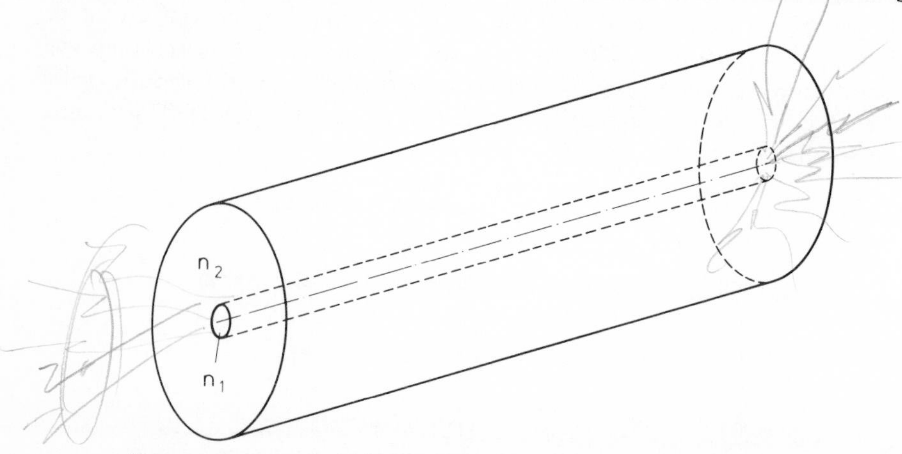

Figure 1.2. Fibre waveguide with a core of higher refractive index n_1 and a cladding of lower refractive index n_2

with lower refractive index (Figure 1.2). Such a fibre acts as an open optical waveguide. The electromagnetic field is guided only partially within the core region, whereas outside the core the electromagnetic field is evanescent in a direction normal to propagation. Among the electromagnetic field modes there is one, namely the HE_{11} which has no cutoff wavelength.[29] Only this HE_{11} mode can propagate in a fibre for wavelengths greater than the highest cutoff wavelength of the other modes. For a given core refractive index n_1, a cladding refractive index n_2 and the core diameter d, the vacuum cutoff wavelength λ_c for monomode operation is given by[29]

$$\lambda_c = \frac{\pi d}{2.405}(n_1^2 - n_2^2)^{1/2} \tag{1.3}$$

For $d = 5\ \mu m$, $n_1 = 1.5$ and n_2 smaller than n by 0.3%, for example, we obtain $\lambda_c = 0.758\ \mu m$. The bandwidth of an optical-fibre transmission link is limited by the fibre dispersion. The dispersion is usually measured by the broadening of a light pulse propagating across the fibre. In the monomode regime the dispersion is caused mainly by the wavelength dependence of the refractive index of the fibre material. The pulse dispersion depends on the wavelength and the spectral width of the optical source. For an optical source with a centre wavelength of 0.85 μm, single-mode fibres exhibit a pulse dispersion of 80 ps per kilometre fibre length and per nanometre spectral width of the optical source.[30] In the 1.27 μm wavelength region, monomode fibres exhibit a dispersion minimum[31] which is shifted into the wavelength region between 1.3 and 1.4 μm due to the geometric dispersion of the

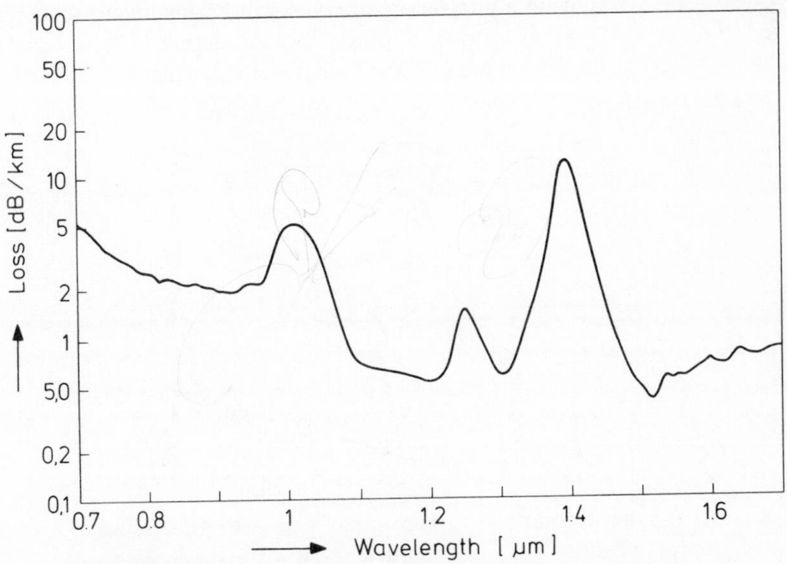

Figure 1.3. Loss spectrum of a single-mode silica fibre[33].

fibre guide.[32] With a germanium-doped silica glass fibre at the wavelength of 1.29 μm, a pulse broadening of only 4 ps nm^{-1} km^{-1} has already been measured.[33] There has been a great effort in the last ten years to reduce the fibre losses. When fibre attenuation values of only 20 dB/km were achieved with silica fibres by Kapron, Keck and Maurer in 1970, the barrier against the application of optical fibres in communication links was broken through.[34] At the beginning of 1978 the best measured attenuation values for silica-based monomode fibres were under 2 dB/km at 0.85 μm wavelength and under 0.5 dB/km at 1.3 μm wavelength.[33,35] Figure 1.3 shows the loss spectrum of a single-mode silica fibre. Compared with other light waveguides, optical fibres are inexpensive to manufacture. Since curvatures in the order of centimetres are allowed, the cabling of the fibres and the installation of fibre cables is as simple as that for electrical cables.

So far we have only referred to the monomode fibre waveguide. The monomode fibre exhibits minimum dispersion and therefore the highest transmission bandwidth. However, since optical fibres are a very inexpensive transmission medium, they also became interesting for lower-bandwidth applications. For these purposes fibre types which allow the propagation of many modes are of special interest. Figure 1.4 shows the three types of fibres mainly used in optical communications. Apart from the monomode fibre, these are the step-index multimode fibre and the graded-index multimode fibre. Whereas the monomode fibre exhibits core diameters in the order of 2 to 8 μm the step index multimode fibre has core diameters in the order of 50 μm. A large number of modes can propagate in such a fibre. Usually, many modes are excited by a light source at the fibre end, and due to fibre bendings mode conversion occurs along the fibre. The different group velocities of the modes yield a considerable broadening of transmitted light pulses, so that the

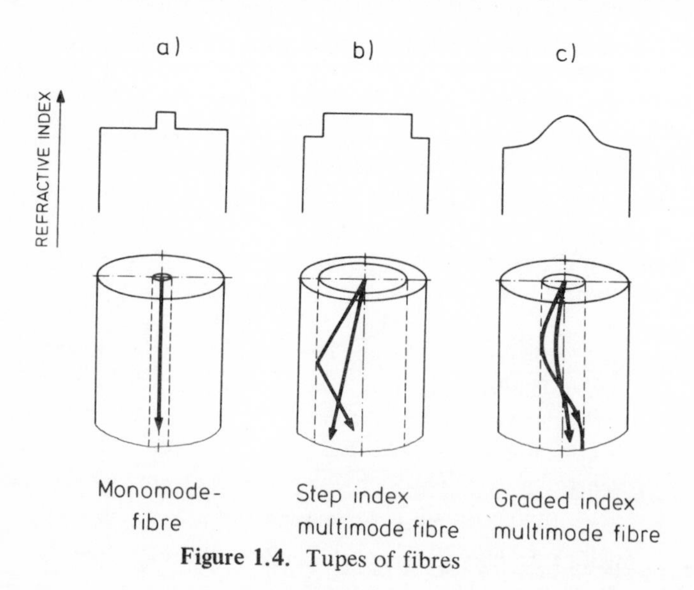

Figure 1.4. Tupes of fibres

bandwidth length product is in the order of 10 to 100 MHz km. For low-bandwidth optical links the multimode fibre has two advantages compared with the monomode fibre due to the higher core diameter. Firstly, the multimode fibre allows the use of incoherent optical sources, which could only be coupled with extremely low efficiency into monomode fibres. Secondly, the multimode fibre imposes lower tolerance requirements on fibre connectors. With a step-index multimode fibre consisting of a phosphosilicate core and a borosilicate cladding, an attenuation of 0.47 dB/km was achieved at 1.2 μm wavelength.[36]

The graded-index fibre exhibits no step-index change but a parabolic refractive index profile with its maximum in the fibre axis.[37,38] The graded index fibre also supports a great number of modes. Compared with the step-index multimode fibre it has the advantage of a low intermode dispersion. This is achieved by an appropriate choice of the refractive index profile which minimizes the group velocity difference of the different modes. This behaviour of graded-index multimode fibres can be easily understood using concepts of geometric optics. Figures 1.4b and 1.4c show the paths of light rays in the fibre cores. In the step-index fibre the light rays are straight within the core and are totally reflected at the core–cladding interface. Since the light velocity is constant within the core, rays intersecting a higher angle with the fibre axis consequently have a smaller velocity component in the fibre direction. In graded-index fibres the light rays follow curved paths. Rays running close to the fibre axis have a shorter path, but they pass through a region with a higher refractive index and therefore lower group velocity, compensating for the shorter path-lengths. Graded-index fibres use a parabolic index profile. The power law index for minimum dispersion depends on the fibre material. With a boron oxide–doped silica-graded index fibre with a power law index of 1.77 Cohen and coworkers[39] achieved an intermodal dispersion of only 170 ps km^{-1} and with a laser light source of 907.5 nm wavelength and approximately 2.5 nm spectral width, a total pulse spreading of 260 ps km^{-1}. An intermodal dispersion of only 150 ps km^{-1} was obtained with a germanium and boron oxide–doped silica fibre.[40] If a laser light source is also used with the best graded-index fibres, gigabit rates can be transmitted across distances of several kilometres.

Let us now turn again to the light sources for optical communications. In the last 20 years optical sources have also undergone a development similar to the optical transmission media from large and expensive devices to small and simply manufacturable ones. In the early stages of laser communication research, helium–neon, argon ion, CO_2 gas lasers and the neodymium doped YAG solid-state laser were interesting candidates for the optical sources in high-capacity optical communication links.[41] The helium–neon laser was the first highly developed CW laser and supplied output powers up to 100 mW at the 0.633 μm wavelength. The CO_2 laser has a high optical output power up into the kilowatt region and oscillates at the long wavelength of 10.6 μm. The long wavelength yields a low photon noise when cooled detectors are used. The CO_2 laser therefore is well suited for long-distance outer-space applications. For fibre-optic applications, gas lasers are in general too large, too expensive and suffer from low efficiency and the need for external modulation. Neodymium compound solid-state lasers are of some interest as sources for fibre-optic communications. When Danielmeyer and Weber[42] succeeded in 1972 in growing stoichiometric neodymium ultraphosphate crystals, a laser material with an optical gain 60 times higher than that of neodymium-doped YAG was available and the fabrication of small-dimension lasers became possible. Since then many stoichiometric neodymium laser materials with emission wavelengths between 1.0477 and 1.0641 μm, and in one case 1.32 μm, have been investigated.[43] In 1975 Burrus and Stone grew single-crystal neodymium-doped YAG fibres.[44] Using such a fibre 0.5 cm long and 80 μm in diameter, a room-temperature CW laser pumped from one end by a GaAs superluminescent diode was realized.[45] It has been estimated that with optimized mirrors an optical output of 1 mW could be achieved. Neodymium compound lasers have the advantages of a narrow spectral width and an emission wavelength close to the dispersion and attenuation minima of optical fibres. However, the need for an external pump and external modulation complicates their application.

Today the semiconductor injection laser is the most promising coherent light source for optical-fibre communications. Its main advantages are simple construction, small dimensions, high efficiency and direct modulation capability up into the GHz range. The semiconductor injection laser utilizes stimulated emission due to the recombination of carriers injected across a semiconductor *p–n* junction in forward direction. This principle was first suggested by Basov *et al.* in 1961.[46] Unfortunately, the germanium semiconductor material proposed by Basov is not suitable for this purpose, since as an indirect semiconductor it has too low an optical transition probability. Also in 1961, Bernard and Duraffourg derived the laser condition for semiconductor lasers.[47] Then, in 1962, lasing action was achieved with semiconductor injection lasers by three different research groups.[48–50] The first semiconductor injection lasers were made from gallium arsenide in the

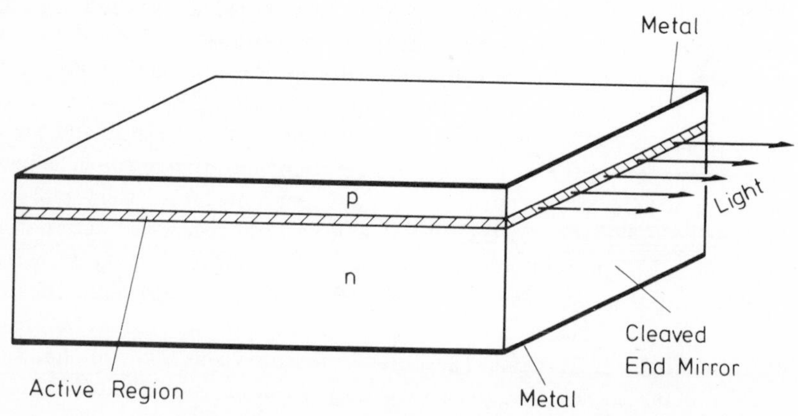

Figure 1.5. Semiconductor injection laser

form of a parallelepiped with a planar diffused *p–n* junction perpendicular to two opposite ends of the semiconductor crystal (Figure 1.5). If a current in forward direction is impressed, radiative carrier recombination occurs near the junction plane. The semiconductor material also exhibits optical gain for a sufficiently large injection current. Since the semiconductor crystal has a refractive index greater than that of air, the cleaved end faces of the crystal act as mirrors, so that the radiation is generated and amplified within a Fabry–Perot cavity. At a certain threshold level the round-trip gain exceeds the bulk and mirror losses for a certain mode and the laser starts to oscillate. The first injection lasers exhibited threshold current densities of up to 10^5 A cm^{-2} at 300 K so that room-temperature CW operation was impossible. In 1963, Kroemer suggested heterostructures in which the active region of the *p–n* junction was followed by a semiconductor layer with a higher band gap and a lower refractive index in order to provide better carrier and optical confinement, and hence to reduce the threshold current density.[51] In 1970 Hayashi and Panish built single-heterostructure lasers with room-temperature threshold current densities of 10^4 A cm^{-2} [52] and double-heterostructure lasers with a room-temperature threshold current density of only 1600 A cm^{-2}.[53,54] GaAs and $Ga_xAl_{1-x}As$ heterostructures fabricated by liquid-phase epitaxy were used in these cases. The narrow-gap active GaAs layer of the double heterostructure laser with a thickness considerably below 1 μm is bounded by two wide-gap layers of $Ga_xAl_{1-x}As$. The threshold current of injection lasers could be further reduced by the introduction of stripe geometry. Figure 1.6 shows the first stripe-geometry laser.[55] The active *p*–GaAs layer is sandwiched by two $Ga_xAl_{1-x}As$ layers. The stripe is etched in a thin SiO_2 layer deposited on the semiconductor crystal and forms a window for the metal contact. As a consequence, only the part of the active region under the stripe is pumped. With a laser length of 400 μm and a stripe width of 13 μm, Ripper and coworkers achieved threshold currents as low as 300 mA at room temperature.[55] Furthermore, with stripe-geometry lasers single

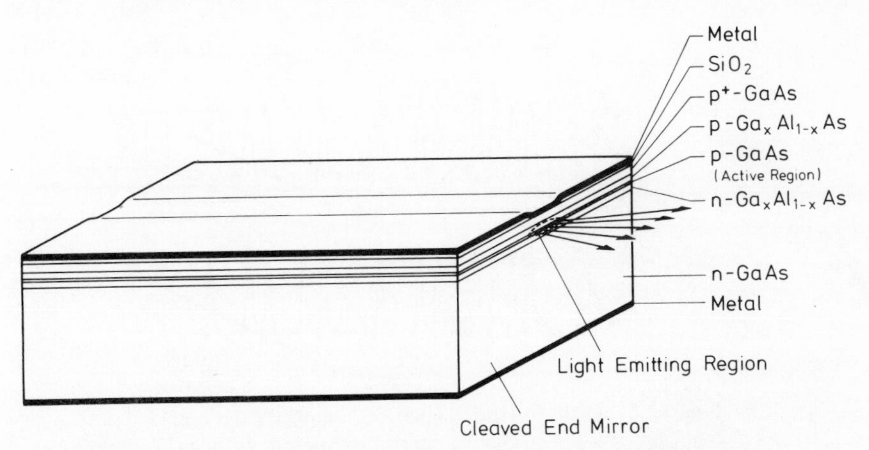

Figure 1.6. Stripe geometry double-heterostructure injection laser

tranverse-mode operation and single-frequency operation became feasible, whereas first injection lasers exhibited a broad multimode spectrum. Single transverse-mode operation is necessary for a good coupling efficiency into monomode fibres. The more severe demands of single-frequency operation must be met if a low pulse dispersion in single-mode fibres is required.

The first lasers for CW room-temperature operation exhibited lifetimes in the order of minutes only. Much technological effort has been undertaken to increase the laser lifetime to several 10^5 hours for CW room-temperature operation with light output powers of several milliwatts.[56] When CW lasers became available at the beginning of the 1970s, experimental investigation into the direct modulation capability at high modulation frequencies and high duty cycles started.[57] For high bit-rate digital communications from several 100 Mbit/s to the Gbit/s range, lasers must exhibit a narrow spectral bandwidth, no modulation distortions, and no high spectral broadening due to direct modulation.[58] Furthermore, the injection laser must not exhibit spontaneous fluctuations in the light intensity when it is biased above threshold.[59] The attempt to meet all these requirements led to the development of a variety of stripe-geometry structures, for example, the proton-bombarded,[60] the zinc-diffused stripe,[61] the stripe mesa,[62] the buried hetero-structure,[63] the channelled substrate planar,[64] the transverse junction stripe[65] and the v-groove structure.[66] In some cases a direct modulation capability of up to 2 Gbit/s has been achieved.[58,63,64] Figure 1.7 shows the light output signal of a v-groove laser modulated with a 1 Gbit/s return-to-zero pcm signal.

The quarternary semiconductor material $Ga_xIn_{1-x}As_yP_{1-y}$ is now of growing interest for injection laser fabrication. Lasers with an emission wavelength in the 1.1 to 1.5 μm region can be realized with this material.[67] In this wavelength region one can take advantage of the dispersion and attenuation minima of optical fibres. The dynamic behaviour of $Ga_xIn_{1-x}As_yP_{1-y}$ injection lasers is identical to that of gallium arsenide injection lasers.[68,69]

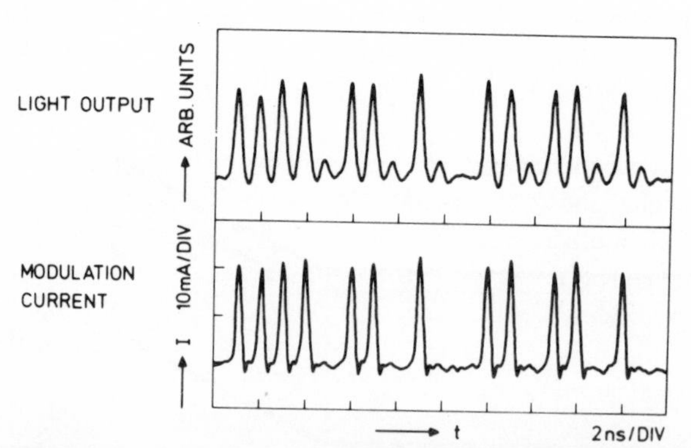

Figure 1.7. Light output signal and modulation current signal of a v-groove laser modulated with a 1 Gbit/s return-to-zero signal[66]

Apart from the semiconductor injection laser, another kind of semiconductor light emitting device, namely the light-emitting diode (LED), has gained importance as the light source for optical-fibre communications. In the LED, light is generated in the same way as in the laser diode, but since no optical feedback is introduced the LED produces incoherent light. The first gallium arsenide LED suitable for optical communications was made in 1962 by Keyes and Quist.[70] The spectral bandwidth of gallium arsenide LEDs is typically 300 Å i.e. higher by a factor of at least 20 than that of lasers. LEDs emit light into many spatial modes. Since the number of modes which can be coupled into a fibre approximately equals the number of modes which can propagate in the fibre, adequate coupling efficiency into the fibre can be achieved only if a multimode fibre is used. LEDs for fibre-optic communication must have a small light-emitting area of high radiance. Examples of such diodes are the Burrus diode[71] and the edge emitter diode.[72] The Burrus diode emits light from a small circular spot of approximately 50 μm diameter perpendicular to the junction plane, whereas the edge-emitter diode has a geometry similar to a stripe-contact laser and emits the light in the direction of the junction stripe. The superluminescent diode is a modified edge-emitter diode with a stripe long enough so that the spontaneously emitted light is considerably amplified by the stimulated processes, but it uses no optical feedback.[73] A lower spectral width (typically 100 Å) and a lower risetime (2 ns) can be achieved with superluminescent diodes than with ordinary LEDs.[74]

In the fibre-optic receiver, a photodetector with a sufficiently wide bandwidth and a high sensitivity is required. Semiconductor photodiodes meet these requirements for fibre-optic communications and have the additional advantages of small dimensions and simple construction.[75,76] The electron–hole pairs created in the depletion layer of a photodiode by absorption of photons give rise to a photocurrent. Since the photocurrent is proportional to the incident optical power, the photodiode acts as a quadratic detector. Silicon photodiodes are suitable for wavelengths up to 1 μm whereas germanium photodiodes can be used up to 1.5 μm. Figure 1.8a shows the cross section of a silicon PIN photodiode. The intrinsic region has a width in the order of 10 to 20 μm.

Avalanche photodiodes were first reported by Johnson in 1965[77] and allow a considerable signal amplification due to avalanche carrier multiplication when reverse biased between 100 and 300 V.[75–78] Figure 1.8b shows the cross section of a silicon avalanche photodiode. The electron–hole pairs are created in the drift region. Avalanche multiplication occurs in the depletion layer of the p–n^+ junction. Gain–bandwidth products greater than 200 GHz have been achieved with silicon avalanche photodiodes.[79] Unfortunately, due to the statistical nature of the avalanche process, the noise increases more strongly than by the square of the signal gain. If the signal amplification factor is M, the noise amplification equals M^{2+x}, where x depends on the diode. For silicon avalanche photodiodes, x usually equals between 0.2 and 0.5. An optimum gain for the photodiode can be determined by taking into account the noise of the following amplifier stage. The optimum design

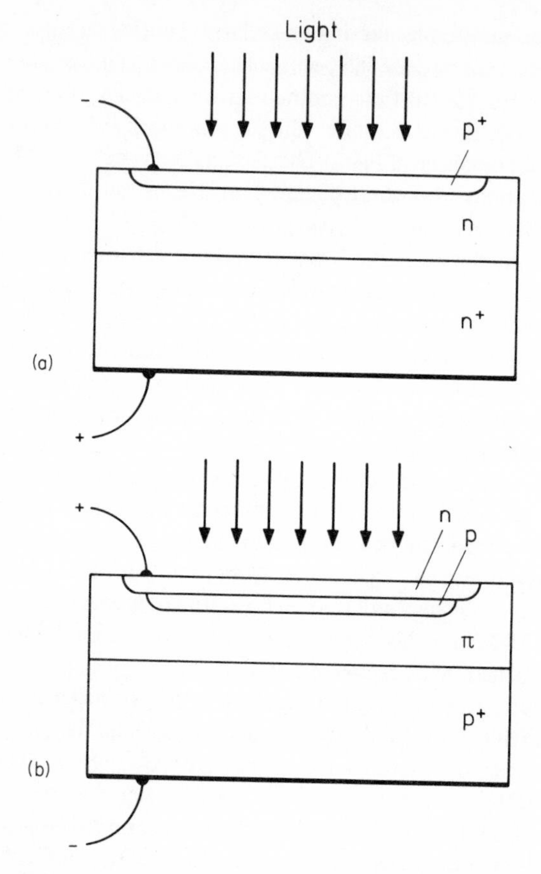

Figure 1.8. Cross sections of: (a) a silicon PIN photodiode; (b) a silicon avalanche photodiode

of the front-end amplifier has been studied by Personick.[80] Avalanche photodiodes have been developed using germanium[81] and InGaAsP[82,83] for the longer-wavelength region. However, germanium avalanche photodiodes suffer from a much higher dark current and a higher excess noise ($x \approx 1$) than silicon photodiodes. InGaAsP avalanche photodiodes are superior to germanium devices but at present are not competitive with silicon avalanche photodiodes. Therefore for the longer-wavelength region (1.1–1.3 μm), a combination of a PIN photodiode with a gallium arsenide field-effect transistor seems to be advantageous.[84] Since photodiodes are quadratic detectors, optical heterodyning to improve the detection sensitivity is possible in principle.[75] However, the spectral width and frequency stability of semiconductor injection lasers are not sufficient to use this method.[58]

Fibre-optic communication links exhibit a number of advantages compared with conventional cable transmission links. The bandwidth and repeater separation are higher than with coaxial cables. Furthermore, fibres are immune to inductive interference, and exhibit no crosstalk. Fibre cables also have a smaller diameter and a lower weight than coaxial cables.

In fibre-optic communications, analog as well as digital modulation is possible. However, for analog modulation a higher signal-to-noise ratio at the receiver is required and the linearity of injection lasers decreases at higher modulation frequencies.[58] For these reasons, analog links are limited to lower bandwidths and lower distances than digital links.

Figure 1.9 shows the block diagram of a typical digital fibre-optic link. The laser driver directly modulates the injection laser with the pcm signal. The APD is followed by the front-end amplifier, a filter for linear signal processing and noise bandwidth reduction, a nonlinear signal processing circuit and phase lock loop for clock extraction and the baseline regenerator and decision detector for signal regeneration.[85] Table 1.1 gives the data of some fibre-optic digital transmission experiments. The possibility of transmitting gigabit rates over fibre links has also stimulated the development of gigabit electronic circuits, necessary for transmitters and receivers.[87,96,97] For low-capacity, short-distance digital fibre-optic links with data rates up to 15 Mbit/s the electronic circuits for the transmitter as well as for the receiver have been monolithically integrated, each on a single chip.[98]

In future, fibre-optic links will find application in telephone networks. In a study by the British Post Office, the economic aspects of fibre-optic links for junction networks with bit rates of 2, 8 and 34 Mbit/s and trunk networks with 140, 280 and 565 Mbit/s were investigated.[99] As a result, the bit rate of 140 Mbit/s would most likely find application for trunk networks, followed by 8.448 Mbit/s (and 2.048 Mbit/s) for junction networks. Fibre-optic links for medium bit rates of

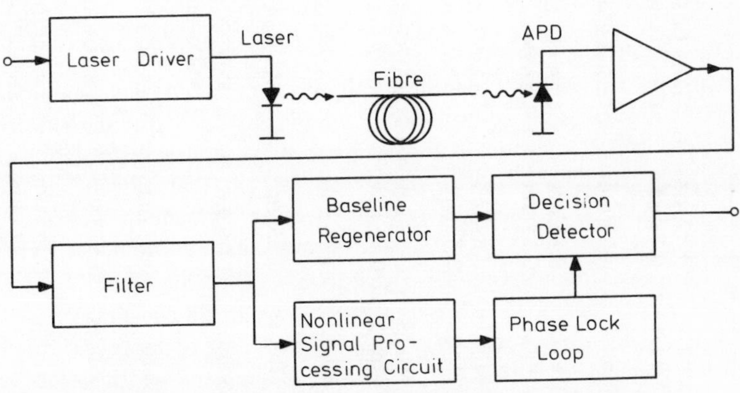

Figure 1.9. Block diagram of a digital fibre-optic link

Table 1.1 *Experimental digital fibre-optic transmission links*

Bit rate	Optical wavelength	Light source	Fibre	Detector	Power at the receiver for error rate 10^{-9}	Reference
10 Mbit/s	0.82 μm	GaAs-LED	6 km graded index	Si-APD	– 63.5 dBm	86
10 Mbit/s	0.85 μm	GaAs/GaAlAs-LD	13 km graded index	Si-APD	– 63 dBm	
32 Mbit/s	1.27 μm	InGaAsP/InP-LD	53.3 km graded index	Ge-APD	– 43.1 dBm	87
50 Mbit/s	0.85 μm	GaAs-LED or		Si-PIN-diode	– 41.5 dBm	88
50 Mbit/s	0.85 μm	GaAs/GaAlAs-LD		Si-APD	– 56.6 dBm	
100 Mbit/s	0.83 μm	GaAs/GaAlAs-LD	12 km graded index	Si-APD	– 50 dBm	89
400 Mbit/s	0.83 μm	GaAs/GaAlAs-LD	8 km graded index	Si-APD	– 38 dBm	
159 Mbit/s	0.85 μm	GaAs/GaAlAs-LD	6 km graded index	Si-APD	– 47 dBm	86
159 Mbit/s	0.85 μm	GaAs/GaAlAs-LD	8 km graded index	Si-APD	– 44 dBm	
400 Mbit/s	0.85 μm	GaAs/GaAlAs-LD	5.9 km single mode	Si-APD	– 38 dBm	90
800 Mbit/s	0.85 μm	GaAs/GaAlAs-LD	7.3 km single mode	Si-APD	– 35.5 dBm	91
800 Mbit/s	1.05 μm	$LiNdP_4O_{12}$-laser $LiNbO_3$-modulator	0.8 km single mode	Ge-APD	– 32 dBm	92
1 Gbit/s	0.83 μm	GaAs/GaAlAs-LD	1.6 km single mode	Si-APD	– 30.8 dBm	85,93
1.12 Gbit/s	0.813 μm	GaAs/GaAlAs-LD	3 km graded index	Si-APD		94
100 Mbit/s	1.293 μm	GaInAsP/InP-LD	11 km single mode	Ge-APD	– 39.9 dBm	95
400 Mbit/s	1.293 μm	GaInAsP/InP-LD	11 km single mode	Ge-APD	– 36 dBm	
800 Mbit/s	1.293 μm	GaInAsP/InP-LD	11 km single mode	Ge-APD	– 33.3 dBm	
1.2 Gbit/s	1.293 μm	GaInAsP/InP-LD	11 km single mode	Ge-APD	– 29.1 dBm	

LD laser diode
LED light emitting diode

8 to 280 Mbit/s are already sufficiently developed and will soon be able to compete economically with coaxial cables. A Japanese study by the Nippon Telegraph and Telephone Public Corporation also comes to the conclusion that as a first step 32 and 100 Mbit/s digital fibre-optic links using graded-index fibres will be introduced for short- and medium-haul trunks with heavy traffic.[100] Large-capacity, long-haul digital transmission systems, for example at 400 Mbit/s and using graded-index or monomode fibres, will follow.

Field trials have been performed in order to evaluate lightwave technology and to test cabling techniques, splicing techniques and fibre-optic equipment under field conditions. A 44.7 Mbit/s fibre-optic digital transmission system was put into operation by the Western Electric and Bell Laboratories in Atlanta in 1976.[101] Similar field trials are being performed in England at 8 and 140 Mbit/s by the British Post Office[86] and in Berlin, Germany, at 34 Mbit/s by the Deutsche Bundespost.[102,103]

Besides the introduction of fibre-optic links in existing telephone networks plans exist also for fibre-optic digital wide-band networks with a full integration of data, voice, video telephone, and broadcasting services including radio and television. A model of such an integrated network, comprising fibre-optic links from 140 to 560 Mbit/s, is now being built in the Heinrich Hertz Institute in Berlin.[104]

In the near future, fibre-optic analog systems will find application in radio and television broadcast distribution systems. In analog systems the subscriber terminals are simple and inexpensive. Since there are problems at present with the linearity of optical sources, analog systems operate at low bandwidths and over short distances. In the Japanese Hi–OVIS project an individual fibre is used for every television channel. The subscriber is connected to the network by two fibres (one for upstream and one for downstream video transmission).[105] In a Canadian project, two television channels and one audio channel are transmitted over a single fibre by analog intensity modulation with frequency-division multiplexed signal. The programmes can be selected via a return line.[106]

A further application for fibre-optic communication links is for high-reliability information transmission within electric power systems.[107] The insensitivity of optical links to electromagnetic interferences in this case is advantageous. For the same reason, optical fibres are also of interest for inter-module and inter-system wiring in electronic systems. In complex electronic circuits, ground loops can be avoided by the use of fibre links. For lower bit rates, monolithic integrated receiver and transmitter modules can be mounted directly on printed-circuit boards.[98] A data bus, that is a single transmission line carrying a multiplexed signal stream fed into the line by different transmitters and distributed over different receivers, can also be realized with fibre optics.[108–110] Coupling of the optical transmitters and receivers to the fibre can be performed by optical star couplers[110] or optical T-couplers.[111] Transparent photodiodes can be inserted into the transmission path to tap off a signal from a fibre.[112]

1.2 PHYSICAL ASPECTS OF OPTICAL COMMUNICATIONS

Electromagnetic radiation in the optical waveband differs from the electromagnetic radiation in the radio-frequency band not only in frequency but also in its statistical properties. The reason for this is that the quantum nature of the electromagnetic radiation plays an important role in the optical region. The information carried by a light wave is in general limited by quantum noise and not by thermal noise. In this section, therefore, we shall briefly discuss the implications of the statistical properties of coherent and incoherent radiation as carriers of information.

When Planck derived his formula for black-body radiation in 1900 he assumed that only integer multiples of discrete energy quanta can be exchanged when radiation and matter interact.[113] The energy quantum ϵ is related to the radiation frequency f by

$$\epsilon = hf \tag{1.4}$$

where h is Planck's constant. Using heuristic arguments Einstein suggested in 1905 that radiation really consists of independent energy quanta.[114] By 1927, the development of modern quantum theory was largely completed. In quantum theory the concept of quanta includes radiation as well as matter. Although quantum theory cannot be explained on the basis of classical physics, analogies to classical physics are helpful in order to develop a clear concept of the phenonoma. For example, an electromagnetic wave confined in a resonator can exhibit discrete frequency values only. This quantization of frequency values is a classical phenomenon. Now in quantum theory the electromagnetic field undergoes a second quantization, namely the quantization of the energy in each mode. Each resonator mode is occupied by a discrete number of radiation quanta, called photons, having an energy ϵ according to Equation 1.4.[115] Matter, on the other hand, also exhibits wave properties according to quantum theory. If a particle is spatially bounded, the matter-wave assigned to it can also have discrete frequency values only, and therefore, according to Equation 1.4, only discrete energy values are possible. To sum up, it can be said that the first quantization, i.e. the quantization of frequency and energy values, is the contribution from the wave picture to the quantum picture whereas the second quantization, i.e. the particle quantization of wave fields, is the contribution from the particle picture to the quantum picture of physics.

The quantum behaviour of radiation must be taken into consideration when $hf > kT$, since in this case the quantum fluctuations dominate over the thermal fluctuations. For a vacuum wavelength of 0.85 μm for example, $hf/kT = 56.5$ at $T = 300$ K, whereas in the radio-frequency region at a wavelength of 3 cm we obtain $hf/kT = 1.6 \times 10^{-3}$. The detection of light by a photodiode is a discrete process since the creation of an electron–hole pair is effected by the absorption of a photon. If I is the light intensity incident on the photodiode, and η is the quantum efficiency (i.e. is the ratio of the number of generated electron–hole pairs to the number of incident photons), then within a period T the average number of electron–hole pairs $\langle n \rangle$ generated by the incident light is given by

$$\langle n \rangle = \frac{\eta I T}{hf} \tag{1.5}$$

The number of photons detected in a certain period of time is discrete, whereas the light intensity is a continuous quantity. According to quantum theory, the light intensity is a measure for the probability of detecting a photon within a small interval of time. Now let us consider a light beam of constant intensity incident on the photodiode. We would expect a constant photon-detection probability for such a light beam. Since no interaction between photons occurs due to the linearity of the electromagnetic wave equation, the detection probability for a photon is statistically independent of the number of photons previously detected. We would therefore expect that the probability $p(n)$ of detecting n photons in a time T obeys the Poisson distribution[116]

$$p(n) = \frac{\langle n \rangle^n \exp(-\langle n \rangle)}{n!} \tag{1.6}$$

Figures 1.10a and 1.10b show the Poisson distributions for $\langle n \rangle = 10$ and $\langle n \rangle = 1000$. The analogue to the classical electromagnetic sine wave in quantum theory is the single-mode coherent state of the light wave.[117,118] From Glauber's theory of optical coherence it follows that the photon-detection process obeys Poisson statistics for a monochromatic coherent light wave and an arbitrary counting interval. The fluctuations in a light wave of constant intensity are called 'particle fluctuations'. The Poisson distribution exhibits the mean-square deviation

$$\overline{\delta n^2} \equiv \langle n^2 \rangle - \langle n \rangle^2 = \langle n \rangle \tag{1.7}$$

Since the particle fluctuation is proportional to the square root of the light intensity, we would expect the signal-to-noise ratio caused by the particle fluctuations to be inversely proportional to the light intensity.

Incoherent light is emitted by independent atoms and there is no phase correlation between the emission processes. Incoherent light is the optical analogue to noise in the radio-frequency band; like noise, incoherent light has an exponential intensity distribution

$$p(I(t)) = \frac{1}{\langle I \rangle} \exp(-I(t)/\langle I \rangle) \tag{1.8}$$

where $I(t)$ is the instantaneous intensity and $\langle I \rangle$ the average intensity.[118,119] By instantaneous intensity we mean the light intensity averaged over only a few cycles of the light oscillations. Averaging the Poisson distribution over the distribution $p(I(t))$ yields

$$p(n) = \frac{1}{\langle I \rangle} \int \exp\left(-\frac{I}{\langle I \rangle}\right) \frac{\left(\frac{\eta I T^n}{hf}\right) \exp\left(-\frac{\eta I T}{hf}\right)}{n!} \, dI = \frac{\langle n \rangle^n}{(1 + \langle n \rangle)^{n+1}} \tag{1.9}$$

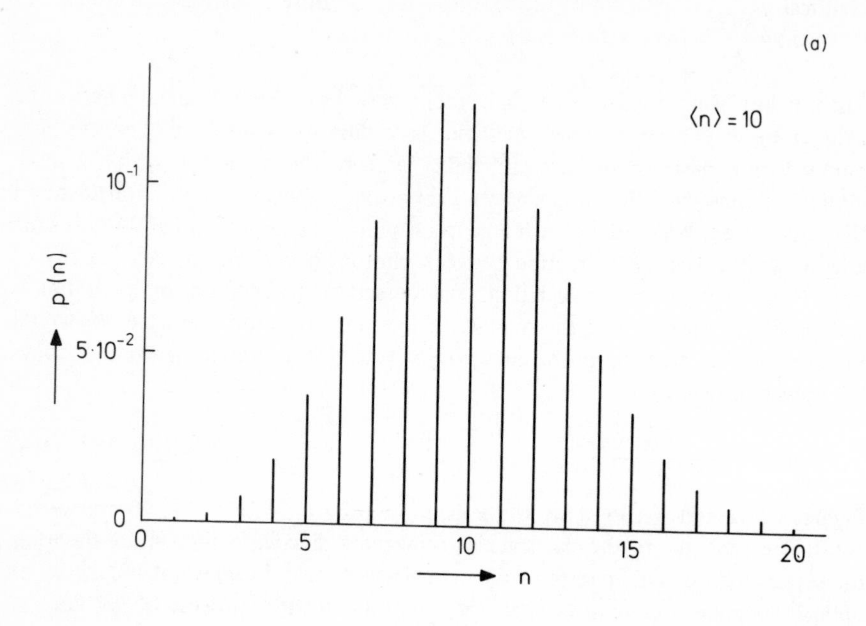

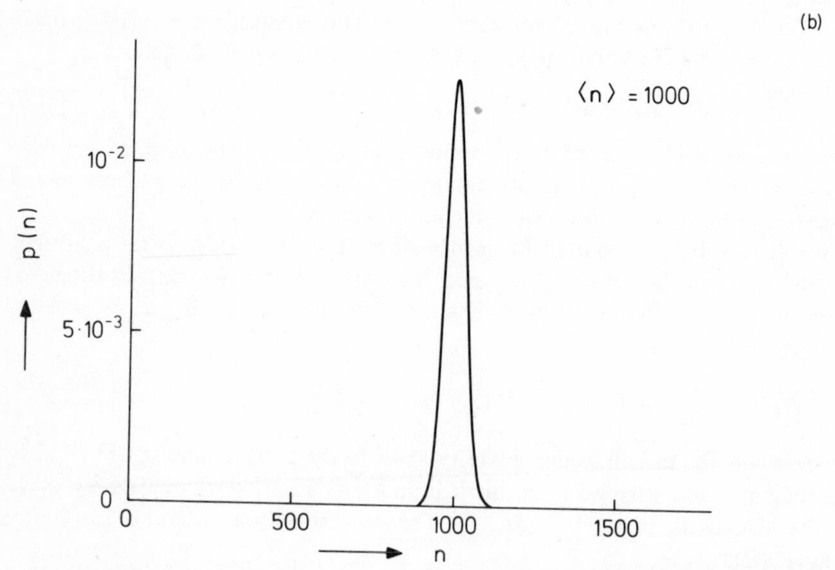

Figure 1.10. Poisson distributions $p(n)$ for: (a) $\langle n \rangle = 10$; (b) $\langle n \rangle = 1000$

where we have used Equations 1.5, 1.6 and 1.8. This probability distribution is identical with the probability distribution for the Bose–Einstein distribution for thermal light. The corresponding mean-square deviation is

$$\overline{\delta n^2} = \langle n \rangle + \langle n \rangle^2 \tag{1.10}$$

The instantaneous fluctuations of incoherent light are proportional to the light intensity and are of the same magnitude. Figure 1.11 shows the photon-count probability distributions for incoherent light for $\langle n \rangle = 10$ and $\langle n \rangle = 1000$. However, if the light intensity is averaged over a certain interval of time, the fluctuations of this averaged intensity are smaller than the fluctuations of the instantaneous intensity. This can be compared with the measurement of Gaussian noise intensity

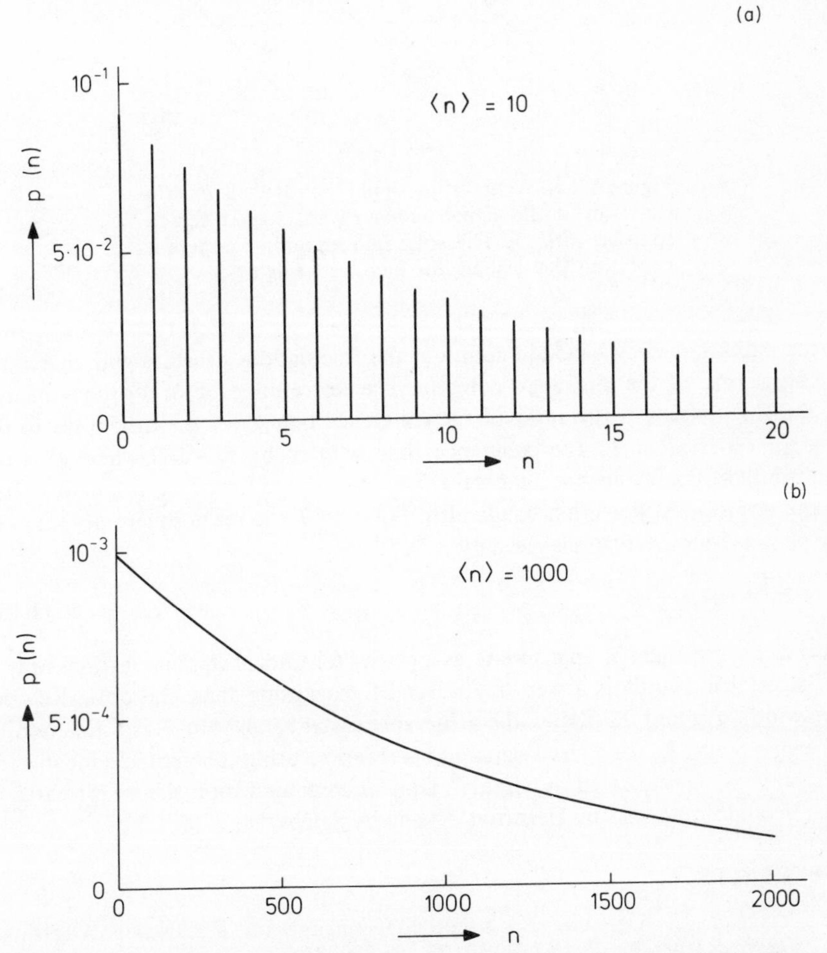

Figure 1.11. Probability distributions $p(n)$ for the instantaneous fluctuations of incoherent light for: (a) $\langle n \rangle = 10$; (b) $\langle n \rangle = 1000$

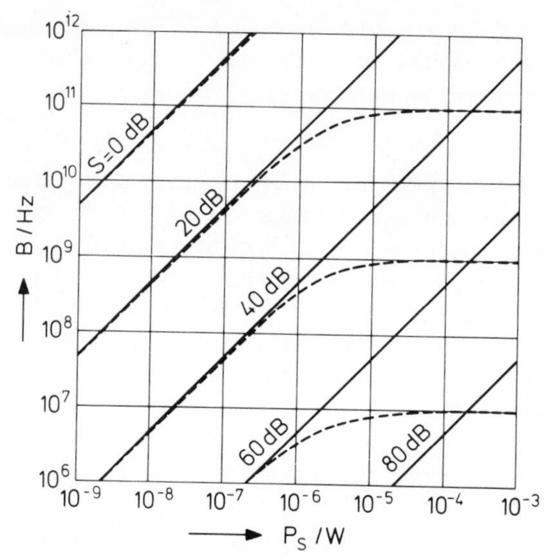

Figure 1.12. Achievable signal bandwidth B as a function of the signal power P_S for a given signal-to-noise ratio S. The solid line represents coherent light, and the dashed line incoherent light[10]

in the radio-frequency band. In addition, the fluctuations decrease with increasing averaging time of the square-law detector. The determining factor for the reduction of the fluctuations is the ratio of the coherence time τ_c of the light wave to the time of observation T. The coherence time is given by $\tau_c = 1/B$ where B is the bandwidth of the incoherent light wave. The observation time T corresponds to half of the inverse post detection bandwidth. For $\tau_c \ll T$ the mean-square deviation of the photon-count statistics is given by[118,119]

$$\overline{\delta n^2} = \langle n \rangle + \frac{\tau_c}{T} \langle n \rangle^2 \tag{1.11}$$

Thus incoherent light is appropriate as a carrier for optical communications only if the signal bandwidth is lower by orders of magnitude than the optical carrier bandwidth. Figure 1.12 shows the achievable signal bandwidth B as a function of the signal power P_S for a given signal-to-noise ratio S using coherent and incoherent light sources, as calculated by Grau.[10] Optical communication theory is treated in great detail in the books by Helstrom[120] and by Saleh.[121]

REFERENCES

1. V. Aschoff, 'Optische Nachrichtenübertragung im Klassischen Altertum', *Nachrichtentechn. Z.* (NTZ), **30**, 23–28, 1977.
2. K. Steinbuch, Die informierte Gesellschaft, 60–66. Stuttgart, 1966.

3. A. G. Bell, 'Selenium and the photophone', *The Electrician*, 214, 215, 220, 221, 1880.
4. W. S. Huxford and J. R. Platt, 'Survey of near infra-red communication systems', *J. Opt. Soc. Am.* **38**, 253–268, 1948.
5. N. C. Beese, 'Light sources for optical communication', *Infrared Phys.*, **1**, 5–16, 1961.
6. A. L. Schawlow and C. H. Townes, 'Infrared and optical masers', *Phys. Rev.*, **112**, 1940–1949, 1958.
7. T. H. Maiman, 'Stimulated optical radiation in ruby', *Nature*, **187**, 493–494, 1960.
8. A. Javan, W. R. Bennett and D. R. Herriott, 'Population inversion and continuous optical maser oscillation in a gas discharge containing a He–Ne mixture, *Phys. Rev. Lett.*, **6**, 106–110, 1961.
9. H. E. Rowe, 'Amplitude modulation with a noise carrier', *Proc. IEEE*, **52**, 389–395, 1964.
10. G. Grau, 'Temperatur- und Laserstrahlung als Informationsträger', *Arch. Elektron. Übertragungstech.* (AEÜ), **18**, 1–4, 1964.
11. T. Elder and J. Strong, 'The infrared transmission of atmospheric windows', *J. Franklin Inst.* **255**, 189–208, 1953.
12. R. Kompfner, 'Optical communications', *Science*, **150**, 149–154, 1965.
13. S. E. Miller and L. C. Tillotson, 'Optical transmission research', *Appl. Optics*, **5**, 1538–1549, 1966.
14. R. Gruss, 'Übertragung von Laserstrahlung durch die Atmosphäre', *Nachrichtentechn. Z.* (NTZ), **22**, 184–192, 1962.
15. A. R. Kraemer, 'Free-space optical communications', Signal, **1977**, 26–32.
16. S. E. Miller, 'Directional control in light-wave guidance', *Bell Syst. Tech. J.* **43**, 1727–1739, 1964.
17. G. Goubeau and F. Schwering, 'On the guided propagation of electromagnetic wave beams', *IRE Trans. Atennas and propagation* **AP–9**, 248–255, 1961.
18. G. Goubeau and J. R. Christian, 'Some aspects of beam waveguides for long distance transmission at optical frequencies', *IEEE Trans. Microw. Theory Techniques*, **MTT–12**, 212–220, 1964.
19. D. Marcuse and S. E. Miller, 'Analysis of a tubular gas lens', *Bell Syst. Tech. J.* **43**, 1759–1782, 1964.
20. C. C. Eaglesfield, 'Optical pipeline: a tentative assessment', *Proc. IEE*, **109B**, 26–32, 1962.
21. E. A. J. Marcatili and R. A. Schmeltzer, 'Hollow metallic and dielectric waveguides for long distance optical transmission and lasers', *Bell Syst. Tech. J.*, **43**, 1783–1803, 1964.
22. D. Hondros and P. Debye, 'Elektromagnetische Wellen an dielektrischen Drähten', *Annalen d. Physik*, **32**, 465–476, 1910.
23. O. Schriever, 'Electromagnetische Wellen an dielektrischen Drähten', *Annalen d. Physik*, **63**, 645–673, 1920.
24. H. H. Hopkins and N. S. Kapany, 'A Flexible fibrescope, using static scanning', *Nature*, **173**, 39–41, 1954.
25. A. C. S. van Heel, 'A new method of transporting optical images without aberrations', *Nature*, **173**, 39, 1954.
26. K. C. Kao and G. A. Hockham, 'Dielectric-fibre surface waveguides for optical frequencies', *Proc. IEE*, **113**, 1151–1158, 1966.
27. A. Werts, 'Propagation de la lumière cohérente dans les fibres optiques', *L'Onde Électrique*, **46**, 967–980, 1966.

28. M. Boerner, 'Mehrstufiges Übertragungssystem für in Pulscodemodulation dargestellte Nachrichten', DBP Nr. 1 254 513, 1966 (French Pat. Nr. 1 548 972; Brit. Pat. Nr. 1 202 418; US Pat. Nr. 3845–293).
29. E. Snitzer, 'Cylindrical dielectric waveguide modes', *J. Optical Soc. America,* **51**, 491–498, 1961.
30. D. Marcuse, 'Review of monomode fibres', *Proc. 3rd Europ. Conf. Optical Communication,* 60–65, München (September 14–16 1977).
31. J. W. Fleming, 'Material dispersion in lightguide glasses', *Electron. Lett.,* **14**, 326–328, 1978.
32. T. Kimura and K. Daikoku, 'A proposal on optical fibre transmission systems in a low-loss 1.0–1.04 μm wavelength region', *Opt. and Quant. Electron.,* **9**, 33–42, 1977.
33. N. Niizeki, 'Single mode fiber at zero-dispersion wavelength', *Digest of the Topical Meeting on Integrated Optics and Guided-Wave Optics*, (MB 1–1–4, Salt Lake City (Jan. 16–18 1978).
34. F. P. Kapron, D. B. Keck and R. D. Maurer, 'Radiation losses in glass optical waveguides', *Appl. Phys. Lett.,* **17**, 423–245, 1970.
35. A. Kawana, T. Miyashita, M. Nakahara, M. Kawachi and T. Hosaka, 'Fabrication of low-loss single-mode fibres', *Electron. Lett,* **13**, 188–189, 1977.
36. M. Horiguchi, 'Spectral losses of low-OH-content optical fibres', *Electron. lett.,* **12**, 310–312, 1976.
37. D. Gloge and E. A. J. Marcatili, 'Multimode theory of graded-core fibres', *Bell Syst. Tech. J.,* **52**, 1563–1578, 1973.
38. R. Olshansky and D. B. Keck, 'Pulse broadening in graded-index optical fibres', *Appl. Opt.,* **15**, 483–491, 1976.
39. L. G. Cohen, G. W. Tasker, W. G. French and J. R. Simpson, 'Pulse dispersion in multimode fibres with graded B_2O_3–SiO_2 cores and uniform B_2O_3–SiO_2 cladding', *Appl. Phys. Lett.* **28**, 391–393, 1976.
40. J. P. Hazan, J. J. Bernard and D. Kuppers, 'Medium-numerical-aperture low-pulse-dispersion fibre', *Electron. Lett.,* **13**, 540–542, 1977.
41. J. E. Geusic, W. B. Bridges and J. I. Pankove, 'Coherent optical sources for communications', *Proc. IEEE,* **58**, 1419–1439, 1970.
42. H. G. Danielmeyer and H. P. Weber, 'Fluorescence in neodymium ultraphosphate', *IEEE J. Quant. Electron.,* **QE–8** 805–808, 1970.
43. P. Möckel, 'Optically pumped miniature solid-state lasers from stoichiometric neodymium compounds', *Frequenz* **32**, 85–90, 1978.
44. C. A. Burrus and J. Stone, 'Single-Crystal fiber optical devices: a Nd: YAG fiber laser', *Appl. Phys. Lett.,* **26**, 318–320, 1975.
45. J. Stone, C. A. Burrus, A. G. Dentai and B. I. Miller, 'Nd: YAG single-crystal fiber laser: Room-temperature cw operation using a single LED as an end pump', *Appl. Phys. Lett.,* **29**, 37–39, 1976.
46. N. G. Basov, O. N. Krokhin and Y. M. Popov, 'Production of negative-temperature states in p–n junctions of degenerate semiconductors', *Sov. Phys.–JETP,* **13**, 1320–1321, 1961.
47. M. G. A. Bernard and G. Duraffourg, 'Laser conditions in semiconductors', *Phys. Status Solidi,* **1**, 699–703, 1961.
48. R. N. Hall, G. E. Fenner, J. D. Kingsley, T. J. Soltys and R. O. Carlson, 'Coherent light emission from GaAs junctions', *Phys. Rev. Lett.,* **9**, 366–368, 1962.
49. M. I. Nathan, W. P. Dumke, G. Burns, F. H. Dill and G. J. Lasher, 'Stimulated emission of radiation from GaAs p–n junctions, *Appl. Phys. Lett.,* **1**, 62–64, 1962.

50. T. M. Quist, R. H. Rediker, R. J. Keyes, W. E. Krag, B. Lax, A. L. McWhorter and H. J. Zeiger, 'Semiconductor maser of GaAs', *Appl. Phys. Lett.*, **1**, 91–92, 1962.
51. H. Kroemer, 'A proposed class of heterojunction injection lasers, *Proc. IEEE,* **51**, 1782–1783, 1963.
52. I. Hayashi and M. B. Panish, 'GaAs-$Ga_xAl_{1-x}As$ heterostructure injection lasers which exhibit low thresholds at room temperature', *J. Appl. Phys.*, **41**, 150–163, 1970.
53. M. B. Panish, I. Hayashi, and S. Sumski, 'Double-heterostructure injection lasers with room-temperature thresholds as low as 2300 A/cm^2, *Appl. Phys. Lett.*, **16**, 326–327, 1970.
54. I. Hayashi, M. B. Panish, P. W. Foy and S. Sumski, 'Junction lasers which operate continuously at room temperatures, *Appl. Phys. Lett.* **17**, 109–111, 1970.
55. J. E. Ripper, J. C. Dyment, L. A. D'Asaro and T. L. Paoli, 'Stripe-geometry double heterostructure junction lasers: mode structure and cw operation above room temperature', *Appl. Phys. Lett.*, **18**, 155–157, 1971.
56. I. Hayashi, 'Recent progress in semiconductor lasers – cw GaAs lasers are now ready, for new applications', *Appl. Phys.*, **5**, 25–36, 1974.
57. T. L. Paoli and J. E. Ripper, 'Direct modulation of semiconductors lasers', *Proc. IEEE,* **55**, 1457–1465, 1970.
58. G. Arnold and P. Russer, 'Modulation behaviour of semiconductor injection lasers', *Appl. Phys.*, **14**, 255–268, 1977.
59. G. Arnold, F.-J. Berlec and K. Petermann, 'Investigations on repetitive self-pulsations in injection lasers', *Proc. 4th Europ. Conf. Optical Communication*, 346–351 Genova (September 12–15 1978).
60. J. C. Dyment, L. A. D'Asaro, J. C. North, B. I. Miller and J. E. Ripper, 'Proton-bombardment formation of stripe-geometry heterostructure lasers for 300 K cw operation', *Proc. IEEE,* **60**, 726–728, 1972.
61. H. Yonezu, I. Sakuma, K. Kobayashi, T. Kamejima, M. Ueno and Y. Nannichi, 'A GaAs–$Al_xGa_{1-x}As$ double heterostructure planar stripe laser', *Jap. J. Appl. Phys.*, **12**, 1585–1592, 1973.
62 T. Tsukada, R. Ito, H. Nakashima and O. Nakada, 'Mesa-stripe-geometry double-heterostructure injection lasers', *IEEE J. Quantum Electron.*, **QE–11**, 418–420, 1973.
63. M. Maeda, K. Nagano, I. Ikushima, M. Tanaka, K. Saito and R. Ito, 'Buried-heterostructure lasers for wideband linear optical sources', *Proc. 3rd Europ. Conf. Optical Communication*, 120–122, München (Sept. 14–16 1977).
64. Nakamura, K. Aiki, J. Umeda, N. Chinone, R. Ito and H. Nakashima, 'Single-transverse-mode low-noise (GaAl)As lasers with channeled substrate planar structure', *Proc. Int. Conf. Integrated Optics and Optical Fiber Communication*, 83–86, Tokyo (July 18–20 1977).
65. M. Nagano and K. Kasahara, 'Dynamic properties of transverse junction stripe lasers', *IEE J. Quantum Electron.*, **QE–13**, 632–637, 1977.
66. P. Marschall, E. Schlosser and C. Wölk, 'A new type of diffused stripe geometry injection laser', *Supplement to the Proc. 4th Europ. Conf. Optical Communication*, 94–97, Genova, (September 12–15, 1978).
67. J. H. Hsieh, J. A. Rossi and J. P. Donelly, 'Room temperature CW operation of GaInAsP/InP double-heterostructures diode lasers emitting at 1.1 μm, *Appl. Phys. Lett.*, **28**, 709–711, 1976.
68. K. Oe, S. Ando and K. Sugiyama, '1.3 μm CW operation of GaInAsP/InP DH diode lasers at room temperature', *Jap. J. Appl. Phys.*, **16**, 1273–1274, 1977.

69. S. Akiba, K. Sakai and T. Yamamoto, 'Direct modulation of InGaAsP/InP double heterostructure lasers', *Electron. Lett.*, **14**, 197–198, 1978.
70. R. J. Keyes and T. M. Quist, 'Recombination radiation emitted by gallium arsenide', *Proc. IRE*, **50**, 1822–1823, 1962.
71. C. A. Burrus and R. W. Dawson, 'Small-area high-current-density GaAs electroluminescent diodes and a method of operation for improved degradation characteristics', *Appl. Phys. Lett.*, **17**, 97–99, 1970.
72. M. Ettenberg, K. C. Hudson and H. F. Lockwood, 'High-radiance light-emitting diodes', *IEEE J. Quant. Electron.*, **QE–9**, 987–991, 1973.
73. T.-P. Lee, C. A. Burrus and B. J. Miller, 'A stripe-geometry double-heterostructure amplified-spontaneous-emission (superluminescent) diode', *IEEE J. Quantum Electron.*, **QE–9**, 820–828, 1973.
74. M.–C. Amann and W. Harth, 'Superluminescent diode as light source in optical fibre systems', *Proc. 3rd Europ. Conf. Optical Communication*, 148–150, München (September 14–16 1977).
75. H. Melchior, 'Demodulation and photodetection techniques', *Laser Handbook* (ed. F. T. Arecchi and E. O. Schultz-Dubois), Amsterdam 1972, 725–835.
76. H. Melchoir, M. B. Fisher and F. R. Arams, 'Photodetecters for optical communication systems', *Proc. IEEE*, **58**, 1466–1486, 1970.
77. K. M. Johnson, 'High-speed photodiode signal enhancement at avalanche breakdown voltage', *IEEE Trans. Electron. Devices*, **ED–19**, 55–63, 1965.
78. R. P. Webb, R. J. McIntyre and J. Conradi, 'Properties of avalanche photodiodes', *RCA Review*, **35**, 234–278, 1974.
79. K. Berchtold, O. Krumpholz and J. Suri, 'Avalanche photodiodes with a gain-bandwidth product of more than 200 GHz', *Appl. Phys. Lett.*, **26**, 585–587.
80. S. D. Personick, 'Receiver design for digital fiber optic communication systems – Part I and II', *Bell Syst. Tech. J.*, **52**, 843–886 1973.
81. T. Shibata, Y. Igarashi and K. Yano, 'Passivation of germanium devices (III) – Fabrication and performance of germanium planar photodiodes', *Rev. Electrical Communication Laboratories*, **22**, 1069–1077, 1974
82. C. E. Hurwitz and J. J. Hsieh, 'GaInAsP/InP avalanche photodiodes', *Appl. Phys. Lett.*, **32**, 487–489, 1978.
83. M. Ito, T. Kaneda, K. Nakajima, Y. Toyoma, T. Yamaoka and T. Kotani, 'Impact ionisation ratio in $In_{0.73}Ga_{0.27}As_{0.57}0.43$', *Electron. Lett.*, **14**, 418–419, 1978.
84. D. R. Smith, R. C. Hooper, I. Garrett, 'Receivers for optical communications: a comparison of avalanche photodiodes with PIN–FET hybrids', *Opt. and Quant. Electron.*, **10** 293–300, 1978.
85. J. Gruber, M. Holz, P Marten, R. Petschacher, P. Russer and E. Weidel, 'A 1 Gbit/s fibre optic communications link', *Proc. 4th Europ. Conf. Optical Communication*, 556–563, Genova, (September 12–15, 1978).
86. R. W. Berry, D. J. Brael and I. A. Ravenscroft, 'Optical fiber system trials at 8 Mbit/s and 140 Mbit/s', *IEEE Trans. Communications*, **COM–26**, 1020–1027, 1978.
87. T. Ito, K. Nakagawa, K. Aida, K. Takemoto, and K. Suto, 'Non-repeatered 50 km Transmission experiment using low-loss optical fibres', *Electron. Lett.*, **14**, 520–521, 1978.
88. P. K. Runge, 'An experimental 50 Mb/s Fiber optic PCM repeater', *IEEE Trans. Communications*, **COM–24** , 413–418, 1976.
89. S. Sugimoto, K. Minemura, T. Yomase, Y. Oolagiri and R. Ishikawa, '100 Mb/s 12 km and 400 Mb/s 8 km fiber transmission experiments', *IEEE J. Quant. Electron.*, **QE–13**, 180, 1977.

90. T. Ito, S. Machida, T. Izawa, T. Miyshita and A. Kawana, 'Optical-transmission experiment at 400 Mb/s using a single-mode fibre', *Trans. IECE Japan,* **E 59**, 19–20, 1976.
91. K. Nawata, S. Machida and T. Ito, 'An 800 Mbit/s optical transmission experiment using a single-mode fiber', *IEEE J. Quant. Electron.*, **QE–14**, 98–103, 1978.
92. T. Kimura, S. Uehara, M. Samuvatari and J.–I. Yamada, '800 Mbit/s optical fiber transmission experiment at 1.05 μm', *Proc. Int. Conf. Integrated Optics and Optical Fiber Communications*, 489–492, Tokyo, 1977.
93. R. Petschacher, J. Gruber and M. Holz, 'Error rate measurement on a 1 Gbit/s fibre optic communications link', to be published in *Electron. Let.*.
94. C. Baack, G. Elze, B. Enning and G. Heydt, '1.12 Gbit/s regenerator experiment for an optical transmission system', *Frequenz,* **32**, 151–153, 1978.
95. J. I. Yamada, M. Samuwatari, K. Asatani, H. Tsuchiya, A. Kawana, K. Sugiyama and T. Kimura, 'High-speed optical pulse transmission at 1.29 μm wavelength using low-loss single-mode fibers', *IEEE J. Quant. Electron.* **QE–14** 791–800, 1978.
96. B. G. Bosch, 'Gigabit Electronics – A Review', *Proc. IEEE*, **67**, 340–379, 1979.
97. J. Gruber, P Marten, R. Petschacher and P. Russer, 'Electronic circuits for a high bit rate digital fiber optic communication systems', *IEEE Trans. Communications,* **COM–26**, 1088–1098, 1978.
98. W. M. Brown, D. C. Hanson, T. Hornak and S. Garvey, 'System and circuit considerations for integrated industrial fiber optic data links', *IEEE Trans. Communications,* **COM–26**, 976–982, 1978.
99. W. J. Murray, 'Economic planning and system parameter objectives for optical fibre transmission systems in the British Post Office', *Proc. 4th Europ. Conf. Optical Communication*, 502–509, Genova (September 12–15, 1978).
100. K. Okura and M. Ejiri, 'Optical fiber system applications aspects in future NTT networks', *IEEE Trans. Communications,* **COM--26**, 968–974, 1978.
101. The whole issue **57** no. 6, 1717–1895, 1978. of the *Bell Syst. Tech. J.* deals with the Atlanta fiber system experiment.
102. H. Liertz, H. Goldmann and D. Schicketanz, 'Laying and properties of an optical wave guide test link in Berlin', *Proc. 4th Europ. Conf. Optical Communication*, 583–587, Genova (September 12–15, 1978).
103. E. Adler, H. Haupt and W. Zschunke, 'A 34 Mbit/s Optical field trial system', *Proc. 4th Europ. Conf. Optical Communication*, 588–596, Genova (September 12–15 1978).
104. U. Haller, W. Herold and H. Ohnsorge, 'Problems arising in the development of optical communication systems', *Appl. Phys.*, **17**, 115–122, 1978.
105. T. Nakahara, H. Kumamanu and S. Takeuchi, 'An optical fiber video system', *IEEE Trans. Communications,* **COM–26**, 955–961, 1978.
106. E. H. Hara, 'Conceptual design of a switched television distribution system using optical-fiber waveguides', *IEEE Trans. Cable Television,* **CATV–2**, 120–130, 1977.
107. F. Aoki, K. Ando, M. Nishida, Y. Ueno, S. Hiuoshita and M. Ishikawa, 'Practical use of optical fiber communications for electric power companies', *Proc. Int. Conf. Integrated Optics and Optical Fiber Communication*, 485–488, Tokyo (July 18–20 1977).
108. I.Reese, 'Fibre optic data bus for control – a reality', *Control Engineering*, 43–48, July 1977.
109. H. W. Giertz, V. Vucina and L. Ingre, 'Experimental fibre optical databus', *Proc. 4th Europ. Conf. Optical Communication*, 641–644, Genova (September 12–15 1978).

110. M. K. Barnoski, 'Data distribution using fiber optics', *Appl. Optics*, **14**, 2571–2577, 1975.
111. A. F. Milton and A. B. Lee, 'Optical access couplers and a comparison of multiterminal fiber communication systems', *Appl. Optics*, **15**, 244–252, 1976.
112. J. Müller, W. Eickhoff, P. Marschall and E. Schlosser, Transparent photodiodes for optical transmission systems', *Proc. 3rd Europ. Conf. Optical Communication*, 173–175, München (September 14–16 1977).
113. M. Planck, 'Zur Theorie des Gesetzes der Energieverteilung im Normalspectrum', *Verhandlungen der Deutschen Physikalischen Gesellschaft*, **2**, 237–245, 1900.
114. A. Einstein, 'Über einen die Brzeugung und Verwandlung des Lichtes betreffenden heuristischen Gesichtspunkt', *Annalen der Physik*, **17**, 132–148, 1905.
115. See for example: P. A. M. Dirac, *The Principles of Quantum Mechanics* (14th edn.), Oxford, 1958.
116. That the above assumptions yield the Poisson distribution is shown for example in: W. B. Davenport and W. L. Root, *An Introduction to the Theory of Random Signals and Noise*, New York, 1958, 115–117.
117. R. J. Glauber, 'Coherent and incoherent states of the radiation field', *Phys. Rev.*, **131**, 2766–2788, 1963.
118. A simple treatment of quantum optics is given in: R. Loudon *The Quantum Theory of Light*, Oxford, 1973.
119. L. Mandel, 'Fluctuations of photon beams and their correlations', *Proc. Phys. Soc.*, **71**, 1037–1048, 1958.
120. C. W. Helstrom, *Quantum Detection and Estimation Theory*, New York, 1976.
121. B. Saleh, *Photoelectron Statistics*, Berlin, 1978.

Optical Fibre Communications
Edited by M. J. Howes and D. V. Morgan

CHAPTER 2

Optical source devices

R. C. GOODFELLOW and R. DAVIS

2.1 INTRODUCTION

The ever-widening field of applications of fibre-optic communications has provided a driving force for the development of a large range of devices. The applications spectrum extends at the simplest end from automobile links which will need to be of a few metres length, handling up to perhaps kilobit data rates and costing pennies, to the hundreds of Mbit/sec, tens of kilometre links presently planned by telecommunications companies which will be very much more expensive. Many links with gigabit data rates could be made using many different independent wavelength channels in one fibre but applications do not yet appear to have emerged for such advanced systems. However, the capability is there.

Selection of the best source for a particular application is difficult at present because of the pace of developments in fibres and detectors as well as in the sources themselves. The paper by Kao and Hockham[1] which triggered fibre-optical communications, viewed the GaAs injection laser as the best prospective source. Some of the sources which now exist which would be suitable for fibre-optic systems are shown in Figure 2.1. The GaAs laser underwent several fundamental developments before its maturity towards a viable source. The first of these was the implementation of the double-heterojunction concept of Kroemer[2] and Alferov and Kazarinov[3] in a GaAlAs/GaAs laser structure by Panish *et al.*[4] This allowed continuous operation at room temperature which is almost essential in real systems. However, the double-heterostructure concept has also been applied to several other material systems and although presently less well established, (GaIn)(AsP)/InP, (GaAl)(AsSb)/GaAsSb, and other compound III–V lasers which operate continuously at room temperature over a range of wavelengths have been reported. But the appearance of the low-loss fibres now available has removed the necessity for the very high radiance properties of lasers and incoherent LED sources now also have a very real and useful role to play in the fibre-optic scene. The carrier confinement and waveguiding properties of heterojunctions assume less importance in LEDs, so homojunction and single-hetero-

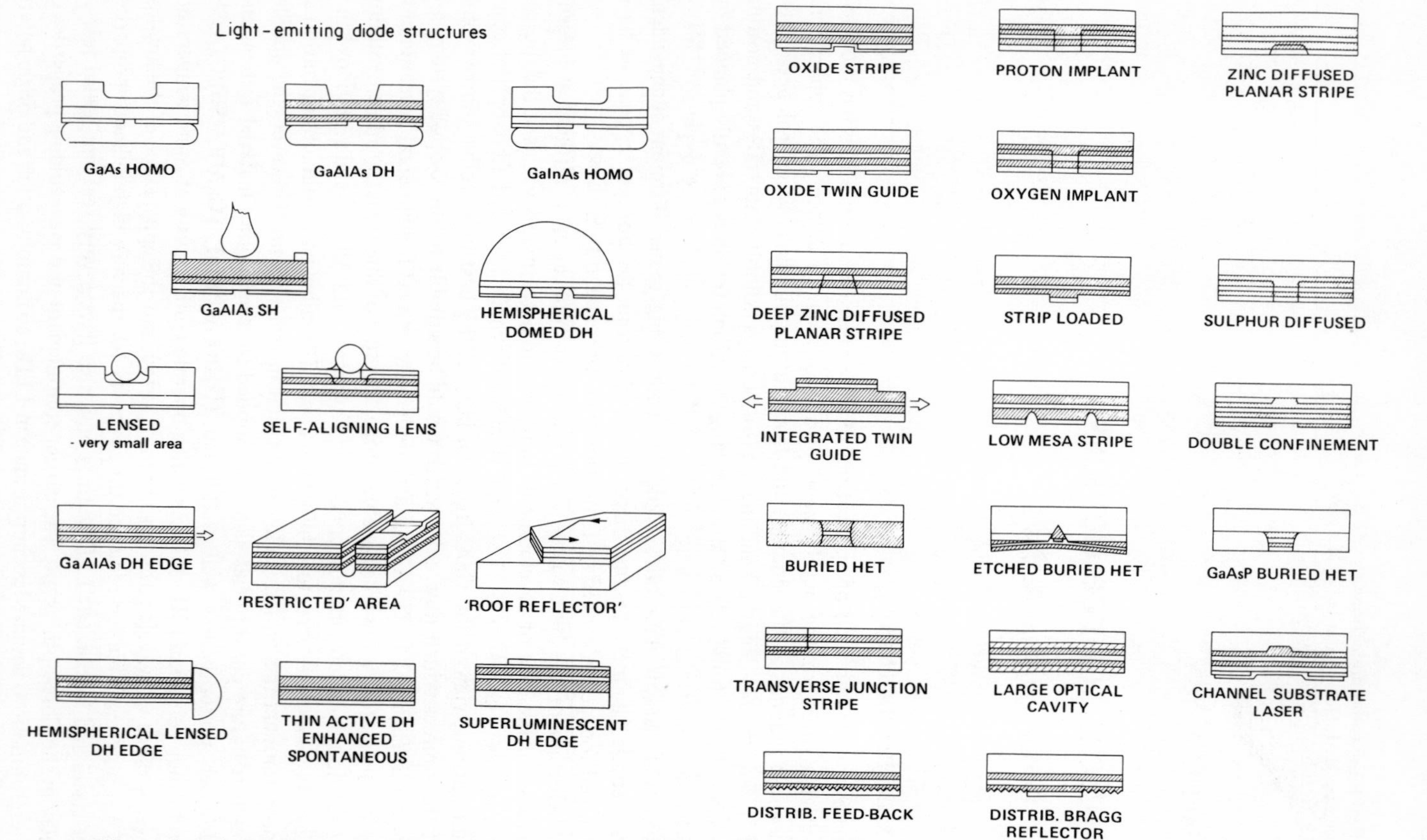

Figure 2.1. A pictorial summary of the LED and laser structures which have been proposed as fibre-optic sources

junction structures compete with double heterojunctions here. The simplest LED perhaps is the GaAs zinc-diffused LED which emits at 0.9 μm but this competes quite strongly with the more complicated but considerably more efficient GaAlAs LEDs, because of its simplicity and reliability.

The practical zinc-diffused GaAs LED is usually made in a so-called surface-emitting configuration but an alternative configuration allied to that adopted for lasers, the edge-emitting geometry, has been used with DH structures and is also well established at present. The debate as to which LED configuration is best suited to fibre communications is presently under way and examples of each type are commercially available for fibre-optic applications.

The disadvantages of incoherent LED sources for many applications arise from their broad spectral widths. As no optical resonant cavity exists for wavelength selectivity, the emitted photon energy depends on the distribution of energies of the recombining electrons and holes which is generally between 1 and 2 kT. Here k is the Boltzmann constant and T the absolute temperature at the junction. This spread in photon energy means that different regions of the emission spectrum propagate in a fibre at different velocities and results in a fibre material dispersion limitation on bit rate and bandwidth in real systems. The velocity spread depends on the value of the material dispersion for the core glass of the fibre which falls to zero in doped silica around 1.25–1.3 μm (and at much longer wavelengths in the newer-proposed crystalline halide glass fibres). Material dispersion considerations have provided a convincing argument for the development of LED sources for the 1.25–1.3 μm region and GaInAsP/InP high-radiance LEDs for this wavelength region have now come to the fore for wide-bandwidth, long fibre systems.

A main advantage of injection sources is that they can be directly modulated simply by varying the drive current applied. However, the transient response, wavelength, linewidth, noise and other device parameters vary, in some devices quite markedly, when the drive current is changed. This provides a justification for using an external optical switch or external optical modulator in series with the source, so that stability of the source can be maintained as far as possible. The direct modulation properties of injection sources are then almost disadvantageous. With such external modulators, gas lasers and crystal lasers which are difficult to modulate directly can be considered. Burrus[5] has described neodymium YAG fibre lasers which are end-pumped by GaAlAs DH LEDs, and emit continuously at 1.06 or 1.3 μm, and a lithium neodymium pentaphosphate crystal laser pumped by a GaAlAs laser or krypton gas laser has also been reported. The latter laser has been operated in an 800 MBit/sec fibre-optic link by Kimura *et al.*[6] Such crystal lasers can be made with very well defined narrow spectra. The disadvantages of using such lasers are that wavelength selectivity is limited to the available lasing transitions of the neodymium ion and total system efficiency taking in the crystal laser, the pump source and the modulator, does not compare favourably with that of present injection sources.

The fact that crystal lasers have been developed at all for fibre communications reveals the need for sources with very narrow spectra and well-defined centre wavelengths and this has been a target set in the development of injection-laser sources. One approach towards this specification has been to use distributed feedback (DFB) to give an injection laser cavity with a very narrow spectral linewidth, and CW GaAlAs DFB lasers which operate continuously have been reported. This development leads on to the integration of several sources onto one chip because of the planar nature of the DFB laser and the removal of the need for cleaved facets. However, the DFB approach has not been widely adopted for fibre source applications and the more straightforward cleaved Fabry–Perot cavity-structure lasers are considerably more prominent.

For many fibre communications systems reliability is an all-important factor and much of the device selection will be concentrated on the degradation characteristics of the devices. We include here a review of the degradation mechanisms pertaining to injection sources. It appears that similar degradation mechanisms apply both to LEDs and lasers but material deterioration in a small region of the active layer gives rise to a much greater loss in output in a laser than it does in an LED. Kobayashi[7] has demonstrated improved reliability by making a double-heterostructure device with its p–n junction remote from the active layer, which has the effect of gettering defects from the active layer, resulting in decreasing losses with time, and this is likely to result in very high-reliability devices for the future.

In this chapter we aim to consider some of the fundamentals, the structures, the degradation, operation and preparation of a wide range of LED and laser sources for fibre-optic communications systems.

2.2 MATERIAL CONSTRAINTS

2.2.1 What emission wavelengths are needed?

It is the fibre characteristics which mainly influence the choice of source wavelength. Fibre losses vary widely with wavelength and so also do their bandwidths. A graded-index fibre generally will only have the optimum profile for the design wavelength and so will probably have poorer characteristics such as a smaller acceptance cone, or a smaller bandwidth outside the design band of wavelengths. Furthermore, the bandwidth limitation imposed by material dispersion when using broad-linewidth sources can be quite severe if the source wavelength is far from the zero material-dispersion wavelength for the fibre. In Figure 2.2a we show the loss characteristics of a state-of-the-art low-loss fibre.[8,9] The germanium-doped fibres have less than 5 dB/km loss over the band 0.6–1.8 μm and beyond, and the phosphorus-doped fibres between 0.8 and 1.7 μm with an absorption peak of around 10 dB/km amplitude between about 1.25 and 1.45 μm.

In Figure 2.2b we show the refractive index versus wavelength characteristics for various doped silica fibres.[10] The zero-dispersion wavelength ranges around 1.26–1.31 μm depending on dopant and doping concentration. Fortunately, this co-

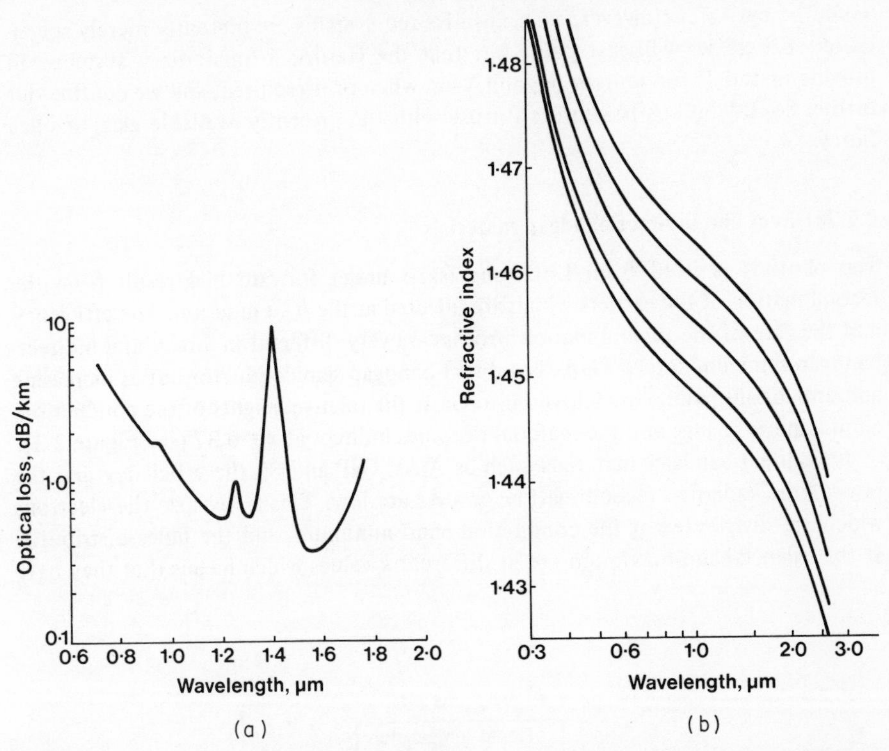

Figure 2.2. (a) The wavelength dependence of attenuation of a state-of-the-art doped silicon fibre (1978).[8] (b) The refractive index variation with wavelength for different fibre materials showing an inflexion at which wavelength the material dispersion falls to zero.

incides with a low loss region in these fibres. Sources which emit around 1.3 μm will therefore allow best exploitation of the fibre characteristics, although for many less demanding applications any source wavelength between 0.6 and 1.8 μm might be accommodated.

For the future, halide materials are presently being actively investigated for use as fibres in place of doped silica.[11,12] The halide materials such as TlBr and $ZnCl_2$ have absorption edges far into the infra-red, typically between 10 and 25 μm, and have zero-dispersion wavelengths between 1 and 10 μm. Limiting loss in such fibres will not be set by the fundamental absorption as in silica but by impurities and scattering. By operating at long wavelengths the Rayleigh scattering (which is proportional to λ^{-4}) could be very small, and Pinnow[11] has projected losses of less than 10^{-3} dB/Km for such fibres in the proposed operating wavelength region around 4 μm. Here the loss could be low, material dispersion could be small if the best material was utilized, room-temperature detectors could be used (whereas at longer wavelengths the leakage current in detectors becomes excessive), and the system would be eye-safe. However, such far-infra-red systems are presently merely speculation so here we will just point out that the GaInAsSb quaternary system will provide materials for sources beyond 4 μm when or if required, and we confine our further considerations to sources for use with the presently available glass or silica fibres.

2.2.2 Direct and indirect bandgap materials

The photons emitted from LEDs and lasers under forward bias result from the recombination of the carriers which are injected at the *p–n* junction. The efficiency and the rate of the recombination process are very different in direct- and indirect-bandgap semiconductors. GaAs is a direct-bandgap semiconductor but as increasing amounts of aluminium are alloyed into GaAs the relative heights of the conduction-band minima change and the material becomes indirect at $x = 0.37$ (see Figure 2.3).

In indirect-bandgap materials such as AlAs, GaP and Si, the efficiency and the rate of the radiative recombination process are low. This is because the electrons which are distributed at the conduction-band minimum, and the holes distributed at the valence-band maximum are at different *k*-values which means that they have

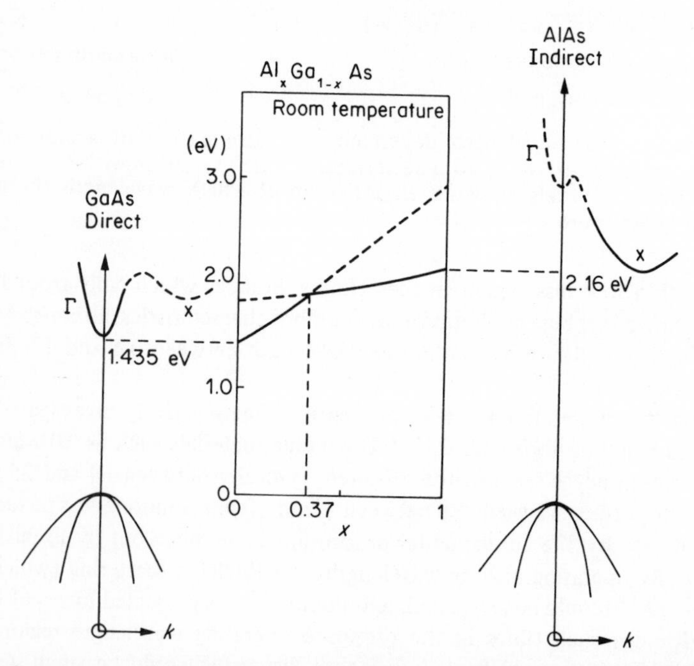

Figure 2.3. Direct and indirect semiconductors – $Al_xGa_{1-x}As$. Changes from a direct to an indirect material with increasing x. Derived from data of Casey and Panish[63]

different average momenta. Each recombination transition must therefore involve the emission or absorption of a momentum- (or k-) balancing phonon for momentum conservation. This momentum-balancing requirement is the reason for the low rate, and because alternative non-radiative recombination processes can occur it also explains the low efficiency of the radiative process. Indirect-bandgap materials tend to be used for the passive regions of devices where no recombination takes place and the low radiative efficiency is not important here.

The radiative efficiency is much higher and can approach 100% in direct-bandgap semiconductors such as GaAs, InP, GaSb and ternary and quaternary alloys of these compounds. In these materials the minimum of the conduction band occurs at the same k-value as the maximum in the valence band so that the electrons and holes have an overlapping range of k-values. In such cases, k- or momentum-conservation is satisfied by the high-probability 'vertical' electron–hole recombination processes and phonons are not necessarily involved.

Direct-gap materials are used for the highly efficient LEDs useful for fibre optics and, as the stimulated recombination necessary for laser action is also a 'vertical' process which has high probability in direct-gap materials, they are also chosen for lasers.

2.2.3 Band–band and band–acceptor recombination

Electron–hole spontaneous recombination in direct bandgap materials takes place at a rate R depending upon the numbers of electrons and holes available, i.e.,

$$R \propto np \qquad \text{or} \qquad R = Bnp \tag{2.1}$$

The constant B is the recombination coefficient and depends on the nature of the transition. If the hole is captured on an acceptor atom, it is localized in space so Δx is small, whereas Δx is large for mobile holes. The Heisenberg uncertainty principle states that $\Delta k \Delta x > 1$ so a small Δx implies a large Δk and vice versa. It follows that recombination via an acceptor is favoured over the band–band process because k-conservation is more readily attained. Rees and Milne[13] calculated the recombination coefficient for transitions via acceptors, B_A, for germanium, zinc and magnesium in GaAs and obtained values of 2.75×10^{-10}, 5.19×10^{-10} and 6.8×10^{-10} cm^3 s^{-1} respectively, whereas the band–band recombination coefficient B_V (for mobile electrons and hole recombination) is typically $1–2 \times 10^{-10}$ cm^3 s^{-1} in GaAs. The effective spontaneous lifetime for the radiative recombination τ depends on the number of acceptors ionized, p_V, and un-ionized, p_A, as

$$\tau = (B_A p_A + B_V p_V)^{-1} \tag{2.2}$$

The ratio p_A/p_V decreases as the acceptor level becomes shallower so at normal temperatures it is not possible to realize fully the benefits of the larger recombination constant of the shallow acceptors, and the effective recombination constant approaches the band–band value B_V.

For a fast radiative response we need to increase R, but as we have just discussed, B can only be increased by a relatively small amount. Now

$$R = \frac{\Delta n}{\tau}$$

by definition, where Δn is the excess electron concentration. We can therefore write Equation 2.1 as

$$\frac{\Delta n}{\tau} = Bnp \tag{2.3}$$

Often $p \approx N_A$ so we can say that the lifetime is inversely proportional to the acceptor concentration. If, conversely, we use undoped material and operate at very high current densities so that $\Delta n \approx \Delta p$ which is large, the lifetime can be made very small and a fast modulation rate achieved.

2.2.4 Heterojunctions

When double-heterojunction structures were first used to provide potential barriers in lasers, the threshold currents for lasing reduced by a factor of around 100. This resulted because injected carriers could now be confined and so the active-layer thickness was no longer related to the minority-carrier diffusion length. Relatively small currents could then produce the required quasi-fermi level separation (degree of population inversion) needed for gain. Furthermore, a smaller value of gain could produce lasing because of the smaller optical losses arising from the better waveguiding due to the dielectric discontinuity at the two hetero-interfaces and also due to the transparency of the layers cladding the active layer.

The heterostructure is also useful for LEDs because light can be extracted with almost no absorption loss through the cladding wide-bandgap layers either normally to the active layer as in Burrus LEDs[14] or a separate waveguide can be formed to make edge-emitting LEDs.

The heterojunction interface must comprise a very small density of defects such as misfit dislocations or inclusions for effective reflection of carriers without large losses due to nonradiative recombination at the interface states. The size similarity of the gallium and aluminium atoms makes Ga_xAl_{1-x} systems very useful because the lattice parameter is almost constant over the whole system. High-quality heterojunctions have been made in the GaAlAs system with aluminium concentration steps of 80% (atomic) which corresponds with potential barrier heights of about 25 kT.

The band diagrams for abrupt and graded multilayer heterojunctions are shown in Figure 2.4. In the case of the abrupt interfaces, the spikes or notches in the conduction band arise because GaAs and GaAlAs have different electron affinities or conduction-band positions with respect to the vacuum level. Charge must be transferred across the heterojunction on contact to preserve electrical neutrality or to

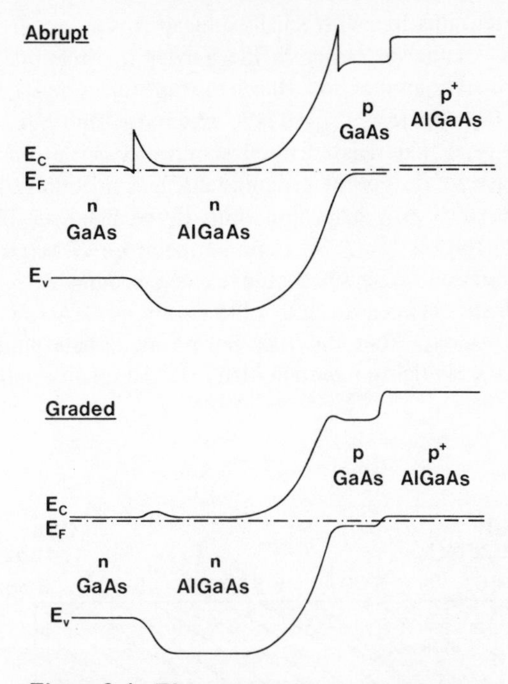

Figure 2.4. The energy-band diagrams of double heterostructures having abrupt and graded interfaces between layers[15]

bring the two fermi levels into line. The depletion of charge at the boundary reflects as a notch in the conduction band. These notches are potential barriers which theoretically can be as large as 15 kT.[15,16] In practice, the heterojunction is graded, albeit over a small distance, and this results in a reduction in the notch height. The natural tendency for the heterojunctions grown for injection lasers and LEDs to be sufficiently graded that the notches have little effect on device operation is very fortuitous and the heterojunctions are usually represented as step-like potential barriers with the notches often omitted.

Electrical properties of notch-free heterojunctions

With no applied bias voltage, the fermi levels of the two materials must be common when they are in contact. This determines how the bandgap difference is accommodated between the conduction and valence bands. The steps in the conduction and valence bands are barriers to the flow of carriers from the narrow to the larger bandgap material. The proportion of carriers which are able to surmount a barrier of height Δ is determined by the Boltzman factor $e^{-\Delta/kT}$.

At an n–p heterojunction with small-bandgap n-type material, the barrier for electron injection is enhanced whereas the barrier for hole injection is similar to that for the equivalent homojunction. Hence the hole injection efficiency is enhanced for this situation (by exp $[Eg_2 - Eg_1]/kT$, where Eg_1 and Eg_2 are the band gaps for the n- and p-type materials). Effective minority-carrier confinement can be achieved by the use of isotype heterojunctions, and injection of one carrier type only can be achieved at p–n heterojunctions if the bandgap differences are sufficiently large with respect to kT. At room temperature $kT \approx 0.025$ eV. For room-temperature laser action, $\Delta Eg/kT$ should exceed around 8.[17] Here ΔEg is the difference in bandgaps between the active and confining GaAlAs layers.

It follows now to ask, 'What materials can we make heterojunctions with?' For optical sources it is essential to maintain high material quality with a low density of

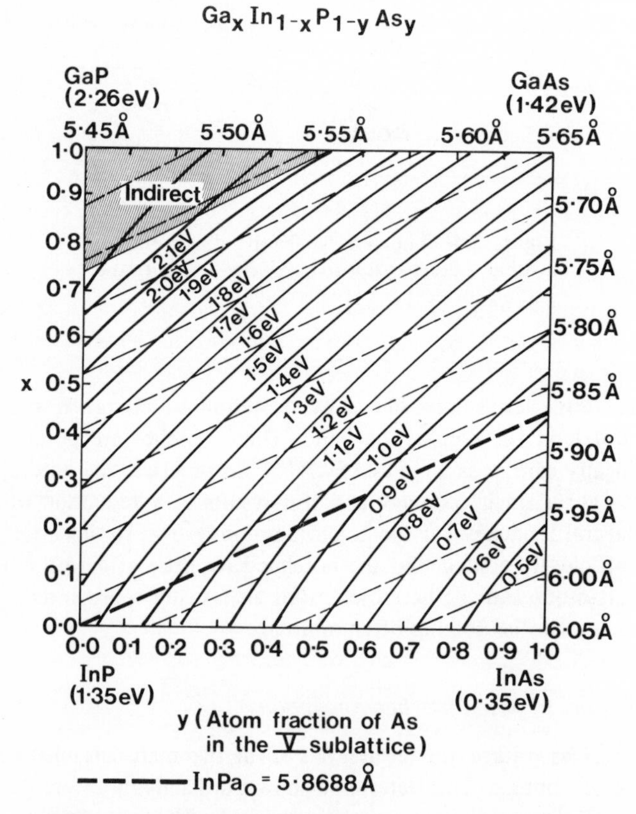

Figure 2.5. Diagram showing the relationship between lattice parameter (broken lines), bandgap (solid lines) and material composition (x and y axes) for the quaternary material $Ga_x In_{1-x} P_{1-y} As_y$ [20]

dislocations so that nonradiative recombination via interface states and states associated with dislocations is kept to a minimum. If the lattice parameter of the component materials of a heterojunction differ, the difference must be accommodated in the lattice by means of misfit dislocations. These misfit dislocations are needed even if the composition is gently graded from one value to another, although then the dislocations can be distributed through the graded layer. The only III–V materials for which the lattice remains essentially constant with composition x are the aluminium/gallium alloys, i.e. $Ga_xAl_{1-x}As$, $Ga_xAl_{1-x}P$, $Ga_xAl_{1-x}Sb$. If we wish to use other alloys we have therefore to go to quaternary alloys so that we can control both the lattice parameter and the band gap and then there are several III–V contenders. The most advanced are $Ga_xAl_{1-x}As_ySb_{1-y}$ and $Ga_xIn_{1-x}As_yP_{1-y}$. Up to now the GaAlSbAs system has been somewhat limited because of the need to prepare a substrate with the correct lattice parameter, and it has been necessary to grow a graded or stepped composition layer of $GaAs_ySb_{1-y}$ in which y is increased gradually to the value required for the double heterostructure. The value of y is then kept constant throughout the heterostructure. Such structures were produced by Sugiyama and Saito[18] and Nahory *et al.*[19] The graded layer accommodates the lattice-parameter change by incorporating the misfit dislocations.

By comparison, the growth of the $Ga_xIn_{1-x}As_yP_{1-y}$ quaternary is much more straightforward as the composition can be chosen so that InP or GaAs substrates can be grown on with lattice-parameter matching. A bandgap-lattice parameter-composition diagram for GaInAsP is shown in Figure 2.5 (after Hsieh *et al.*[20]). Now that the feasibility of quaternary systems has been demonstrated a very large field of new materials can be accessed. Glisson *et al.*[21] and Williams *et al.*[22] have computed the energy-bandgap and lattice-constant contours for most of the Group III–V quaternaries.

2.3 LIGHT-EMITTING DIODES

LED optics

The double-heterostructure (DH) semiconductor laser discussed in Section 2.4 is in many respects an ideal source for fibre-optic communication systems. The light-emitting diode (LED), however, is simple to construct, extremely easy to modulate and has well-defined reliability/degradation characteristics. These properties are attractive to the system designer for applications where a relatively broad emission linewidth is not a disadvantage.

Although minority-carrier injection-induced electroluminescence in LEDs can be of very high quantum efficiency, the measured external efficiency of rectangular shaped LEDs is only a few per cent. This is because only a small fraction of the light is radiated from the active region within the crystal towards the crystal surfaces at less than the critical angle, so the majority of the light generated is trapped by total internal reflection. The power coupled into a medium of low index n_0 from the planar face of an LED crystal of index n_x is approximately

$$P_{in}\, T\, \frac{n_0^2}{4n_x^2}$$

where P_{in} is the power generated internally and T is the transmission factor at the crystal/air interface. This is 1.3% of P_{in} for GaAs into air where $n_x \approx 3.5$ and $n_0 = 1$.

The power P coupled from an LED of radiance R into a fibre with acceptance angle Ω is

$$P = RA\Omega \approx RA\pi\,(\mathrm{NA})^2 \tag{2.4}$$

for small values of numerical aperture NA. A is the smaller of the areas of the emitter and the fibre core cross section. For highest coupled powers the product RA must be maximized.

Burrus-type surface emitters

One simple way of achieving a high radiance is to restrict the emission to a small region within a larger chip (see Figure 2.6a). This approach was first adopted by Burrus and Dawson[14] and high-radiance LEDs of this geometry are commonly termed 'Burrus' emitters. With such designs, low thermal impedances can be achieved at the active region allowing very high current-density operation. With the high internal efficiencies η_{int} usually achieved this results in high-radiance emission at the front surface. If it is assumed that the emission distribution from the active region is isotropic, the external emission distribution, after transformation by refraction from a high- to a low-index medium at the front face, is approximately lambertian (i.e. $I_\theta \approx I_0 \cos\theta$).

The axial radiance R is approximated by

$$R \approx \frac{T\eta_{int} JV}{4n_x^2\pi} \tag{2.5}$$

where T is the transmission factor which may take into account the absorption loss through the crystal and losses at the surface due to reflection at the dielectric/air interface, J is the current density and V is the junction voltage.

If the internal absorption is low and the reflection coefficient at the back crystal face very high, the radiance may be up to twice that predicted by Equation 2.5. This situation is approached when the double-heterojunction configuration is adopted as in Figure 2.6b whereas the relationship of Equation 2.5 would more accurately apply for the simple homojunction configuration of Figure 2.6a, where the internal absorption in the p-layer is relatively high.

With the fibre butt-coupled to the emitting area of the LED as shown, there is obviously no advantage in making the emitting area larger than the fibre core area

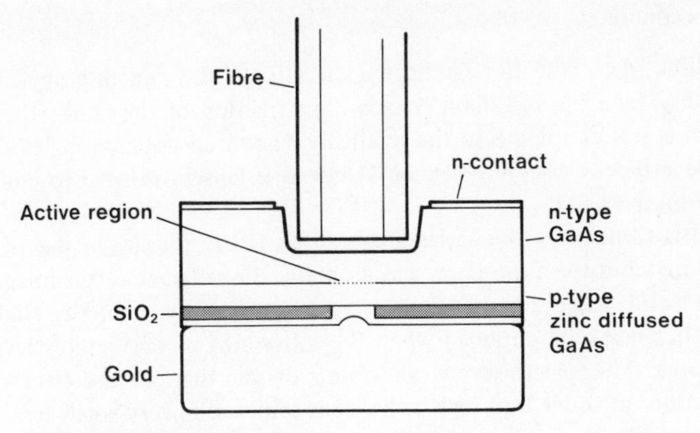

Figure 2.6.(a) Schematic section of the small-area, high-radiance gallium arsenide homojunction LED

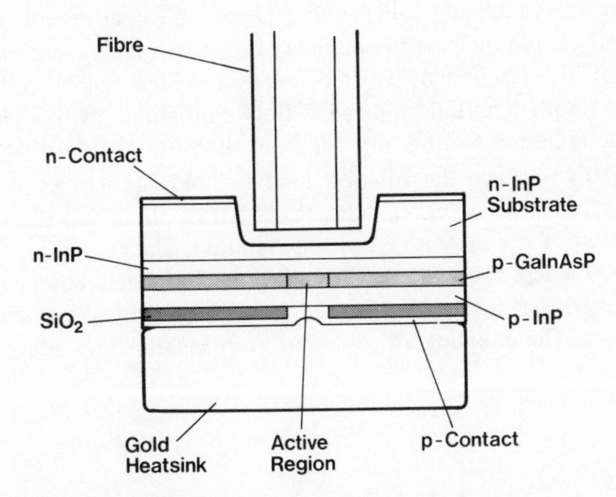

Figure 2.6.(b) Schematic of a double heterostructure high radiance LED structure. Internal absorption is small in this geometry

as light would only be coupled into the fibre cladding and ultimately be lost or otherwise wasted. With a step-index fibre, the emitting area may therefore be matched to that of the fibre core so that the active region will then have lowest thermal impedance and therefore a high maximum operating current.

With graded-index fibres, the effective acceptance numerical aperture is a maximum on the fibre axis and decreases radially so that increased butt coupling efficiency results with smaller sources but thermal impedance increases with decreasing source area so some compromise dimension must be determined.

2.3.1 Lens coupling

The situation where the fibre is butted directly to the emitting aperture of the device is very far from optimum since a large fraction of the lambertian emission distribution is not coupled into the relatively narrow acceptance angle of the fibre. Much more efficient coupling can be achieved if lenses are used to collimate the emission from the LED.

The aim is to magnify the active area so that firstly the size of the image of the active area matches the fibre core, and secondly the radiance of the image over the acceptance solid angle of the fibre is not significantly reduced. Highest lens-coupling efficiencies are obtained when the active area is considerably less than the fibre-core area. This places severe constraints on the thermal and electrical design of the structure in order that high radiance is achieved in very small-area devices at reasonable drive levels.

Several lens-coupling configurations have been adopted and are summarized in Figure 2.7. In all cases radiation within a larger acceptance solid angle is transformed by the lens to have a smaller solid angle which can be accepted by the fibre. The main limit to the increase in coupling efficiency which can be achieved with lenses is set by the fraction of emitted radiation which can be collected by the lens. With the simple lens and bulb-ended fibre configurations, the lenses collect a fraction of the radiation emitted into air from the front face of the LED. With the sphere-lens configurations, the adhesive used to hold the lens in place acts as an immersion medium with a higher index than that of air and so the lens acts on a larger proportion of the internally emitted radiation. The best coupling efficiencies will be achieved where the LED–lens interface is completely eliminated, as in the integral-lens LED, because the lens can act upon a large proportion of the internally generated power. The coupling efficiency of both separate and integral sphere lens

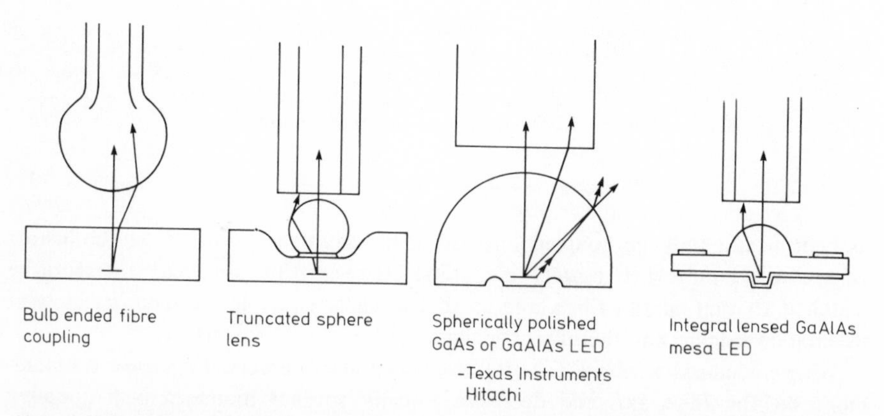

Figure 2.7. Some practical methods by which the power coupling between LEDS and fibres has been increased[24]

configurations has been calculated by Abram, Allen and Goodfellow.[23] Alferov *et al.*[24] have described a GaAlAs DH LED with a novel integral-lens geometry. This was formed by the growth of GaAlAs in near-hemispherical holes in a ⟨111⟩ oriented GaAs substrate which is etched away later in the process.

Improvements in the coupling efficiency compared with the butt-coupling case (with 0.16 NA fibres) of around 3–5 times are typically obtained with bulb-ended fibres, and of 18–20 times with sphere-lens coupling. However, in principle, gains of several hundredfold are feasible with the integral-lens geometry if active regions of just a few micrometres diameter are realized.

From thermodynamic considerations it is not possible to increase radiance by means of lens systems, but magnification without loss of axial radiance at the image can be readily achieved if the acceptance aperture of the lens is sufficiently large. The advantages which can result from sphere-lens coupling depend on the operating conditions one allows for the LED.

Let us assume that the reliability of the LED depends on the junction temperature, and we therefore set a maximum junction temperature T_{max}. For small-area devices the thermal inpedance Z_T of the active area is due mainly to spreading, and is approximated by

$$Z_T \propto \frac{1}{r} \tag{2.6}$$

where r is the radius of the active area.
But

$$Z_T = \frac{(T_{max} - T_0)}{J_{max} \pi r^2 V}$$

where T_0 is the heatsink temperature and T_{max} is the maximum junction temperature and this is proportional to $1/r$ from Equation 2.6.
Hence

$$J_{max} \propto \frac{1}{r} \tag{2.7}$$

which from Equation 2.5 is proportional to the radiance. Hence the radiance is inversely proportional to the radius of the active area.

Maximum coupled power (with this assumption) will be achieved with a high-gain, integral-lens geometry and the smallest active area for which high magnification without significant radiance loss can be achieved. So small active areas can lead to high launch powers.

Let us assume as a second case that reliability is determined by current density, and therefore we set a maximum current density. The maximum coupled power with the Burrus LED geometry is then achieved when the active area matches that of the fibre core and is butt-coupled to it. Here lens coupling with high gains gives a great improvement in efficiency over the butt-coupled case, but no increase in launched power.

In practice, the device lifetime of Burrus emitters does not vary very dramatically with operating current density, and quite acceptable lives are achievable with small area devices operating at around 100 kA/cm^2. We presently are not certain of the real limitations on current density with respect to reliability, but when the excess carrier concentration approaches the condition where population inversion is achieved then the carrier lifetime varies dramatically with drive condition and performance becomes nonlinear at high modulation rates.

The excess carrier concentration can be reduced by increasing the thickness dimension of the active region but then absorption in the layer becomes significant. Also, it is more difficult to collimate the radiation using lenses because the object distance becomes diffuse. The latter effect can be avoided by enclosing the active region within a waveguide so that optically the source is located at the end of the waveguide. Then the collimating lens is focused on the end of the waveguide. This idea leads conceptually to the 'edge-emitting LED'.

Edge-emitting LEDs

In this geometry the edge of an emitting *p–n* junction is viewed as in conventional injection lasers. This is equivalent to a surface emitter with a very thick active region with the significant difference that a waveguide around the active region channels radiation to the emitting face of the device (see Figure 2.8). Current flow is perpendicular to the junction plane and the active area usually is much larger than the emitting aperture. With low self-absorption in the active layer, the edge radiance can far exceed the radiance viewed perpendicular to the junction. Zargar'yants *et al.*[25] demonstrated the efficiency of the edge-emitting geometry using a zinc- and silicon-doped compensated active region in a homojunction structure which incorporated no special waveguiding. Ettenberg, Hudson and Lockwood[26] followed with a GaAlAs double-heterojunction version which utilized the refractive-index difference of GaAs and GaAlAs to guide radiated energy diverging from the junction plane. Internal absorption in such devices reduces the external efficiency and Kressel and Ettenberg[27] showed how the efficiency could be enhanced by using a very short structure. An alternative method of reducing the absorption is to use a very thin active layer and to enclose this within a low-loss waveguide.[28,29] In the last method several advantages were reported for the very thin (0.05 μm) active layers employed. Such thin layers support only a few guided modes to which 'spontaneously emitting electrons are coupled much more strongly than to the unguided ones' as shown theoretically in detail by Wittke.[30] This enhancement of the emission in the guided modes gives a much more sharply peaked far-field emission pattern and gives improved butt coupling to low-NA fibres. The high excess-carrier concentration in the thin active region results in the onset of bimolecular recombination at low currents and hence a short, square-root

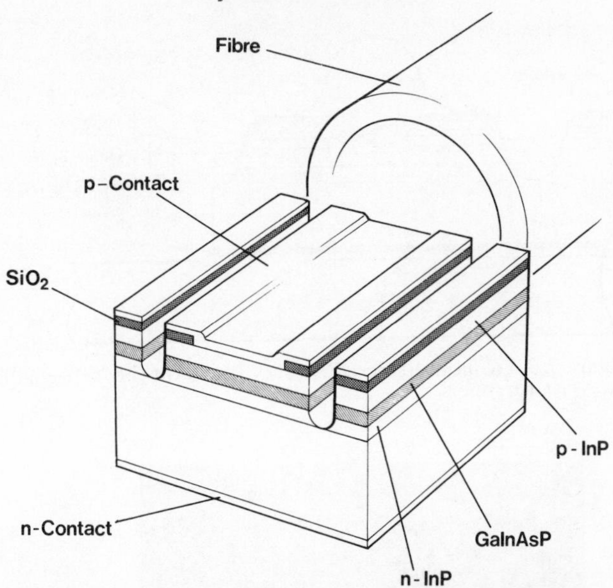

Figure 2.8. Schematic of an 'edge-emitting' LED. Light is generated in the active (GaInAsP) layer and guided to the emitting facet

current density–dependent lifetime with a correspondingly wide bandwidth capability.

Seki[28] assessed microlens coupling with edge emitters and achieved lens gains of around 5 times or so. For high coupling efficiency it is essential to keep the emitting aperture small; hence a narrow waveguide which gives confinement both in the vertical plane by means of the guiding layers and horizontally by means of a ridge mesa or other side-wall reflectors would be an ideal optical configuration.

2.3.2 LED arrays

When using ribbon assemblies made up of several fibres configured accurately within the cable, it is more convenient to couple to arrays of diodes rather than individual devices. Horiuchi *et al.*[31] reported a 'self-aligning lens' LED and later Sugimoto *et al.*[32] described how an array of such devices could be constructed. This is shown in Figure 2.9. This structure uses a 'layer-up' planar geometry which has the advantage that the positive terminal of each diode can be electrically isolated from the others – by mesa etching in the example reported. However, the layer-up geometry and the transverse surface electrodes limit the power dissipation and current drive capability of this structure.

The Burrus geometry and edge-emitting geometries can also be designed and operated in array form by using a combination of mesa etching and proton implan-

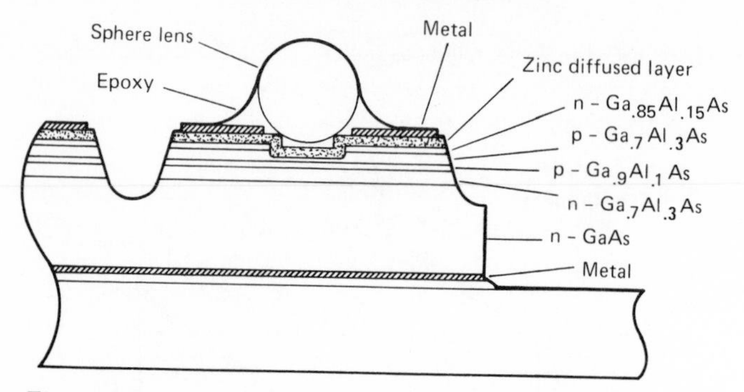

Figure 2.9. A monolithic LED array based on the 'self-aligning lens' DH LED[32]

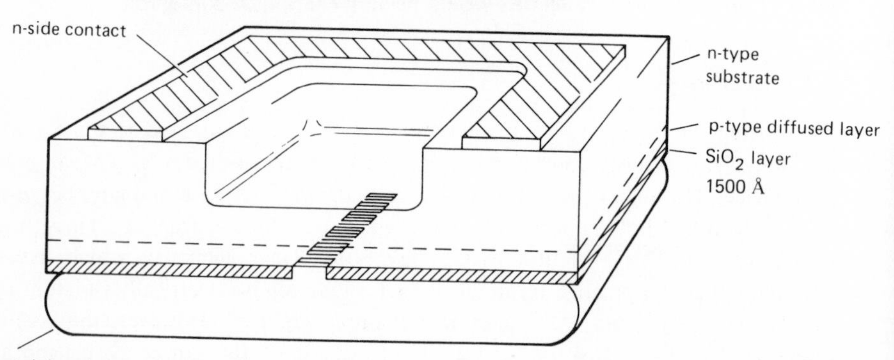

Figure 2.10. An array of high-radiance GaAs homojunction LEDs based on the small-area, etched-well design approach. (Reproduced by permission of Plessey Co. Ltd.)

tation to produce the desired active area isolation. Such an approach has been used in the fabrication of line-surface emitter arrays which have a layer-down configuration for use in shaft encoders, and an example of such a device is shown in Figure 2.10.

An alternative multichannel scheme utilizes several different wavelength sources coupled to a single fibre and the signal is demultiplexed using a prism interference filter or diffraction grating to discriminate the different wavelengths at the receiver. This wavelength multiplexing scheme has been demonstrated both in laser and LED systems,[32,33] and may well provide a relatively simple way of increasing the data rate capability of fibre systems. Arrays of LEDs, each of which emit at a different wavelength, have yet to be realized and would present a formidable materials problem, but the single-wavelength LED array for use with fibre arrays is now practical.

2.3.3 Modulation of high-radiance LEDs

If the first requirement for a fibre-optic source is that it should couple large amounts of power into fibres, surely the next most important is that it must be capable of direct modulation at sufficiently high switching rates. The modulation response $R(\omega)$ under carrier-lifetime limited conditions was calculated by Liu and Smith[34] as

$$R(\omega) = \frac{1}{\sqrt{1 + (\omega\tau)^2}} \tag{2.8}$$

This expression was derived for small-signal conditions but holds quite well for large current swings if:

(a) the carrier lifetime τ can be considered constant; and
(b) if the space-charge capacitance or parasitic junction capacitance is small in comparison with the diffusion capacitance for the range of currents over which the device is modulated.

Harth, Huber and Heinen[35] calculated that this expression holds true also for LEDs where excess carriers are confined close to the p–n junction by heterojunction potential barriers as in single- and double-heterojunction devices. Rees[13] calculated the effects of the electric field present in graded junctions in which the injected minority carriers are confined close to the p–n junction and found significant deviation from the fall-off predicted by Equation 2.8.

The radiative recombination rate R is proportional to the electron and hole densities in the active region. Hence the modulation characteristics depend on the active-region doping level, the heterojunction confinement and the drive level – in fact all of the factors which control the excess carrier concentration in the active region.

The radiative recombination coefficient B varies with material since it derives from the matrix element for the overlap of the electron–hole wave functions. In

GaAs it is approximately 1×10^{-10} cm^3/sec and in InGaAsP is around 3×10^{-10} cm^3/sec for the InP lattice-matched composition which emits at 1.3 μm. These values for B are the authors' best estimates derived from zinc-doped GaAs homojunction LEDs and zinc-doped double-heterostructure GaInAsP/InP LEDs.

The carrier lifetime τ_{rad} is given by Equation 2.3 as

$$\tau_{rad} = \frac{\Delta n}{R} = \frac{\Delta n}{Bnp}$$

If $\Delta n \approx n$ as in undoped or p-type material,

$$\tau_{rad} \approx \frac{1}{Bp} = \frac{1}{B(N_A + \Delta p)} \tag{2.9}$$

The injected carrier density Δp $(= \Delta n)$ is $\Delta p = J\tau_{rad}/ed$ where J is the injected current density and d is the active-layer thickness.

Thus

$$\frac{1}{\tau_{rad}} = BN_A + B\frac{J\tau_{rad}}{qd} \tag{2.10}$$

which is a quadratic whose solution is

$$\tau_{rad} = \frac{edN_A}{2J}\left(\sqrt{1 + \frac{4J}{qdN_A^2B}} - 1\right) \tag{2.11}$$

When J/edN_A^2B is large, that is when the active region doping is low and the operating current density high, the lifetime is

$$\tau_{rad} \approx \sqrt{\frac{qd}{JB}} \tag{2.12}$$

Therefore in a very thin undoped active region, a very short carrier lifetime and hence a wide bandwidth capability can be achieved. However, under large signal conditions harmonic distortion will arise due to the current-dependent lifetime.

When $4J/qdN_A^2B$ is small, the carrier lifetime is given by

$$\tau_{rad} = \frac{1}{N_AB} \tag{2.13}$$

and is independent of the current density. With a doping level $N_A \approx 10^{19}$ cm^3 in a GaAs active region of thickness 1 μm, the lifetime remains essentially constant up to $J = 40$ kA/cm^2.

The modulation response given in Equation 2.8 is optimistic if the space-charge capacitance of the junction is large by comparison with the diffusion capacitance of the diode.

The space-charge or depletion-layer capacitance is usually represented as

$$C_s = \frac{C_0}{(1 - qV/Eg)^n} \tag{2.14}$$

where V is the voltage across the junction and C_0 is the zero-bias capacitance, and E_g is the bandgap in the depleted region. Here the index n is between ½ and ⅓ dependent upon junction parameters. This equation suggests that the capacitance tends to infinity at high forward bias, but several workers, e.g. Poon and Gummel,[36] have estimated that the space-charge capacitance saturates at around $4C_0$ in silicon under heavy forward bias. We assume that a similar saturation occurs in other semiconductor junctions. The diffusion capacitance C_{diff} for a junction of area A forward-biased with a current density J is

$$C_{\text{diff}} \approx \tau_{\text{rad}} A \frac{\mathrm{d}J}{\mathrm{d}V} = \frac{Aq}{kT} J \tag{2.15}$$

where $\omega\tau_{\text{rad}} < 1$, and this is shunted by the diffusion resistance R_{diff} where

$$R_{\text{diff}} = A \frac{\mathrm{d}V}{\mathrm{d}J} \approx \frac{kT}{q} \frac{A}{J} \tag{2.16}$$

again where $\omega\tau_{\text{rad}} < 1$.

At low frequencies the diffusion capacitance is proportional to J and the diffusion resistance is inversely proportional to J. This implies that the characteristic response time (obtained from the product of C_{diff} and R_{diff}) is independent of J and is proportional to τ_{rad}. τ will obviously be longer if the space-charge capacitance exceeds C_{diff}. In a junction structure which has a carrier lifetime of 1 nS, the diffusion capacitance begins to dominate the space-charge capacitance at around 40 A/cm^2. Normal operating current densities are typically 1 to 40 kA/cm^2 so the space-charge frequency-limited current range occurs at very low drive levels only. However, in devices with large passive areas of p–n junction (such as oxide-isolated device structures) the current level at which the space-charge capacitance ceases to limit the response, occurs at current levels which are several times higher than the value calculated above. Lee[37] calculated the delay time and rise time of an LED with a large region of passive p–n junction contributing a large excess space-charge capacitance, for conditions where the current is pulsed from zero to a constant level from a finite source impedance. The delay occurs because no significant diffusion current flows until the space-charge capacitance is charged to the junction turn-on voltage (which is close to the active region bandgap voltage).

It follows from Equation 2.11 that high modulation speeds can be achieved:

(a) by operating thin undoped confined active regions at high current densities; and
(b) by heavily doping the active region.

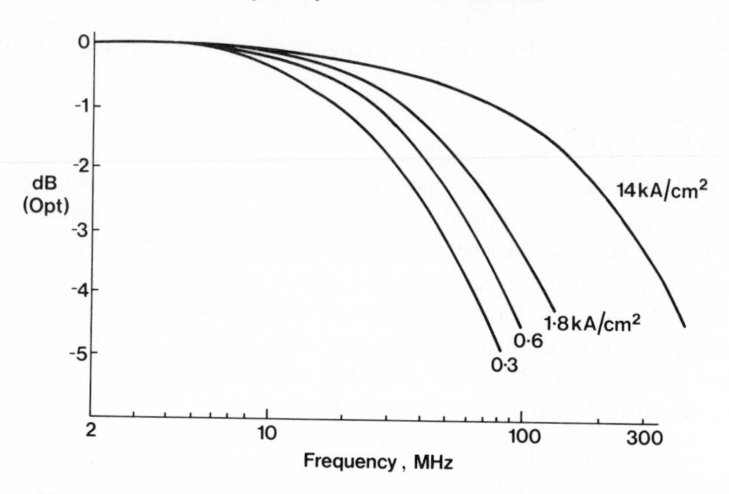

Figure 2.11. Variation of frequency response with drive current density for a GaInAsP/InP high-radiance, small-area LED with 1 μm active layer and 50 μm diameter active area

In both cases, reducing the effect of the space-charge capacitance of passive junction area in the device reduces the current drive required for good frequency characteristics.

Ettenberg *et al.*[29] demonstrated the undoped active-retion approach using an edge-emitter configuration but this also works in surface emitters and Figure 2.11 shows the varying bandwidth with current density on GaInAsP–InP high-radiance LEDs.[38] The low active-region doping results in very high recombination efficiency and high speed at sufficiently high current drive.

Highly doped active regions have given the widest bandwidth characteristics to date. GaAs homojunctions prepared by vapour-phase epitaxy with a very abrupt doping profile at the junction and processed into Burrus-type surface emitters have been modulated well beyond 1 GHz. These fast devices have smaller efficiencies than slower, more lightly doped devices and this is explained as arising from the poorer quality of heavily doped material. The statistical nature of the distribution of the acceptors implies the presence of inclusions and precipitates at which non-radiative recombination processes take place. Nevertheless, devices capable of wideband operation which can launch acceptably high levels of optical radiation into fibres have been made with quite straightforward homojunction geometries.

In Figure 2.12 we show the frequency-response characteristics of some heavily doped homojunction LEDs made in GaAs. Powers of 125 μW have been launched into 0.16 NA in silica 85 μm core fibres at 400 MHz with such devices.

The frequency response of homojunction LEDs can be of the form of B of Figure 2.12 because of emission due to the recombination of holes on the *n*-type

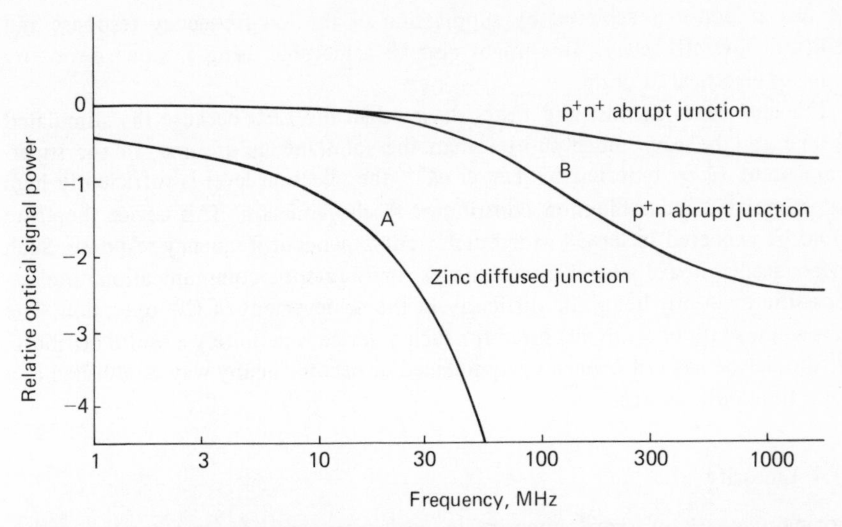

Figure 2.12. Frequency response characteristic of GaAs homojunction LEDs. Curve A shows a device in which electron injection dominates. Curve B shows a device where both electron and hole injection are significant

side of the junction as well as due to electron recombination in the *p*-type material. The characteristic can then be interpreted as made up of two responses of the form of Equation 2.1 with two lifetimes τ_n and τ_p. The low-frequency fall-off of curve B is due to the longer hole recombination lifetime τ_p on the *n*-side of the junction. Increased donor doping on the *n*-side shifts the fall-off to higher frequency and also decreases the hole current density J_h, that is it increases the injection ratio J_e/J_h and increases the wide bandwidth emission component. From simple diode theory,

$$\frac{J_e}{J_h} \approx \frac{D_n N_D L_p}{D_p N_A L_n} \tag{2.17}$$

By employing a heterojunction, this ratio can be modified by exp $(\Delta Eg/kT)$ but bandgap narrowing which occurs at high doping levels in GaAs also tends to enhance the electron injection in high acceptor-doped homojunctions.

An alternative approach to the achievement of a wide bandwidth LED was demonstrated by Heinen, Huber and Harth,[39] who used a short-based diode approach. In their device the active layer was made thin and was contacted directly. In such a device the frequency response is controlled by the nonradiative lifetime τ_{nr}. As the internal efficiency

$$\eta_{int} = \frac{\tau_{nr}}{\tau_r + \tau_{nr}} \tag{2.18}$$

the fast response is achieved by suppression of the low-frequency response and results in low efficiency. This might also be achievable using a 'slow' device by means of electrical filtering.

The laser is an intrinsically faster device than the LED because the stimulated lifetime can be very much shorter than the spontaneous lifetime. In the super-luminescent LEDs reported by Lee *et al.*[40] the injection level is sufficiently high that stimulated recombination contributes to the emission. This device therefore would be expected to have a wide but current-dependent frequency response. Such devices seem to have several disadvantages for fibre-optic communications applications, the main one being the difficulty in the achievement of CW operation. The narrow linewidth of 5 nm measured for such a device is definitely a useful attribute, but the device has not been actively pursued or become in any way established as a useful fibre-optic source.

2.3.4 Linearity

The maintenance of signal integrity in analog transmission systems puts severe constraints on the linearity of the optical source. For example, in frequency-multiplexed systems a high degree of linearity is required in order to minimize interference between the separate channels due to the generation of intermodulation products. Also in baseband TV systems of 'broadcast standard', extremely low amplitude and phase distortions must be maintained. By comparison with most injection lasers, LEDs are intrinsically very linear devices but nevertheless they do exhibit a degree of nonlinearity which varies with LED configuration. Ozeki and Hara[41] measured the second- and third-order intermodulation products of DH Burrus-type GaAlAs LEDs by modulating at three frequencies f_1, f_2, f_3. To determine the linearity of the detector and amplifier the output of three separate LEDs driven at f_1, f_2 and f_3 were combined optically to avoid the mixing in the source. The second and third intermodulation products due to the complete detection system were assessed as −60 dB and −118 dB. The measurement was then carried out with the electrical signals combined on to one LED. The fundamental and intermodulation (IM) products were arranged to be between 1 and 10 MHz so that the signal levels could be compared simply with a spectrum analyser. The performance summary indicates that at 95 mA bias with 75% modulation depth the second and third IM products are −40 and −70 dB of the fundamental respectively.

A linear source which produces very low-level IM products is necessary for transmission of multiple video channels on different carrier signals to minimize interference between channels. However, an assessment of IM products alone is insufficient for high-quality video transmission because no indication of the phase distortion due to the source is obtained. The major TV transmission systems (NTSC, PAL and SECAM) use a signal which comprises a chrominance signal which is modulated both in amplitude and phase, superimposed on a luminance signal.

The phase of the chrominance signal relates to the colour, and the amplitude of the chrominance signal relates to the saturation of that colour. Phase distortion thus disturbs the colour reproduction. Superlinearity or saturation in the light-current characteristic of the LED gives rise to changes in the chrominance amplitude as the luminance is varied, so that distortion also arises in the amplitude of the chrominance signal. The phase- and amplitude distortion are quantified by the parameters differential phase (DP) and differential gain (DG). The magnitudes of DG and DP are defined as the largest changes in the amplitude and phase of the subcarrier component as the luminance is varied between the black and white levels. That is,

$$\text{DG is the maximum value of } \left[\frac{\left.\frac{dP}{dI}\right|_{I_0'} - \left.\frac{dP}{dI}\right|_{I_0}}{\left.\frac{dP}{dI}\right|_{I_0'}} \right] \times 100\% \qquad (2.19)$$

and

$$\text{DP is the maximum value of } [\phi(I_0') - \phi(I_0)] \qquad (2.20)$$

as the currents I_0, I_0' are varied over the working current range of the LED. Here P is the optical power at current I and $\phi(I)$ is the phase delay introduced by the diode at bias I. Asatani and Kimura[42] have related the contributions to DG and DP due to the following factors:

(a) The varying impedance of the LED with bias current.
(b) The injection efficiency which may vary with current under high level injection conditions.
(c) The excess current due to space-charge recombination.
(d) The carrier confinement breakdown which may occur at heterojunctions with small step height ΔEg.
(e) Changes in quantum efficiency with current.
(f) The temperature dependence of radiative recombination rate.
(g) The change in modulation bandwidth with bias.
(h) Changes in absorption coefficient due to increases in the free carrier density in the active layer under injection conditions.

The relative contributions to DG and DP made by factors (a)–(h) are different for homojunction, single-heterojunction and double-heterojunction diodes and also depend on whether the device is driven so that the excess carrier concentration significantly alters the majority carrier concentration in the active layer. This condition is brought about when

$$J \approx \frac{qN_A L_n}{\tau_{rad}} \qquad (2.21)$$

for homojunctions with uniform p-doping and

$$J \approx qN_A d/\tau_{\mathrm{rad}} \tag{2.22}$$

for DH diodes, where N_A is the acceptor doping level in the active region, L_n is the diffusion length for electrons in the active region, τ_{rad} is the carrier lifetime and d is the width of the DH active layer. For 3×10^{18} cm^{-3} p-type, doped active layers,

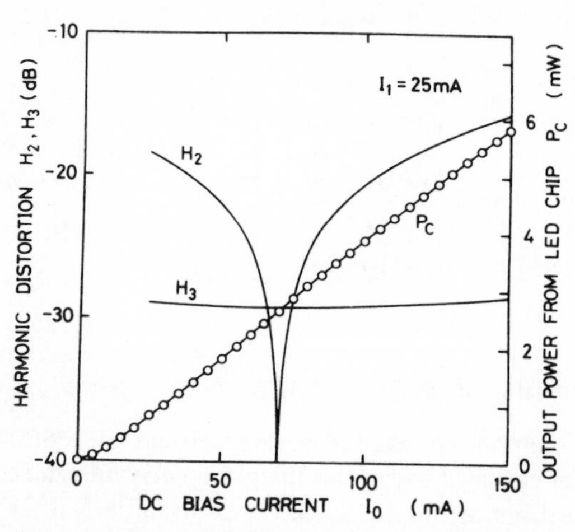

Figure 2.13. (a) Variation of harmonic distortion with bias current for a single heterostructure LED. H_2 and H_3 are the second and third harmonic powers. I_1 is the modulation current

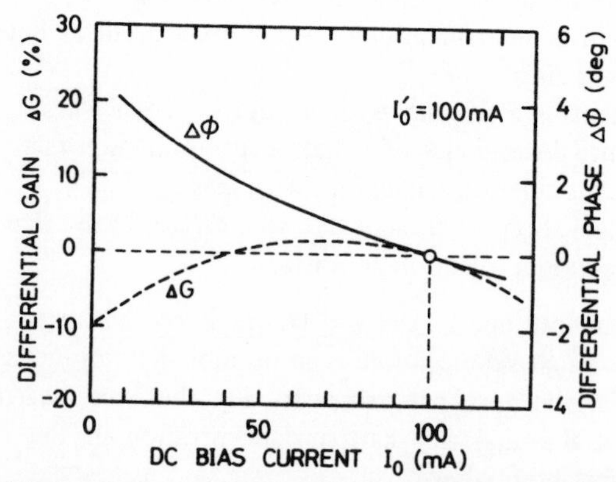

Figure 2.13. (b) Variation of differential gain and phase (ΔG and $\Delta\Phi$) versus d.c. bias current for a constant pulse current I_0' of 100 mA. The LED is an SH GaAlAs small-area device[43]

$J \approx 120$ kA/cm^2 in homojunction diodes with $L_n \approx 2$ μm. Normal operating current densities are $\lesssim 10$ kA/cm^2 so that the DG and DP distortion due to factors (b), (e), (g) and (h) above are likely to be small in devices with heavily doped active regions. If the diode is not modulated to very low bias currents the distortion due to the space-charge recombination current (c) can be avoided. In DH devices with barrier step heights large enough to avoid carrier spillage, (d) can be avoided although the effect of the finite interfacial recombination velocity should be taken into account as pointed out by Asatani and Kimura.[42]

It therefore appears that the major contributions to DG and DP in a well-designed diode will arise from (a), the current-dependent impedance, and (f), the temperature dependence of radiative recombination rate, which determines speed and efficiency.

The space-charge capacitance, which is large in oxide-isolated devices, and is bias-dependent, gives rise to turn-on delays and also gives rise to a non-ideal frequency response characteristic because of its dynamic effect on the current spreading. It will also contribute to DG and DP and needs consideration.

Abe *et al.*[43] measured both the harmonic distortion and DP and DG as a function of bias current for single-heterojunction GaAlAs oxide-isolated LEDs and the results are shown in Figure 2.13.

With the PAL (phase alternate line) colour TV transmission system, total distortions in DG and DP of 25% and 25° can be tolerated over the whole transmission, the large DP being accommodated because of the phase-error cancellation technique in this system. However, typical specification for DG and DP in CCTV links are 10% and 10° in the PAL system and 10% and 5° in the NTSC system. Local links demand maximum distortion of 1% and 1°. The DG of presently available GaAs LEDs ranges up to 20% and the DP up to 10°, so some correction of the device nonlinearity is often necessary.

2.3.5 Compensation of LED distortions

Some of the linearization schemes which have been adopted are shown in Figure 2.14. Ueno and Kajitani[44] reported reductions of 12 dB and 4 dB in second and third harmonic distortion using the negative-feedback approach of Figure 14a. This technique is difficult to implement effectively because of the delay in the feedback circuit and the need for very wideband amplifiers for good phase performance (phase delay $\Delta\phi$ at frequency f is $\Delta\phi = \tan^{-1}(f/f_T)$ where f_T is the unity gain frequency of the amplifier).

The feed-forward compensation scheme described by Strauss and Szentesi[45] (Figure 2.14b) can in principle accommodate all forms of distortion. This scheme is suitable for baseband video transmission, where the efficiency variation due to temperature variations of the junction must be corrected for, and for frequency-multiplexed links where the linearization must function over a wide frequency range. The scheme requires two similar LED sources. The distortion on the optical signal due to the first LED is detected by comparing the output of the monitor with the drive input, and a compensating optical signal derived from this is launched some way down the fibre using a second LED and fibre-branching coupler.

The slightly more simple quasi feed-forward technique described by Strauss and Szentesi[45] is a compromise scheme which involves less fibre 'plumbing'. Here, two closely similar LEDs are necessary. The nonlinearity measured on the output of the first device is used to derive the predistortion signal required to linearize the second LED, which is then coupled to the transmission fibre.

The phase-shift modulation scheme of Strauss and Szentesi[46] illustrated in

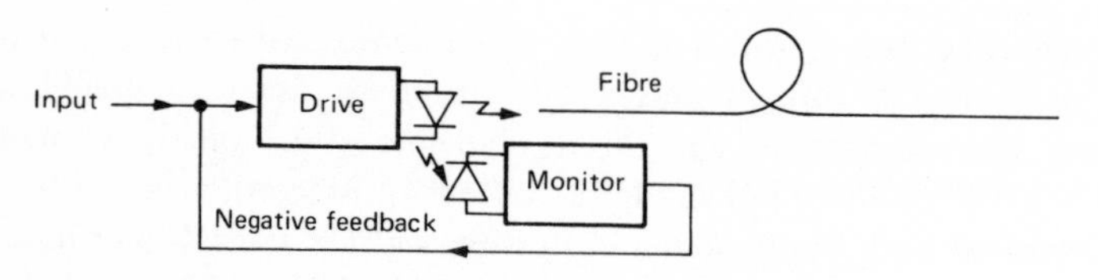

(a) Schematic of negative feedback linearized driver

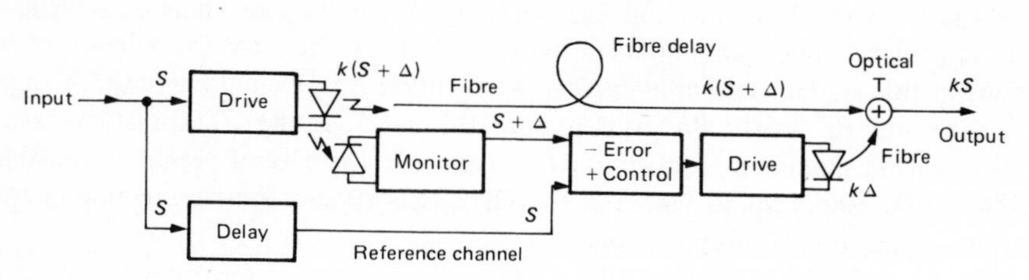

(b) Schematic of feedforward compensation drive scheme (after Strauss and Szentesi[45])

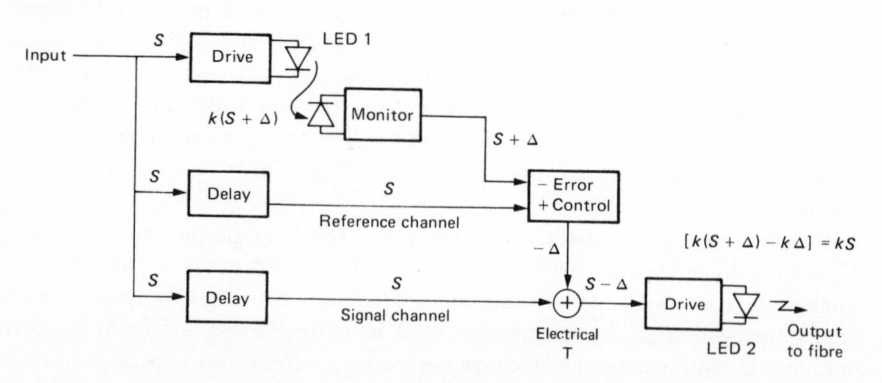

(c) Schematic of quasifeedforward compensation system (after Strauss and Szentesi[45])

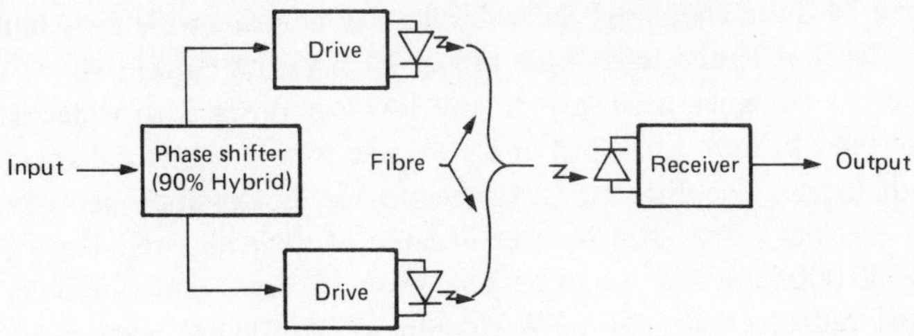

(d) Block diagram of phase shift modulation scheme for 2nd harmonic reduction (after Strauss and Szentesi[45])

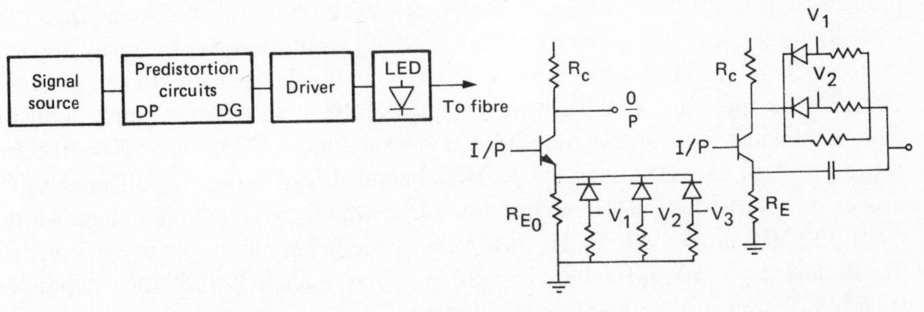

(e) Linearization by the introduction of predistortion—block diagram and DP and DG predistortion circuits (after Asatani and Kimura[47])

Figure 2.14. Summary of linearization schemes for LED drivers.

Figure 2.14d permits the elimination of the second harmonic using two diodes with identical characteristics. Higher-order harmonics can also be eliminated, the required number of sources being doubled for each extra harmonic.

The predistortion circuits of Asatani and Kimura[47] provide the selective compensation of DG and DP for different current regimes and hence linearization can be achieved for large-amplitude modulation. This technique does not require the use of extra optical devices and is probably the most straightforward method. However, it would be somewhat ineffective for baseband video transmission because of the rather slow and accumulative nature of temperature variations at low frequencies, and like all but the negative feedback technique it cannot compensate for the changes in device characteristics which may occur during the operating life.

2.3.6 Practical LEDs for fibre optics

The range of high-radiance LEDs has now become quite comprehensive to meet the large number of proposed applications. For the most basic optical links in which glass or plastic fibres of large NA and large core size are used with modulation bandwidths less than tens of MHz, simple GaAs LEDs with junctions formed by zinc diffusion may be quite adequate. Such devices can be made which will launch a sufficient power level of several tens of μW at just 20 mA drive current.

For longer links it becomes necessary to use low-loss fibres with wider bandwidths, and these tend to have small NA and core size so that higher radiances are demanded. High-radiance, zinc-diffused LEDs (see Figure 2.6a) will probably prove one of the most popular fibre-optic sources because of their high reliability (10^6 hours predicted, 40 000 hours continuous operation to date) and simplicity of operation. Typical radiances around 60 W/sterad/cm^2 at 300 mA over a 50 μm diameter spot are obtained in these devices which will launch around 100 μW into the low-loss silica type fibres. Lensed, smaller-area versions will launch around 0.5 mW at 150 mA into 85 μm core step-index silica fibre of 0.16 NA and 0.25 mW into 60 μm graded-index silica type fibres, which is ample power for many applications, since link length is generally restricted by material dispersion. In silica fibres a 50 nm wide spectrum at 0.9 μm wavelength results in an effective upper limit on fibre bandwidth due to material dispersion of around 125 MHz km. Nevertheless, for short, high data-rate links the 30 MHz bandwidth of GaAs zinc-diffused LEDs proves to be a limitation. Homojunction LEDs which give modulation bandwidths from 200 MHz to beyond 1 GHz have been made using vapour-phase epitaxy.[48,49] These devices, at 400 MHz with 150 mA d.c. bias, launch 125 μW into step-index 0.16 NA, 85 μm core fibre with lens coupling.

Similar homojunction LEDs which emit at wavelengths longer than the 0.9 μm of GaAs devices can be prepared by growing $Ga_{1-x}In_xAs$ instead of GaAs. In practice, the indium fraction x is increased gradually (or sometimes in small incremental steps) until the right fraction (e.g. 16% for 1.06 μm) is reached and the p–n junction is then grown in constant-composition material. GaInAs LEDs for 1.06 μm reported by Mabbit and Goodfellow[50] will launch around 50 μW into silica fibres at 1.06 μm and modulate at 200 MHz (3 dB optical power frequency).

The use of GaAlAs heterostructures in LEDs allows much greater external efficiencies to be achieved because the internal absorption can be reduced to very low levels. In the double-heterostructure GaAlAs LED of Figure 2.15, the 0.82 μm, 1.5 eV photon energy emission is very weakly absorbed in the passive layers on either side of the active layers which have bandgaps around 1.6–1.7 eV. The substrate can be completely and easily removed, leaving a flat emitting surface using a selective etch for GaAs which effectively stops at GaAlAs. Such LEDs are generally designed with $Ga_{1-x}Al_xAs$ active layers with $x \approx 5$–7% giving around 0.82 μm emission which tends to give improved reliability over double heterostructures for which $x = 0$. The 0.82 μm emitters have advantages over 0.9 μm emitters when silica fibres with higher residual water levels are used, for instance the plastic-clad silica type fibres, because absorption is lower at 0.82 μm. 'Dry' fibres are dominated by scattering loss in the 0.8–1 μm region so 0.9 μm is preferable with these.

The radiance of the DH GaAlAs LEDs approaches the theoretical maximum in the devices reported by Burrus and Miller,[51] which are layer-down surface emitters with a similar structure to that shown in Figure 2.15.

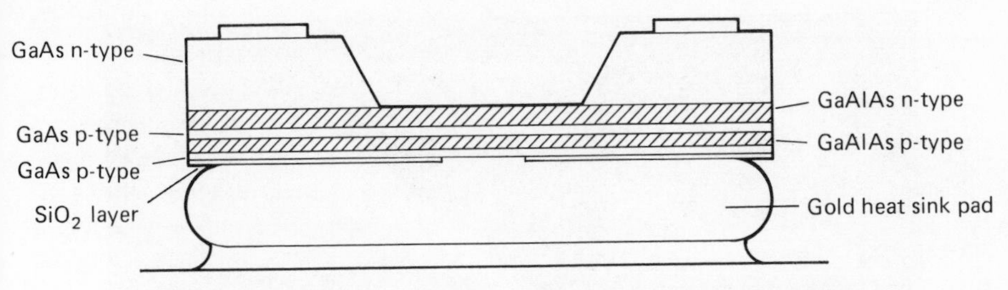

Figure 2.15. The GaAs/GaAlAs double-heterostructure LED

Horiuchi *et al.*[31] have used a novel layer-up surface-emitting DH structure which incorporates a recess which permits location of a sphere lens for improved coupling (see Figure 2.9). This structure has a high efficiency at around 0.82 μm but cannot be driven at such high currents as the layer-down devices. Other workers have adopted edge-emitting configurations and show that similar high launched powers (0.5 mW into 85 μm core, 0.16 NA fibres) can be achieved with such devices. However, the limiting fibre bandwidth due to material dispersion is slightly smaller at 0.82 μm than it is at 0.9 μm and higher-bandwidth systems ($>$10 MHz) tend to be limited by material dispersion rather than insufficient launched power. It is our feeling, therefore, that GaAlAs LEDs at 0.82 μm offer little over the 0.9 μm emitters for most systems. Systems with a great number of connections and splices can be power-limited, consequently they may be the exception where the higher efficiency GaAlAs devices win out.

Longer-wavelength GaInAsP/InP LEDs for the 1.1–1.7 μm band hold much greater promise than devices emitting at 0.9 or 0.82 μm. This is because:

(a) they are double heterostructure devices and therefore highly efficient;
(b) fibre losses can be made much lower in this wavelength band; and
(c) the material dispersion falls to zero at 1.25–1.3 μm in doped silica fibres so wide bandwidth links over long distances can be achieved even with LEDs at this wavelength.

GaInAsP/InP LEDs have now been reported by several groups[52,53,38] (see Figure 2.6b) and are now available as commercial devices.

We show in Figure 2.16 an example of a practical LED, packaged with a fibre tail. This is a GaAs device with a lens for high coupling efficiency, but obviously the package could incorporate any of the GaAs, GaAlAs, GaInAs or GaInAsP/InP device options. We also show alternative packaging configurations which incorporate the device heatsink and fibre-optic connectors.

It is fitting, perhaps, in this discussion of practical LEDs to speculate as to the potential of the different wavelength sources in systems applications. At the present (1979) long-wavelength GaInAsP–InP, 1.2–1.3 μm emitters are just becoming available but the detectors for this wavelength region are somewhat less well

Scanning electron micrograph
of high radiance LED chip

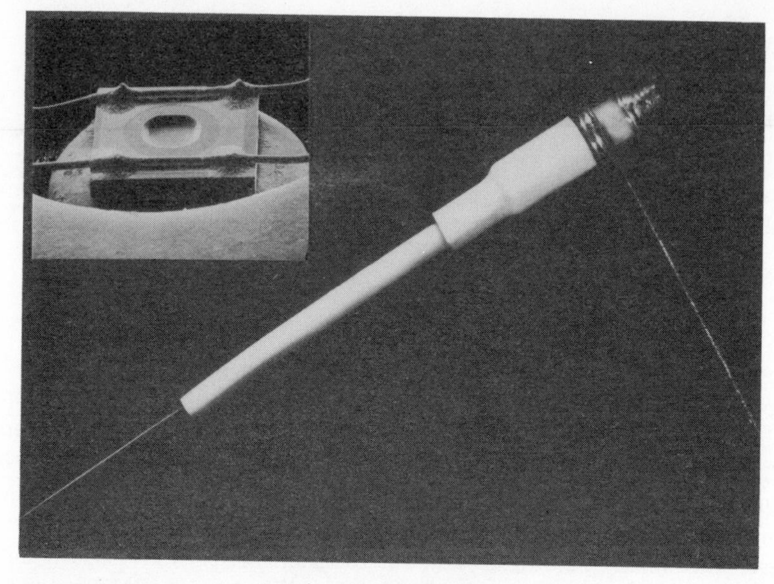

Packaged high-radiance LED
incorporating a Lens and Fibre Tail

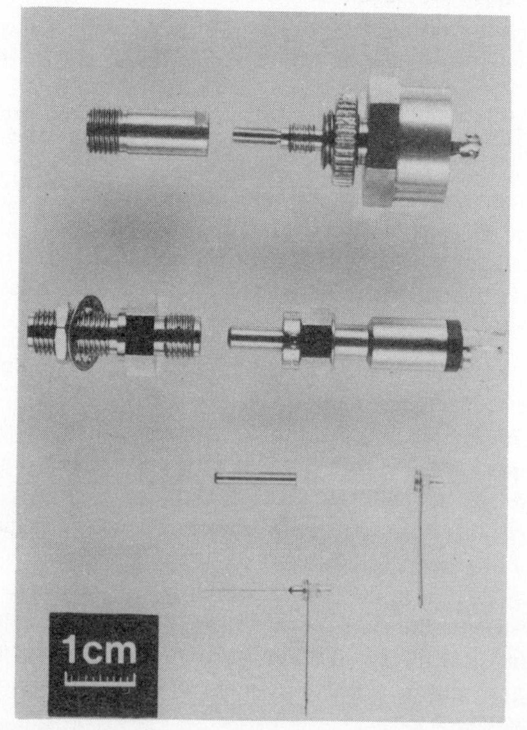

Packaged LED incorporating
connector ferrule,
connector adaptor,
and bulk head
mounting heat sink

Packaged LED with Amphenol
connector compatible components

LED with sapphire ferrule
connector component

LED with fibre tail in
stainless steel sleeve

Figure 2.16. Practical LED configurations (1978). (Reproduced by permission of Plessey Optoelectronics & Microwave Group)

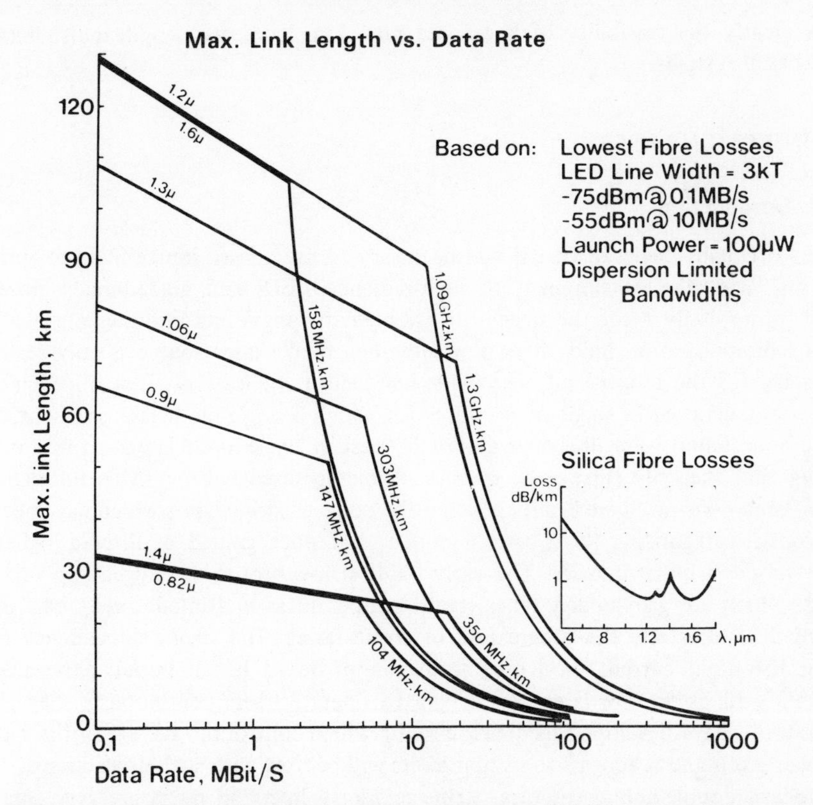

Figure 2.17. The relationship between maximum link length and data rate based on the simple assumption that the system is either power-limited or material-dispersion limited. The basic system parameters assumed are shown. In practice, equalization can be employed to override the fibre bandwidth limitations to some extent, and also the rather peaky loss characteristic of the fibre will tend effectively to narrow the LED spectral linewidth resulting in increased bandwidth. However, the system advantages for the 1.2 and 1.3 μm sources are apparent. This diagram indicates that most of the multimode fibre-optic applications will be fulfilled by LEDs rather than lasers

established. However, we speculate that ultimately the performance of detectors for the longer-wavelength regions will match that of the well-established silicon devices now available. Based on the detectivities presently achieved, we show in Figure 2.17 an estimate of the link-length and bit-rate limits for various wavelength LED sources supposed to have a rather broad emission spectrum of 3 kT width in energy. We have taken as the fibre attenuation lowest fibre losses as reported by Horiuchi and Osanai.[8,9] Fibre bandwidths are the material-dispersion limited bandwidths based on the 3 kT spectral emission width and the curves are based on a launched power level of 100 μW. The graphs clearly show the advantages of operating in the longer wavelength, low-loss, windows of the silica fibre and also show clearly the feasibility of large data rate, long-haul, wavelength multiplexed LED-based systems.

2.4 INJECTION LASERS

2.4.1 Introduction

It was originally believed that it was necessary to use a laser source in fibre-optic systems. With the development of high-radiance LEDs and graded-index fibres which quite easily meet the needs of intermediate length and intermediate bandwidth communication links, it is now apparent that a laser source is only really necessary for the longer-haul, very wide bandwidth applications. Best fibre bandwidths are achieved in single-mode fibres for which a single-mode laser is essential. Noise and stability are likely always to be best in single-mode lasers, so it is our feeling that the only fibre-optic sources ultimately used will be LEDs or single-mode lasers. We have here briefly covered the range of stripe lasers which have been developed, categorizing them as gain-guided, real-index guided or filtered higher-order mode stripe structures. The early oxide-stripe, proton or oxygen-implanted lasers, which are gain-guided types, tend to operate as multimode lasers but are included to illustrate the progression of stripe lasers. The strong dependence of lasing threshold current on temperature can prove to be an initial, expensive, stumbling block to the laser user trying to operate lasers continuously. Several laboratories have developed feedback circuits which control the average or the 1 to 0 power levels and it appears to us that lasers will be used with such drive circuits.

Modern double-heterostructure, stripe-geometry injection lasers are very small but highly complex devices. This complexity has increased through several stages during their development. The first injection lasers, which operated only under short pulse drive at hundreds of kA/cm^2 and at cryogenic temperatures, were simple rectangular chips of gallium arsenide with a planar p–n junction formed by diffusion.

The development of lattice-matched heterojunctions of gallium arsenide/gallium aluminium arsenide to replace the simple p–n junction layers allowed much more efficient pumping of a very thin active layer, reduced the optical absorption and scattering losses and led to laser operation at room temperature.

The idea of restricting the current pumping of the active layer to a narrow central ribbon of the chip, allowed room-temperature CW operation to be achieved at a few hundred milliamperes with a forward voltage around 1.5 V. These developments ensured the future of the injection laser and since that time many further refinements have been made.

The guiding properties of the active lasing stripe, clad above and below with material of lower refractive index and wider bandgap, and on either side by unpumped material, have been extensively studied and the limitations of the first oxide-isolated, stripe-geometry lasers have been overcome to varying extents with the introduction of many newer stripe-geometry structures.

Research in other directions has shown that the GaAs/GaAlAs system is not the only one in which lattice-matched double-heterostructure lasers can be made, although it may be the most simple, and so the range of wavelengths for which CW room-temperature lasers will be available is broadening. The GaInAsP/InP system used by Hsieh *et al.*[20] to make lasers in the 1.3 μm region is becoming particularly well established.

The exact requirements of laser sources are determined by the overall fibre-link specification; laser sources are superior to LED sources in respect to their narrower spectral linewidth, faster switching times and higher coupling efficiency to small-core, low-NA (or single-mode) fibres – characteristics which are necessary for the wider bandwidth, longer repeater length fibre communication links.

The narrowest linewidths, highest coupling efficiencies and lowest noise operation result from single-mode lasers. The attainment of single transverse- and longitudinal-mode lasers has therefore been a major objective of laser development.

In this section on lasers we identify some stripe geometries which have been developed and indicate how optical power in the active layer is guided within the stripe region. We present the established recipes and arguments for the achievement of single longitudinal- and transverse-mode operation and also relaxation oscillation suppression. We briefly outline the distributed-feedback lasers which offer great wavelength stability and selectivity and can be made in a planar geometry which lends itself to integrated optics technology. Finally, we discuss laser mounting, fibre/laser interfacing and laser-drive circuits.

2.4.2 The double-heterojunction laser

To achieve the continuous room-temperature laser operation necessary for fibre communication systems, the rather sophisticated double heterostructure (DH) geometry was developed and this has now been realized in several different semiconducting material combinations.

The basic scheme of a DH laser is shown in Figure 2.18. The changes in bandgap at the top and bottom of the active layer provide potential barriers which confine the injected carriers to the active layer. The number of carriers which leak over these potential barriers is proportional to exp $(-\Delta/kT)$ where Δ is the height of the potential step. Laser action requires that population inversion is achieved. In a semiconductor this occurs when the separation of the quasi-fermi levels describing electrons E_{fn} and holes E_{fp} exceeds the bandgap,[55] i.e.

$$E_{fn} - E_{fp} > E_g \tag{2.23}$$

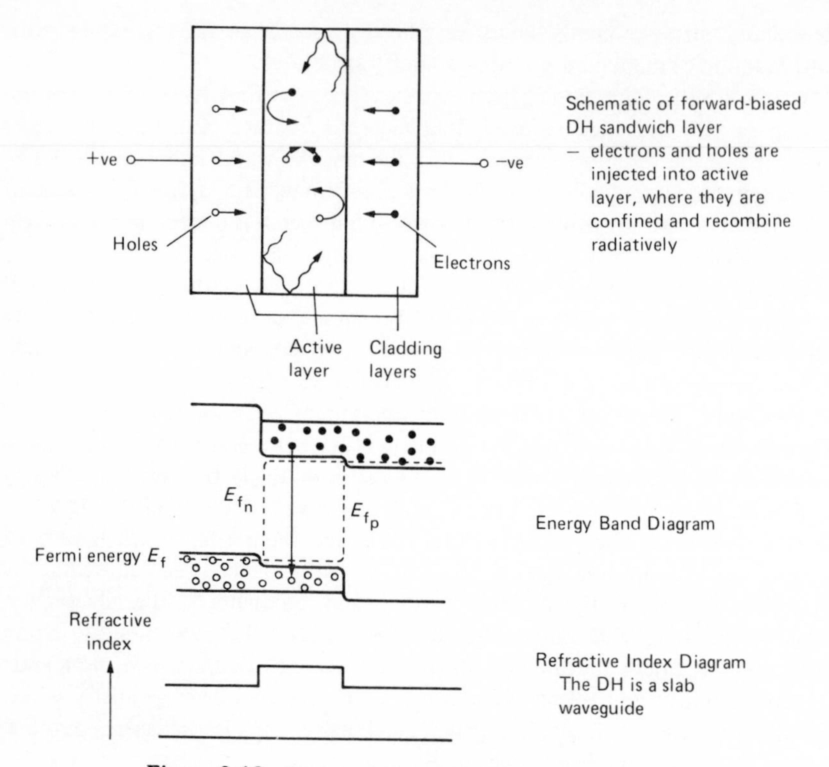

Figure 2.18. Basics of the double-heterojunction laser

In GaAs this laser condition corresponds to a very high excess carrier density of $np \sim 10^{36}\ \text{cm}^{-3}$, whereas $np = n_i^2 \sim 10^{14}\ \text{cm}^{-3}$ in unpumped material. It follows that to achieve the high injected carrier densities necessary for laser action at the low power dissipation required for CW operation it is necessary to restrict the extent of the active layer. The active layer is therefore made thin – usually less than 0.5 μm.

As the refractive index is inversely related to the energy gap in the III–V compound semiconductors, the refractive index of the active layer exceeds that of the cladding layers and so forms the core of a slab waveguide. The active-layer width is therefore a design parameter controlling both the waveguiding and the current density required for achievement of the condition for population inversion.

Lattice matching is generally necessary throughout the double heterostructure for CW operation. This is because the incorporation of dislocations in the active region resulting from the stress and strain due to the lattice mismatch causes non-radiative recombination of the injected carriers. The radiative quantum efficiency in mismatched DH structures is reduced and the lasing threshold current is increased. Lasers have been made using several compound III–V multilayer combinations. These are summarized in Table 2.1.

Table 2.1 Alloy semiconductors from which room-temperature lasers have been fabricated

Active region	Cladding region	Useful λ range (μm)	Min J_{th} (KA/cm^2)	Substrate	Preparation
$Ga_yAl_{1-y}As$	$Ga_xAl_{1-x}As$	0.65–0.9	0.5	GaAs	LPE[56], MBE[57], MOCVD[58]
$Ga_xIn_{1-x}As_yP_{1-y}$	InP	1.1–1.6	0.5	InP	LPE[20] CVD[59] MBE[60]
$GaAs_ySb_{1-y}$	$Ga_xAl_xAs_{1-y}Sb_y$	0.9–1.1	1.9	graded to GaAs	LPE[61]
GaAs	$In_xGa_{1-x}P$	0.9	11.2	GaAs	hydride CVD[62]
$In_yGa_{1-y}As$	$In_xGa_{1-x}P$	0.85–1.1	1.2	graded to GaAs	hydride CVD[62]

LPE = liquid-phase epitaxy
MOCVD = metallorganic chemical vapour deposition

The ternary GaAlAs and quaternary GaInAsP systems are now well advanced. Lasers from these cover the useful fibre window from 0.65 μm to beyond 1.6 μm with a relatively small gap between 0.9 and 1.1 μm, which could possibly be filled by the GaAlAsSb quaternary. However, because this quaternary can not be prepared with a lattice parameter which matches that of a binary III–V substrate, a lattice mismatch must be tolerated. This mismatch is usually accommodated in a graded layer in such a way that the necessary misfit dislocations are distributed and locked in the the grading layer. No such graded-layer structures are required for the other systems in Table 2.1. For the GaAlAs system, almost perfect matching results for any composition because of the almost equal atomic size of gallium and aluminium. For the other systems, lattice matching requires very tight compositional control of the ternary and quaternary layers.

2.4.3 Laser design

A vast majority of the optimization of structure and performance of double-heterostructure lasers up to this date (1978) has been carried out in the GaAlAs system,[63] which has a number of special features compared with other systems. These arise from the ease with which lattice matching can be achieved, from the

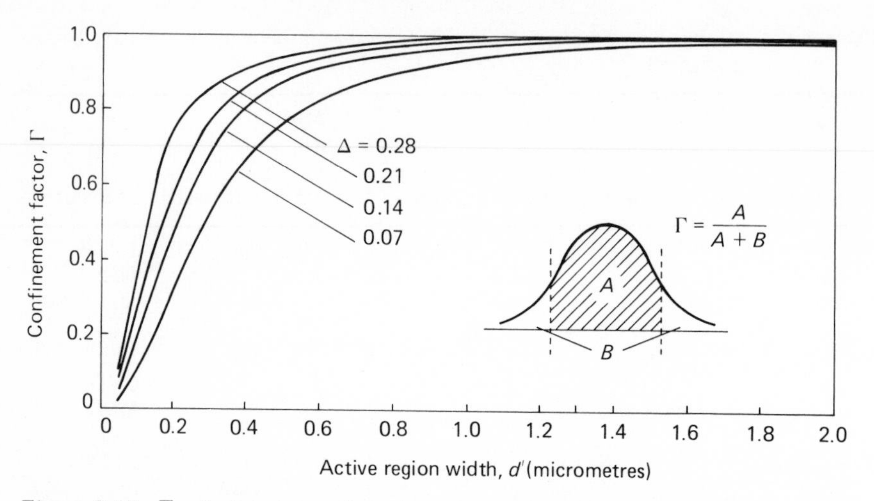

Figure 2.19. Γ calculated as a function of the active layer width for DH diodes with refractive index steps Δ shown, for the lowest-order TE mode. Derived from data of Casey, Panish and Merz[64]

difficulties of making electrical contact to aluminium-rich alloys, and from the ease with which GaAs and GaAlAs can be selectively etched. However, lessons learned in the GaAlAs system should hold well for other systems.

If the active-layer thickness d is less than the diffusion length of injected carriers, the active layer is effectively excited uniformly throughout its thickness. If it is practical to assume that all of the injected minority carriers are confined by the heterojunction then $J_{th} \propto d$ and for the achievement of a low threshold current, d must be made small. However, the active layer is effectively a slab optical waveguide. With d values less than around 1 μm, only the lowest-order transverse mode is guided and then only a fraction of the optical energy Γ (shown in Figure 2.19) is propagated within the active layer.[64] The remainder is guided as evanescent energy in the large energy gap cladding layers.

Obviously when d becomes sufficiently small that Γ is also small, which occurs when d is less than around 0.2 μm, the gain is effectively diluted by Γ and the proportional dependence of J_{th} and d breaks down. In practice it is necessary to consider the modal distributions in both the orthogonal directions perpendicular to the laser axis. We distinguish these directions by referring to modes perpendicular to the junction plane as 'vertical' modes, and those parallel to the junction plane as 'horizontal' modes.

The active-layer thickness which corresponds to minimum threshold current is close to the value which gives maximum beam divergence in the 'vertical' direction because it coincides with the smallest 'vertical' near-field width (see Figures 2.20a and b).[65] Hence a compromise between threshold current and laser/fibre coupling efficiency, which depends on beam divergence, must be made.

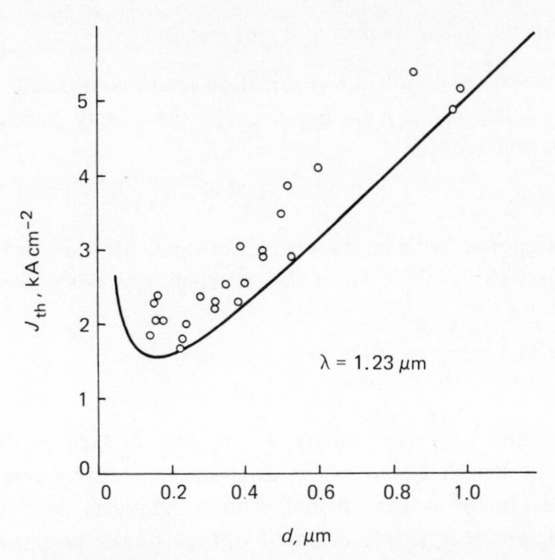

Figure 2.20. (a) Measured threshold current density J_{th} against active layer thickness d for pulsed room-temperature InGaAsP/InP DH lasers[65]

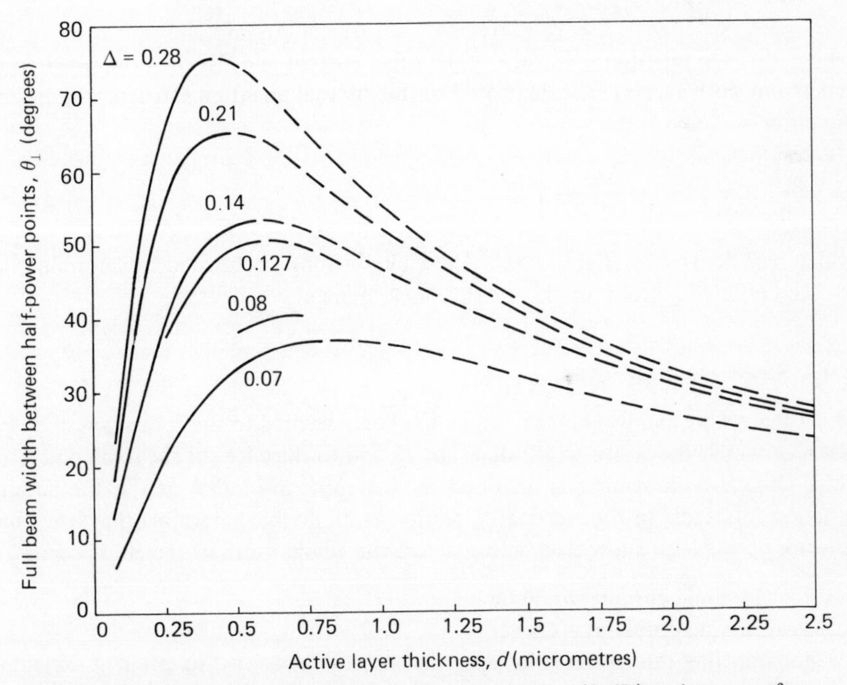

Figure 2.20. (b) Full beam width at the half-power (3 dB) points as a function of active-layer thickness d and refractive index step Δ for lowest order TE mode. The dashed region shows where higher-order modes will exist[64]

Simple relationships of use in analysing laser behaviour

The population-inversion condition in injection lasers is satisfied when the separation of the quasi-fermi levels E_{fn}, E_{fp} exceeds the energy bandgap in the active region, as in Equation 2.23:

$$E_{fn} - E_{fp} > E_g$$

This usually corresponds with an excess carrier concentration in the active region (in GaAs) of around 10^{18} cm^{-3}.[66] Laser action depends on net round-trip gain

$$\Gamma g - (1 - \Gamma)\alpha + \frac{\ln R_1 R_2}{2L} \geqslant 0 \tag{2.24}$$

where L is the length of the laser cavity, R_1, R_2 are the facet reflectivities, g is the linear gain per unit length and α is the attenuation incurred due mainly to free-electron absorption of the evanescent radiation in the layers cladding and surrounding the active region. Γ is the fraction of optical power propagating within the active layer (see Figure 2.19).

The external quantum efficiency is

$$\eta_D = \frac{1}{V}\frac{dP}{dI} \tag{2.25}$$

where V is the junction voltage, I is the drive current and P is the total power output (from both facets). This is related to the internal quantum efficiency η_i by the equation

$$\frac{1}{\eta_D} = \frac{1}{\eta_i}\left[1 - \frac{2\alpha L}{\ln (R_1 R_2)}\right] \tag{2.26}$$

which was derived by Biard *et al.*[67] These expressions can be used to determine the loss and efficiency from simple practical measurements on devices.

2.4.4 Stripe-geometry lasers

Much innovative and ingenious research has been devoted to the formation of stripe lasers in which the active width does not extend to the edges of the semiconductor chip. Optical confinement is provided by the refractive index step at the heterojunction interfaces in the 'vertical' direction in all double-heterojunction lasers but it is not desirable to allow laser action across the whole width of the chip because:

(a) the threshold current would be high;
(b) heat-sinking problems are severe;
(c) the emitting thin rectangular aperture would be unsuited to efficient matching to circular symmetric fibres;
(d) lasing in multiple filaments is difficult to avoid; and
(e) multimode operation in the horizontal plane tends to occur.

Hence stripe structures have been universally adopted for fibre-optic applications, stripe widths ranging between 2 and 30 μm in the modern devices.

The aim has been to incorporate into the stripe structure when it is lasing, just the right balance of guiding so that the laser operates in a single transverse mode. Perhaps the most simple stripe-geometry laser is the deep proton-implanted laser. Here the material on either side of the active volume is rendered high resistance by bombardment of the surface with high-energy (a few hundred keV) protons. Current flows through the undamaged material from the stripe contact to pump the active layer. When the stripe width is comparable with the diffucion length of the injected carriers, carriers are rapidly lost at the edges of the stripe, resulting in increasing threshold current density with decreasing stripe width. This dependence is seen to a lesser or greater extent in all types of stripe laser. Increased radiation losses with decreasing stripe width also contribute to this. In Figure 2.21 we show the dependence of threshold current density on stripe width for several different

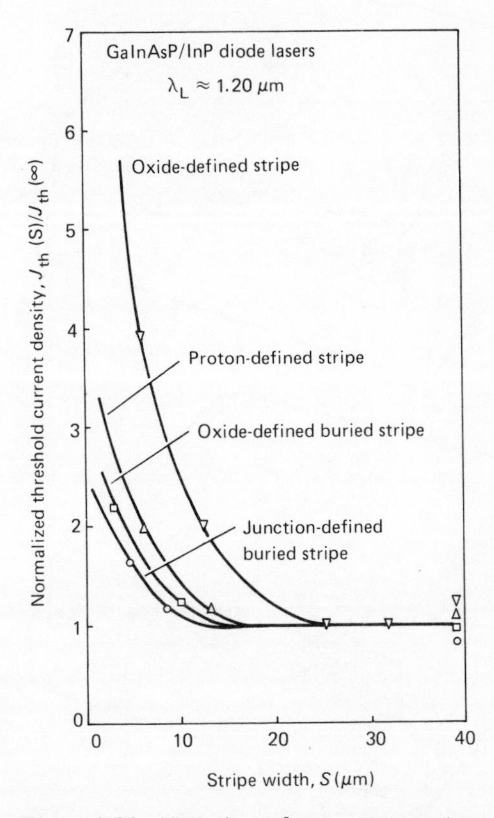

Figure 2.21. Variation of room-temperature pulsed threshold current as a function of stripe width for four types of stripe-geometry DH GaInAsP/InP diode laser[20]

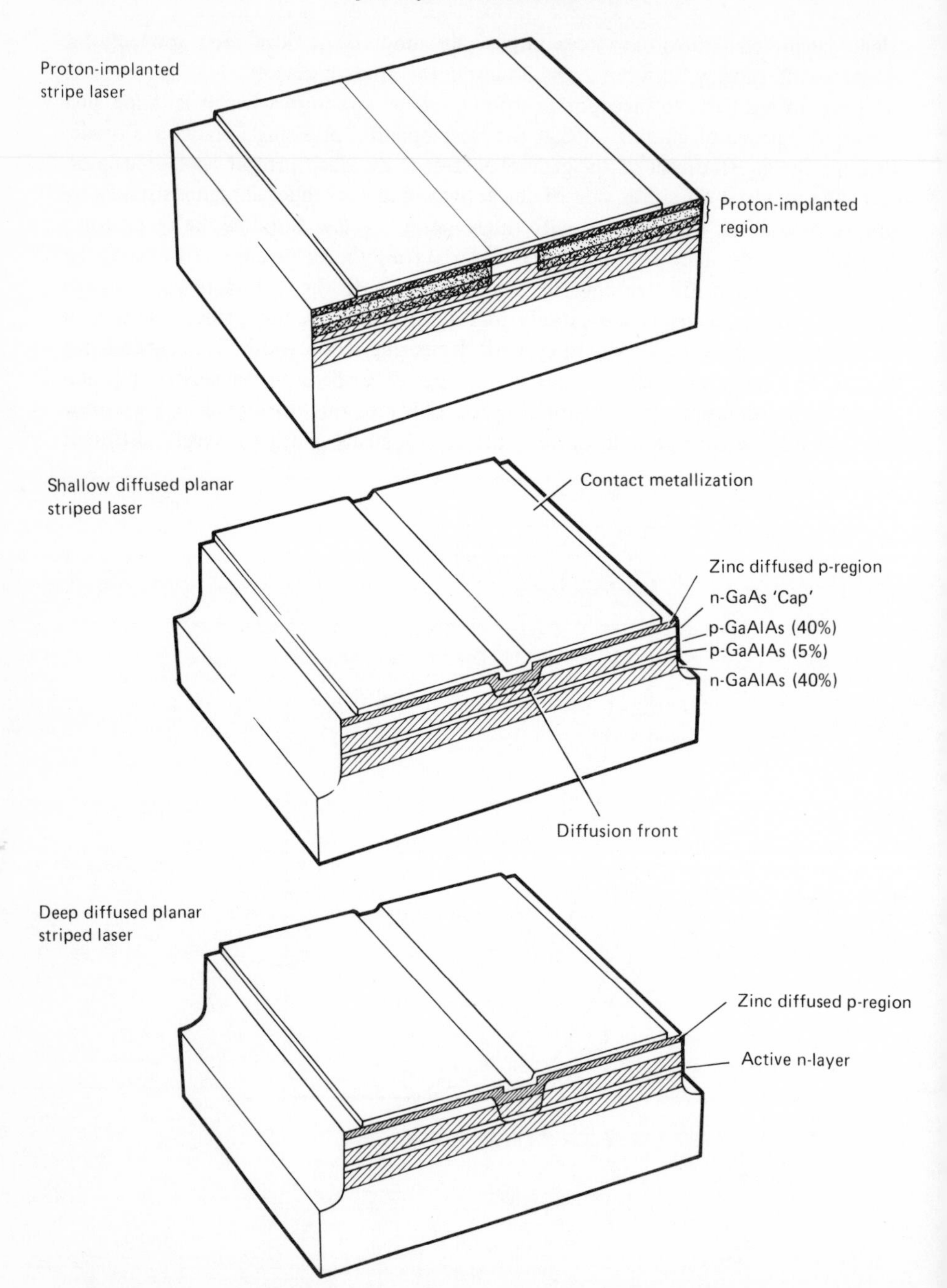

Figure 2.22. Schematics of the proton-implanted, shallow-diffused planar and deep-diffused planar stripe geometry GaAlAs lasers[73]

stripe structures as reported by Hsieh[20] for GaInAsP lasers. Obviously the increase in threshold current with decreasing stripe width strongly depends on the chosen stripe geometry.

In Figure 2.22 we show schematics of the proton-implanted stripe laser and the shallow and deep zinc-diffused planar stripe lasers. These zinc-diffused stripe lasers are developments of the planar structures described by Kobayashi *et al.*[68] but these lasers use a reversed-biased p–n junction to prevent current flow outside the central stripe region. This junction confinement is achieved by reducing the thickness of the GaAs contacting layer in the stripe region and then zinc-diffusing to a depth insufficient to penetrate completely the n-type contact layer outside the stripe.

In the shallow zinc-diffused laser the diffusion front nearly reaches the active layer in the stripe region presenting a low impedance path for current flow to the active stripe which is around 10 μm wide. In the deep-diffused laser, again the diffusion produces a low impedance current path but additionally, what was originally the n-type active layer is converted to p-type in the stripe region so that current spreading is very restricted here. In Figure 2.23 we show the light/current characteristics of some CW laser structures.

The response above threshold can be very linear up to optical power levels of

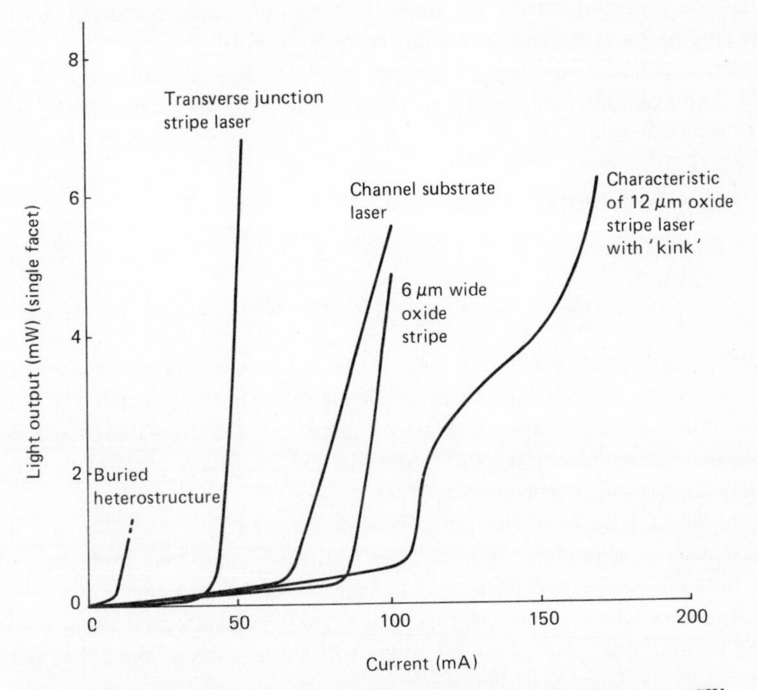

Figure 2.23. The light-current characteristics reported for some CW GaAlAs DH laser

several hundred milliwatts if the laser is very multimode – as would be the case for a deep-diffused planar laser with a wide stripe. However a laser designed for single transverse mode operation has a limited linear region (typically up to ~5 mW) above which kinks appear in the characteristic. The kinks appear when the laser changes from the lowest to higher-order transverse-mode operation. Hence for predictable performance it is important to have an understanding of the confinement properties of the waveguide associated with the stripe.

Schlosser[69] carried out an analysis of the stripe-laser waveguide based on the normalized frequency 'V-value' approach.

Here

$$V = \frac{2\pi a}{\lambda}(\epsilon_1 - \epsilon_2)^{1/2} \tag{2.27}$$

A complex dielectric constant ϵ is employed to allow description of the gain and loss regions where

$$\epsilon = (n + jk)^2 \tag{2.28}$$

with n the real part of the refractive index and k the imaginary part, its sign indicating gain or loss. λ is the free-space wavelength, $2a$ is the effective guide width and suffices 1 and 2 refer to the 'core' and cladding respectively. As for fibres, the approach is powerful, predicting single lowest-order mode operation for V-values below around 2 and multimode operation for large V-values.

The normalized frequency V for the simplest cases of 'step index' rectangular guides with real index steps only, or steps in the imaginary parts of the refractive index only, reduces to

$$V = \frac{2\pi a}{\lambda}(n_1^2 - n_2^2)^{1/2} \qquad \text{where } k_1 = k_2 \tag{2.29}$$

and

$$V = \frac{2\pi a}{\lambda}[(k_2^2 - k_1^2) - 2jn_1(k_2 - k_1)]^{1/2} \qquad \text{where } n_1 = n_2. \tag{2.30}$$

These equations emphasize the two waveguiding situations which occur in stripe lasers. When the refractive index in the stripe exceeds that outside, the guiding is termed 'index guiding'. When this step is absent or negative, the optical confinement arises entirely from the differences in the imaginary parts of the dielectric constant. This is called 'gain guiding'.

Some errors arise from the use of the normalized frequency because this is a two-dimensional simplified approach to a three-dimensional problem and falls down because of the penetration of the electric field into the layers above and below the active layer. The complex index can be corrected by using a weighted average for the index over the multilayers within and outside the active stripe and this approach has been successfully applied by several workers.

When a laser operates in the regime where only a single transverse mode is excited, the light/current characteristic remains quite smooth and continuous and is often quite linear also. However, with increased excitation levels the refractive index and gain profiles can be significantly altered so that the V-value for the stripe is increased. The increase in V may be sufficient for higher-order mode operation, and then kinks are frequently observed in the light/current characteristic as shown for the oxide-stripe laser in Figure 2.23.

Waveguiding considerations account for kink behaviour but the graded nature of the stripe waveguides makes it difficult to quantify the initial pump levels for kinks in the light/current curves. Suematsu and Yamada[70] attempted to predict the range of operation for which single-mode operation could be expected in different types of stripe structure. However, it is difficult to include all of the interacting effects, such as temperature distributions, refractive index changes with pump level, and power density differences along the stripe. Hence such models only provide guidelines for waveguide design. Nevertheless, these analyses highlight the need for selecting the right balance of index- and gain-guiding.

Oxide-isolated stripes

In the oxide-isolated stripe laser shown in Figure 2.24 the active layer is current-pumped in a region wider than the contact stripe by an amount depending on the sheet resistivities of the n- and p-type sides of the junction. This isolation technique has been very widely applied for laser structures but because of the diffuse nature of the activated stripe, analysis of the transverse waveguiding properties of the structure is complex.

Kirkby *et al.*[71] pointed out that there are three main aspects to guiding in oxide-stripe lasers. The lateral fall-off in excess carriers from the centre of the stripe contact gives rise to an increasing refractive index which has a defocusing effect. This is opposed in wide-stripe lasers by the positive guiding arising from the increased index due to enhanced stimulated recombination on axis when operating in the fundamental transverse mode. The third and dominant contribution arises from the imaginary refractive index component which contributes gain-guiding. As the width of the stripe is varied, the relative importance of these guiding factors alters. Paoli[72] reports that the observed output distribution is consistent with an additional contribution to the refractive index profile established by the temperature distribution. Well-behaved lowest-order transverse mode operation has been achieved to optical output power levels of several milliwatts in oxide-stripe lasers. This shows that the opposing gain-guiding and defocusing influences can be balanced over an acceptable dynamic range if care is taken in the laser design.

The proton-isolated laser

The proton-isolated laser structure first reported by Dyment *et al.*[73] is manufactured by subjecting a DH slice, which has been masked by metallic strips, to bom-

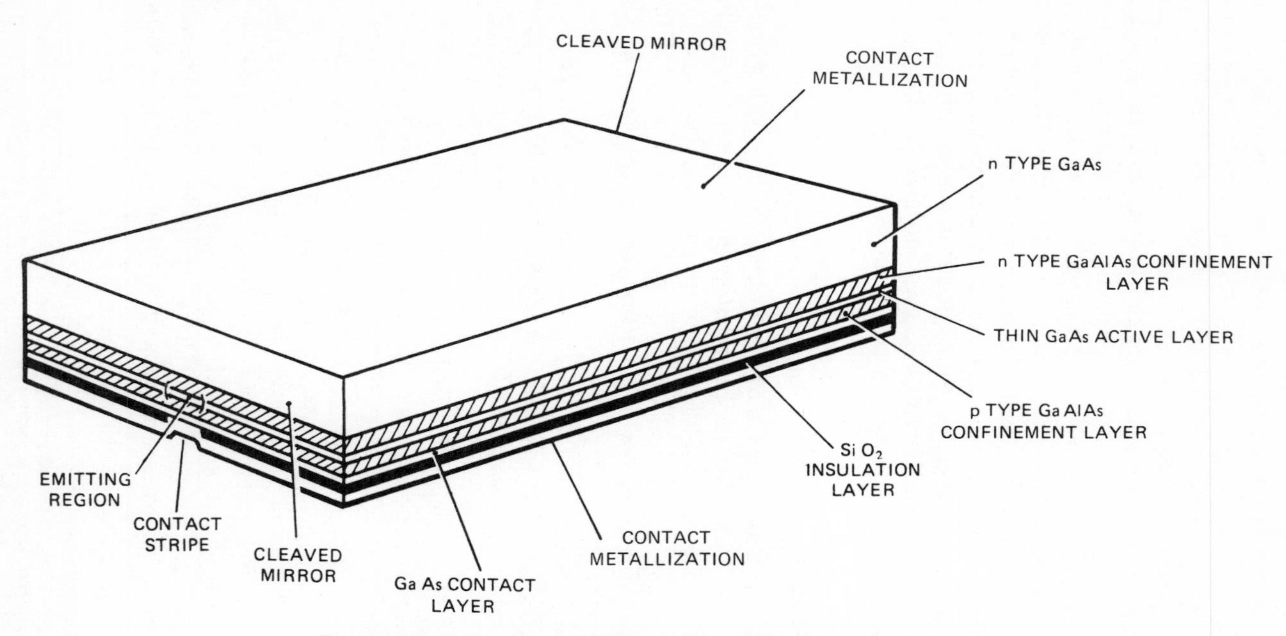

Figure 2.24. Schematic of the oxide-stripe GaAlAs/GaAs laser

bardment with high-energy protons. This design, shown schematically in Figure 2.22, offers better current confinement than the simple oxide-stripe and is also superior thermally, because of the elimination of the SiO_2 glass layer.

Waveguiding is again achieved by a balance of defocusing and focusing influences due to real index variations and gain-guiding but excellent control in manufacture can be achieved by virtue of the reproducibility of the implantation method and because of the abrupt nature of the implanted/unimplanted region boundaries. Steventon *et al.*[74] have shown that good mode control is possible even in shallow implanted structures where the implantation does not penetrate the active layer and that short lasers with 5 μm stripe width can have very low threshold current (~30 mA). Hartman *et al.*[75] have shown that excellent reliability is possible with this approach with more than 30 000 hours continuous operation and projected lives of 10^5 hours at 20°C for their GaAlAs DH lasers. The oxide, shallow diffused planar and the proton-stripe lasers are all examples of predominantly gain-guided lasers.

The transverse-junction stripe laser

The transverse-junction stripe (TJS) laser of Kumabe *et al.*[76] uses rather conventional fabrication techniques but the structure is quite novel. In its latest form the TJS is prepared by growing a double-heterostructure GaAlAs multilayer in *n*-type material on to an insulating substrate. Zinc is then diffused through a stripe mask producing a *p–n* junction as shown in Figure 2.25. The *p–n* junction in the active layer, which has the lowest aluminium composition, turns on at a lower voltage than that in the high-aluminium guide layers and the diffusion front in the insulating substrate obviously presents a high electrical impedance path. Hence in this structure, electrons and holes are injected transversely in the central layer of the DH structure and the heterojunctions confine the injected carriers to this layer. The effective pumped width of this stripe laser is determined by the diffusion length of the injected carriers under conditions of high current density (~50 kA/cm^2).

The optical waveguiding in the horizontal plane (parallel to the heterostructure layers) arises from the built-in doping profile resulting from the zinc diffusion into heavily doped *n*-type material. At the *p–n* junction where the material is exactly compensated, the refractive index is maximum (see the data on the refractive index of doped GaAs of Sell *et al.*[77]). The smooth decrease in compensation on either side of the junction results in a fall-off in refractive index, giving rise to a weak graded-index waveguide which is ideal for transverse mode control.

This laser has excellent characteristics. It has been demonstrated to operate in a single mode (longitudinal and transverse) from −20 to +110 °C, is stable over long periods – a version of the TJS laser has operated for 12 000 hours in a single mode,[78] and threshold currents of 30 mA at 25 °C are obtained.

In its earlier form the TJS laser as described by Namizaki[79] suffered from excessive current leakage which increased dramatically with temperature. This leakage

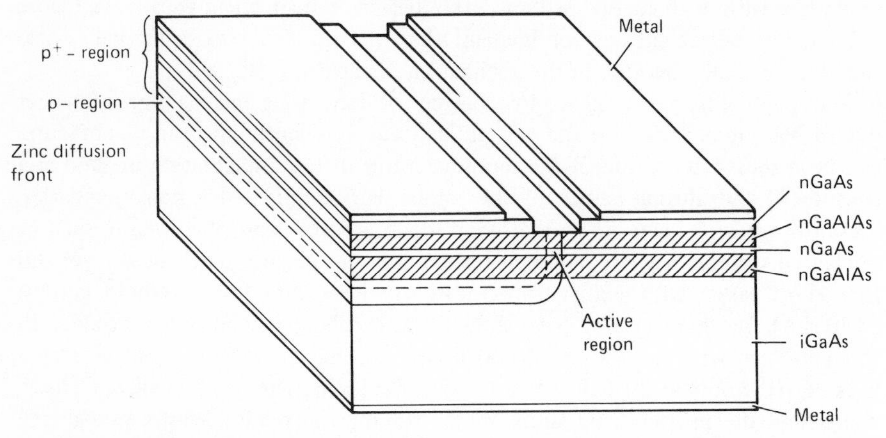

Figure 2.25. Schematic of a 'junction-up' transverse-junction stripe laser[76]

was caused by a large area of slightly forward-biased *p–n* junction which formed part of the stripe isolation. However the use of a semi-insulating substrate in the latest structure[76] has almost completely eliminated this problem and the TJS laser is now one of the most promising laser structures for fibre communications.

The channelled-substrate laser and high-order mode suppression

The channelled-substrate lasers of Aiki *et al.*[80] utilize a different principle to achieve single transverse-mode operation. The structure is formed, as the name suggests, by growing the DH layer structure on to a substrate into which channels have previously been etched (see Figure 2.26c).

In this structure higher-order transverse modes suffer a large propagation loss because the evanescent field of such modes extends over a wider region, so penetrating into the small bandgap and hence absorbing substrate.

Figure. 2.26. Laser structures in which losses increases with mode number. In the case of the channelled-substrate laser this produces a high degree of horizontal transverse mode stability and single longitudinal mode operation is observed. (a) Bent stripe lasers.[81,84] The bend in the stripe filters the higher-order modes preferentially. The kinks associated with higher transverse mode operation are eliminated at the expense of higher threshold current and lower slope efficiency. (b) Angled stripe lasers.[82,83,85] Higher-order transverse mode reflectivity at the facets is more greatly suppressed with increasing angle. This suppresses kinks. (c) The channelled substrate laser.[80] Higher-order transverse modes are attenuated preferentially because of the penetration of the mode field B into the lossy substrate. Excellent mode stability is reported, which results in linear power current characteristics and single transverse and longitudinal mode operation

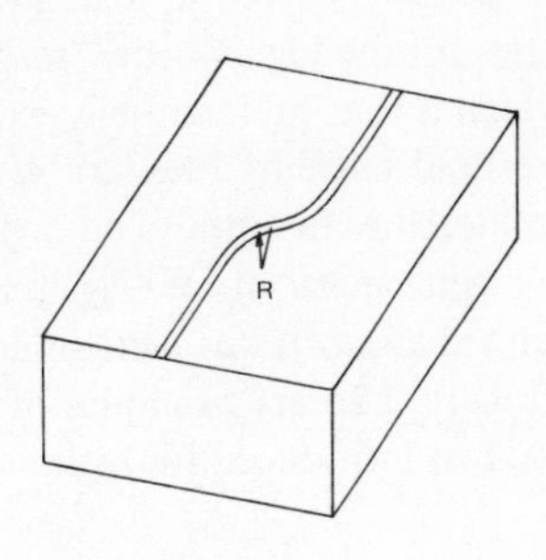
R

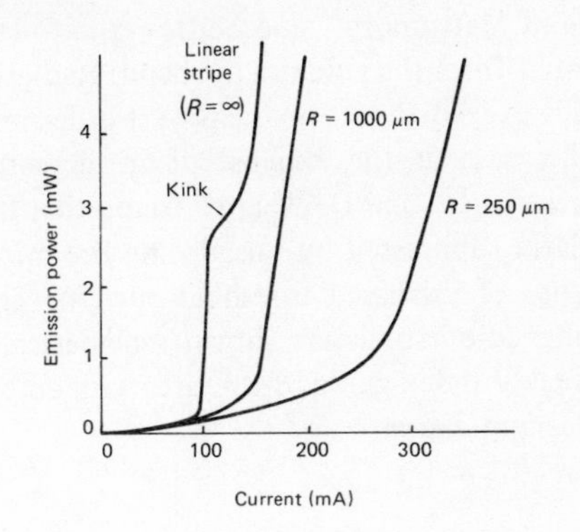
Linear
stripe
($R = \infty$)
R = 1000 μm
Kink
R = 250 μm
Emission power (mW)
Current (mA)
0
1
2
3
4
0
100
200
300

(a)

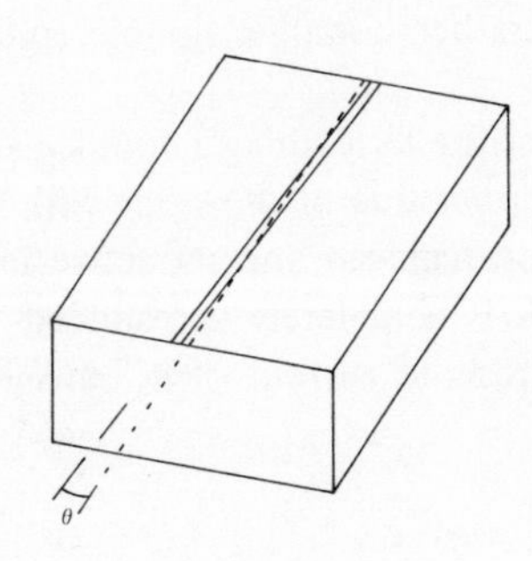
θ

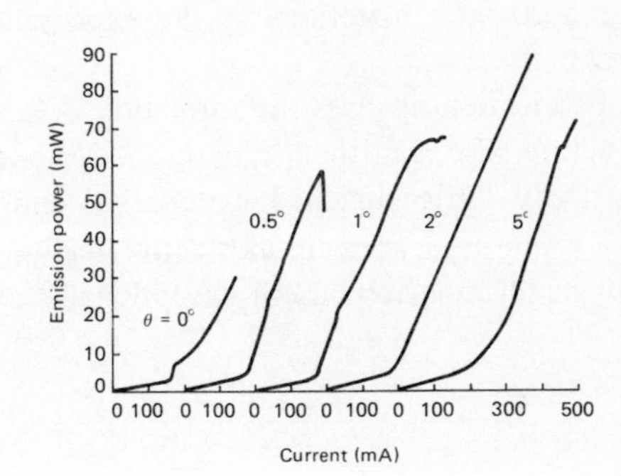
Emission power (mW)
Current (mA)
θ = 0°
0.5°
1°
2°
5°
0 100 0 100 0 100 0 100 0 100 300 500
0
10
20
30
40
50
60
70
80
90

(b)

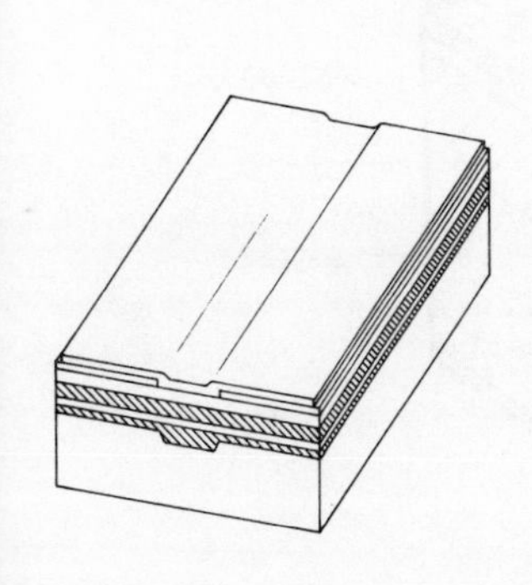

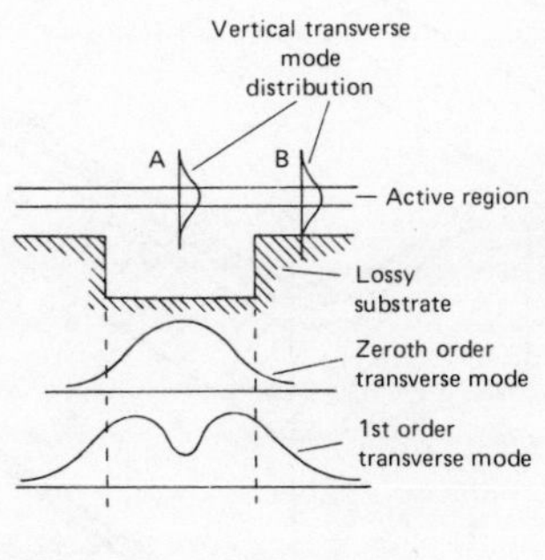
Vertical transverse
mode
distribution
A
B
Active region
Lossy
substrate
Zeroth order
transverse mode
1st order
transverse mode

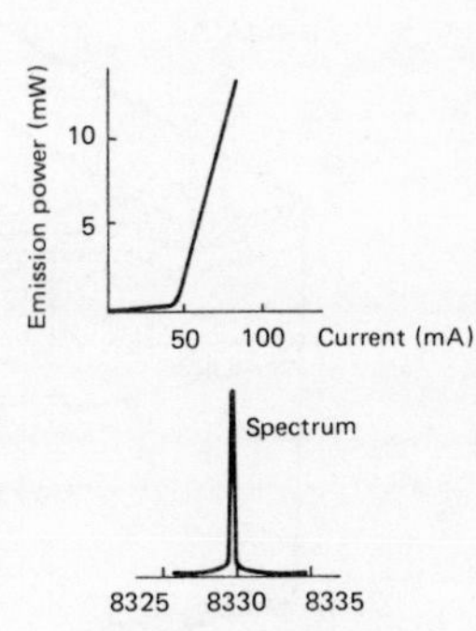
Emission power (mW)
5
10
50
100
Current (mA)
Spectrum
8325
8330
8335
Wavelength (Å)

(c)

A similar effect is achieved in a very different way in the bent-guide laser structures of Matsumoto[81] and Scifres *et al.*[82] Here the stripe is curved so that energy is radiated from the guide at the bend. Radiation losses for the higher-order transverse modes exceed those for the lowest-order mode which is thus preferentially excited, in this case at the expense of an increased threshold current. Frescura *et al.*,[83] Scifres *et al.*[84] and De Waard[85] found that the nonlinearities in output could also be similarly suppressed by misaligning the mirrors by using angled stripes, again at the expense of increased threshold current. The channelled-substrate, bent-guide and misaligned-mirror lasers shown schematically in Figure 2.26 are examples of gain- or weakly index-guided structures with emphasized loss introduced for higher-order mode suppression.

The buried-heterostructure laser

The buried-heterostructure lasers devised by Tsukada[86] and shown in Figure 2.27 achieves complete confinement of the injected minority carriers both in the vertical and horizontal directions at the expense of a rather complex double epitaxial growth process.

Fabrication involves the formation of a mesa stripe by etching a double-heterostructure slice, and then infilling the regions on either side of the stripe with high-resistivity lattice-matched material of appropriate bandgap and refractive index. The result is a filament-like active region which is completely surrounded with wide-bandgap material. An exceptionally low threshold current of 4.5 mA for a

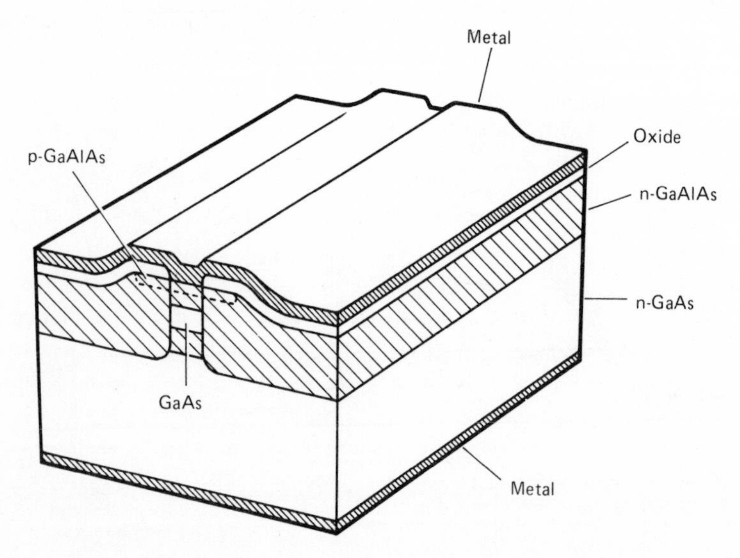

Figure 2.27. Schematic of the buried-heterostructure laser[86]

0.7 μm stripe width has been obtained[87] and good linearity and frequency response characteristics are observed. However, the power output is limited. Chinone *et al.*[88] found that the threshold optical density for catastrophic damage in GaAs lasers is ~1.1 MW/cm^2 or ~11 $mW/\mu m^2$. For reliable operation one must operate considerably below this limit and in practice 1 mW is the maximum CW operating power for the buried-heterostructure lasers which have the small cross section active regions necessary for single-mode operation. The low *V*-value required for a single horizontal mode means that the product of Δn and the guide width must be kept small. For example, with a stripe width of 5 μm for single transverse-mode operation in a step-index guide, Δn must be positive and less than 2×10^{-3}, which demands very tight compositional control of the infill alloy. The reliability of buried-heterostructure lasers has not yet matched that achieved with more straightforward stripe structures. Growth nucleation during the second epitaxy step can impair performance and makes the achievement of low-strain configurations which give best reliability, more difficult. However, the circularly symmetrical emission distribution, the low threshold current, linearity and good modulation characteristics make this real-index-guided structure attractive for single-mode fibre systems and perhaps for integrated optics components.

Longitudinal mode control

The spectra of lasers with the normal cleaved facet Fabry–Perot cavity type structure is usually made up of a series of peaks. The spectrum of such a multimode laser is shown in Figure 2.28a. The bands of energy separated by around 4 Å represent different longitudinal modes and the sub-peaks on each band are due to higher-order horizontal transverse modes. With correct design of the stripe structure, the higher-order transverse modes can be eliminated but most lasers to date have operated in several longitudinal modes as in Figure 2.28b. The TJS laser is presently an exception and operates in a truly single mode (Figure 2.28c), and so too is the channelled-substrate laser.

Ito *et al.*[89] investigated the intensity fluctuations in each longitudinal mode and found that the noise power in each mode *exceeds* the total noise power by greater than 30 dB (see Figure 2.29). This is because fluctuations in the power in one longitudinal mode at any time are largely compensated by related fluctuations in the power excited into other modes. The total noise output will be small only if the correlation between the energy in the different modes is maintained. Dispersion in optical fibres will upset the correlation because of the different propagation velocities of the different modes and so the noise power at the detector will be greater than that at the source. The conclusion is that for lowest noise as well as the attainment of best system bandwidth, the laser must be designed to operate in a *single* longitudinal mode only.

Several approaches have been made towards the control of longitudinal modes. As the longitudinal mode spacing is inversely proportional to the length of the laser

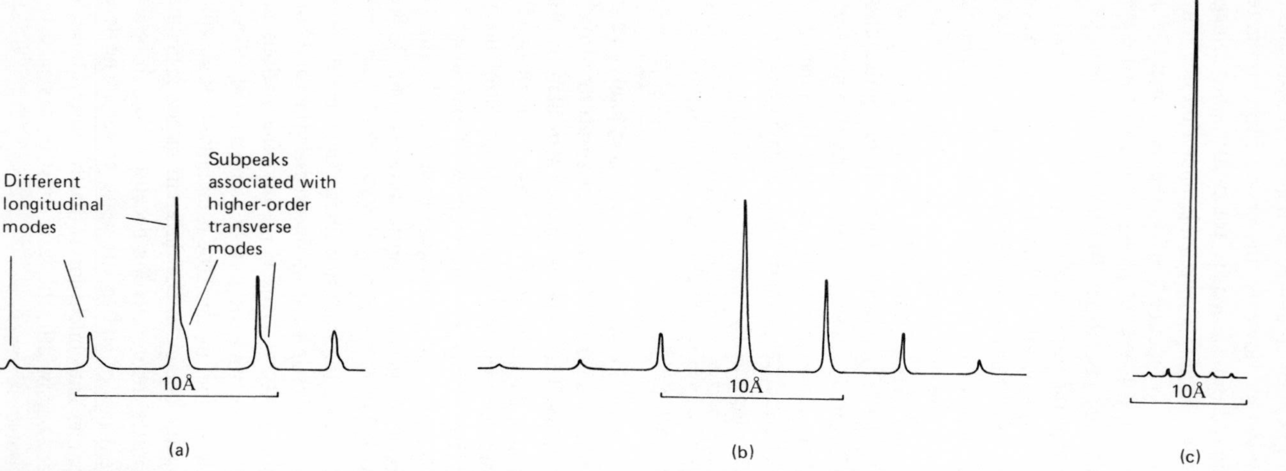

Figure 2.28. Types of spectra obtained from injection lasers. (a) Multitransverse and longitudinal mode. (b) Single transverse multi-longitudinal mode. (c) Single transverse mode

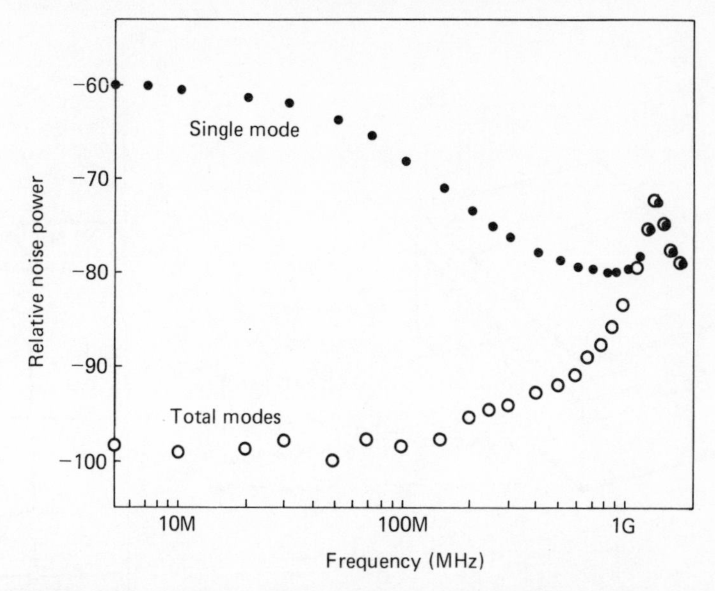

Figure 2.29. Noise power in a 30-kHz bandwidth plotted versus frequency for a single mode, and for all of the output modes of a multi-longitudinal mode laser[89]

cavity, the number of longitudinal modes within the gain wavelength range can be reduced by making the laser cavity very short. Short lasers with cavity lengths of 50 μm have been reported but handling of the chips is difficult and the technique is only marginally successful.

Bent-guide and connected-stripe structure lasers in which the stripe is turned through a right angle or changed in width along the length have been made with the aim of providing longitudinal mode selectivity, also with limited success.

An alternative approach is to use a laser cavity which only has one resonance within the gain band and examples of this are provided by the distributed feedback (DFB) or distributed Bragg reflector (DBR) structures. In such structures the feedback is provided by a periodic perturbation along the active-layer waveguide rather than by reflectors situated at the ends of the waveguide as in the conventional lasers. The periodicity Λ of the disturbance is given by the Bragg condition

$$\Lambda = \lambda/2n \tag{2.31}$$

where λ is the free-space wavelength and n is the effective index of the optical waveguide mode. For a GaAs heterostructure laser operating at 8000 Å with index $n = 3.6$, a periodicity around 1111 Å is required for Bragg reflection. The subject of periodic structures for DBR and DFB lasers has been reviewed in detail by Yariv and Nakamura.[90]

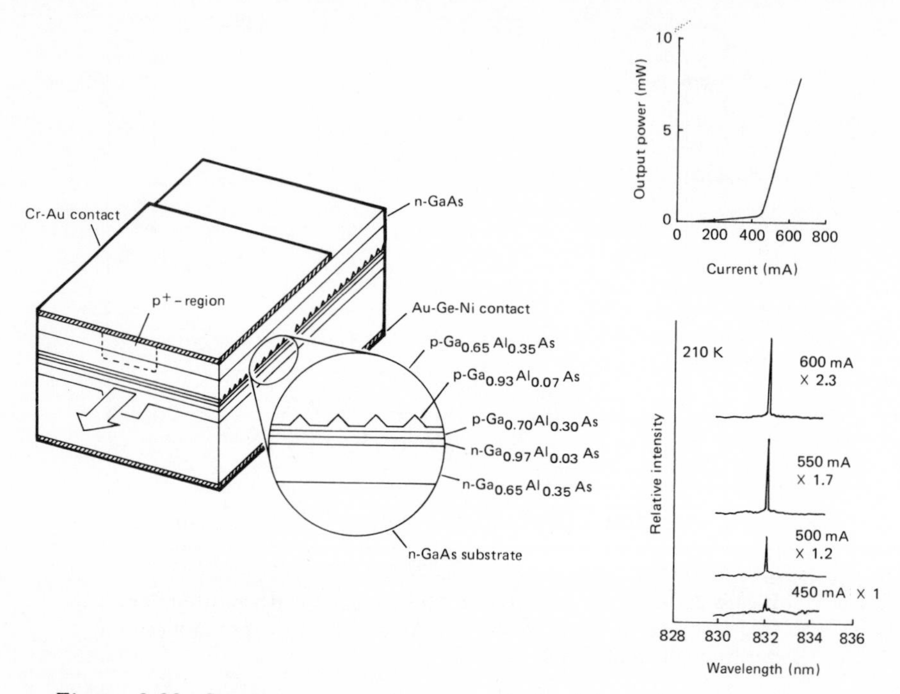

Figure 2.30. Structure, power/current relationship and spectra of the channel-substrate structure D.F.B. laser[91]

The structure and room-temperature spectrum of the channel substrate GaAs–GaAlAs DFB laser of Kuroda *et al.*[91] are shown in Figure 2.30. Fabrication of such devices is quite complex and demands high accuracy in grating periodicity for wavelength selection, and also transverse-mode control is just as important as in more conventional structures. However, the cleaved edges of the device are not part of the resonant cavity in DFB structures and so it becomes feasible to consider the integration of several devices on one chip. In fact this integration of DFB lasers has been demonstrated by Aiki *et al.*[92] Each laser on the chip was designed to, and operated at, a different wavelength and lased at room temperature under pulsed drive. The wavelength selection and control features of the DFB technique will possibly render it attractive for sophisticated wavelength-multiplexing systems of the future. The alternative resonant structure (DBR) in which the grating region is displaced from the active pumped region was demonstrated by Reinhart *et al.*[93]

The molecular-beam epitaxy and vapour-phase epitaxy techniques are amenable to the growth of very large numbers of layers with accurate control of composition and thickness. In 1973 Wang[94] proposed the idea of multilayer Bragg reflectors for waveguide structures. This has now become feasible with the multilayer growth capability and Shellan *et al.*[95] have reported the operation of transverse Bragg

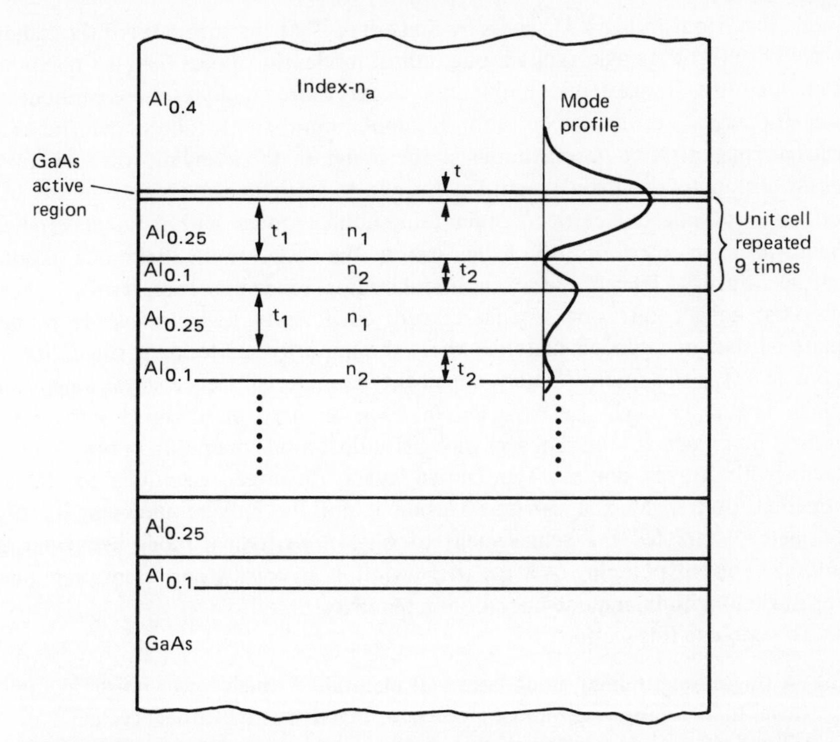

Figure 2.31. The layer geometry of a transverse Bragg reflector laser. The intensity profile of the confined radiation is also shown[95]

reflector injection lasers (see Figure 2.31). The wavelength selectivity of the reflectivity is less than that of the Fabry–Perot cavity formed by the cleaved facets but the technique nevertheless lends a new versatility to laser design which with further development may give rise to improved mode selectivity. However, it is obvious that the DFB approach is a significant departure from the present cleaved-facet geometry whose usefulness has been demonstrated, and we now consider longitudinal mode control in the conventional laser structures.

Longitudinal mode control in lasers with cleaved facets Transverse-junction stripe lasers have stood apart from the other stripe structures in that they naturally oscillate in a single longitudinal mode only. The main difference between this laser and all of the other structures is that the active layer is quite heavily doped *p*-type material. Streifer, Burnham and Scifres[54] considered this fact and have formulated an explanation which should enable most laser structures to be designed to operate in a single mode.

The theory they propose is an adaptation of the explanation of the longitudinal mode behaviour in Nd:YAG lasers by Statz *et al.*[96] At the antinodes of the standing wave pattern of a single excited longitudinal mode, the optical field is a maximum. Consequently, stimulated recombination occurs more rapidly at these antinodes, so that the excess carrier concentration is a minimum there. It follows that the excess carrier concentration is maximum at the nodes in the standing wave where the contribution to the main longitudinal mode is least. In the absence of any axial diffusion of injected carriers, other longitudinal modes within the spectral-gain bandwidth are able to use the higher gain at the nodes of the first mode to obtain sufficient gain to lase, and hence multilongitudinal mode operation results.

Axial carrier diffusion is theoretically sufficiently high in heavily p-doped material and in undoped material where the impurity scattering is small, but it is poor in n-type or intermediately doped p-type material. Hence single longitudinal mode operation might be expected in lasers with p^+ or undoped active layers. Indeed Scifres *et al.*[97] obtained single longitudinal-mode operation in broad-contact lasers with heavily doped p-type active layers. However, recently it has become apparent that high axial carrier diffusion is not the only requirement in stripe-geometry lasers for the achievement of single longitudinal mode behaviour and uniform current pumping over the stripe width is an equally important prerequisite for single longitudinal mode operation in practice.

To conclude this section:

(a) A single longitudinal mode laser will maintain a much lower noise fluctuation level than a multilongitudinal mode laser in a dispersive optical system.
(b) Distributed feedback lasers operate in a single longitudinal mode, are amendable to integration and wavelength multiplexing systems, but use complex technology.
(c) Single longitudinal mode operation can be achieved in cleaved-facet lasers if axial gain variations can be smoothed out, e.g. by encouraging axial excess carrier diffusion by the use of undoped or of heavily acceptor doped active layers.

2.4.5 Modulation

On applying current to a laser, a delay of a few nanoseconds generally occurs before the onset of lasing because of the time taken to establish sufficient excitation. A decaying relaxation oscillation of the optical output power with time is usually observed immediately after the lasing onset because of the coupling between the excess carrier concentration n and the photon flux P. This can be described if a single mode only is confined by the simple coupled rate equations[98]

$$\frac{\mathrm{d}n}{\mathrm{d}t}=\frac{J}{dq}-\frac{n}{\tau_n}-gP \tag{2.32}$$

$$\frac{\mathrm{d}P}{\mathrm{d}t}=gP-\frac{P}{\tau_s}+\frac{Fn}{\tau_n} \tag{2.33}$$

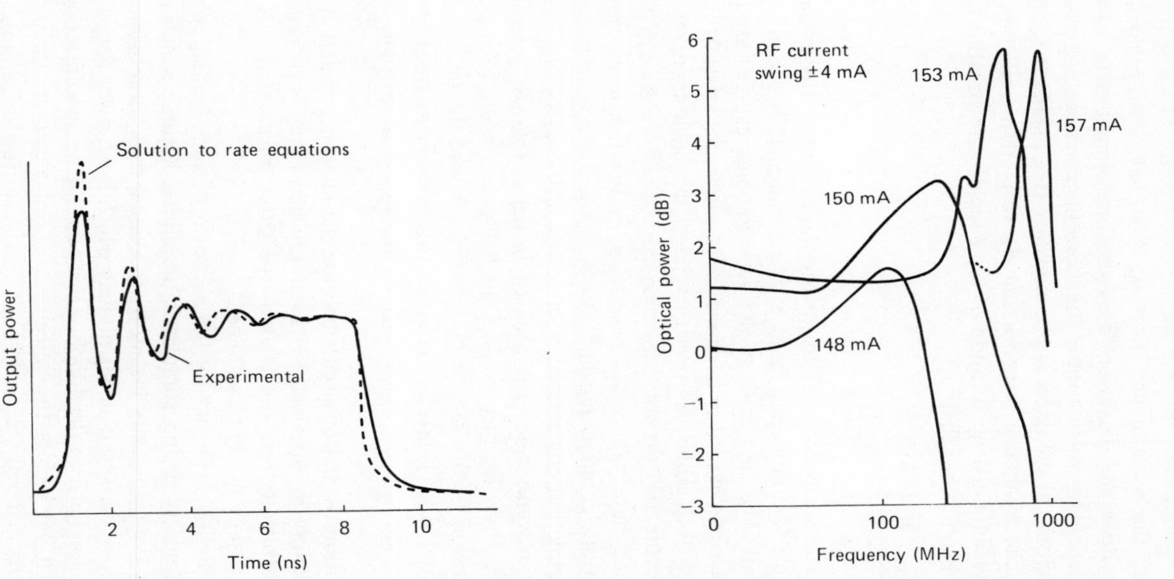

Figure 2.32. Transient pulse response and frequency response characteristics of stripe geometry GaAlAs lasers. (a) Measured transient response of an oxide stripe laser d.c. biased at 0.9 threshold, compared with theoretical model.[133] (b) Frequency response of a 5 μm wide deep zinc diffused stripe laser with a Ge doped active region (10^{18} cm^{-3})

where g is the gain and is approximately linearly dependent on the excess carrier concentration, d is the active region width, q is the electronic charge, τ_s the stimulated lifetime, τ_n the spontaneous lifetime and F is the fraction of the spontaneous emission radiating into the lasing mode. These equations have been generalized by Petermann[99] and by Buus and Davidson[100] so that multilongitudinal mode behaviour can be predicted. By solving these equations the transient pulse response and the frequency response of the lasers can be explained (see Figures 2.32a and b).

The resonance in the frequency response and the relaxation oscillation in the pulse response can be obtained by solution of the coupled rate equations, which give under small signal conditions that

$$f_r = \frac{1}{2\pi}\left\{\frac{1}{\tau_n\tau_s}\,\frac{(I_F - I_t)}{I_t}\right\}^{1/2} \tag{2.34}$$

where I_F is the bias or final current and I_t is the threshold current. These relaxation effects are not desirable for fibre-optic systems where minimal distortion of transmitted data is required. The first spike in the pulse response can be very narrow under high current drive and has led to the use of laser sources for assessment of the frequency response of fibre cables. However, several of the single transverse-mode narrow-stripe lasers fortunately do not show such marked relaxation oscillations in the pulse response and have suppressed resonances in the frequency responses which extend well into the gigahertz region.[87]

The improved modulation characteristics of narrow-stripe lasers which operate in the lowest-order transverse mode has been explained as follows: generally the stripe is pumped over a width which exceeds the half-power width of the optical field. Stimulated emission occurs most rapidly on axis where the optical electric field of the fundamental mode is greatest, leaving a reservoir of carriers at the stripe sides which can diffuse towards the axis smoothing out the lateral distributions (see Figure 2.33).

The normal relaxation oscillation occurs because the photon population builds up to a level which cannot be sustained by the steady injection of carriers. In the narrow-stripe case, diffusion from the sides of the stripe means that this initial photon population can more nearly be sustained.

It is clear from Equation 2.34 that for the highest resonance frequency τ_n and τ_s should be minimized. τ_s is the stimulated or photon lifetime which can be reduced by increasing the laser-cavity loss or by increasing the transmission factor of the laser facets. τ_n is the spontaneous lifetime which for heavily doped active layers is inversely proportional to the donor or acceptor doping concentration. The diffusing species which determines the uniformity in lasing gain are the minority carriers which in p-type material are electrons. Electrons diffuse more rapidly than holes (the ratio of the diffusion coefficient for electrons and holes in GaAs is around 50–100 according to Streifer, Burnham and Scifres[101]), and they will therefore more effectively smooth out any axial depletion of the excess carrier population than would holes. Summarizing, for best frequency response characteristics

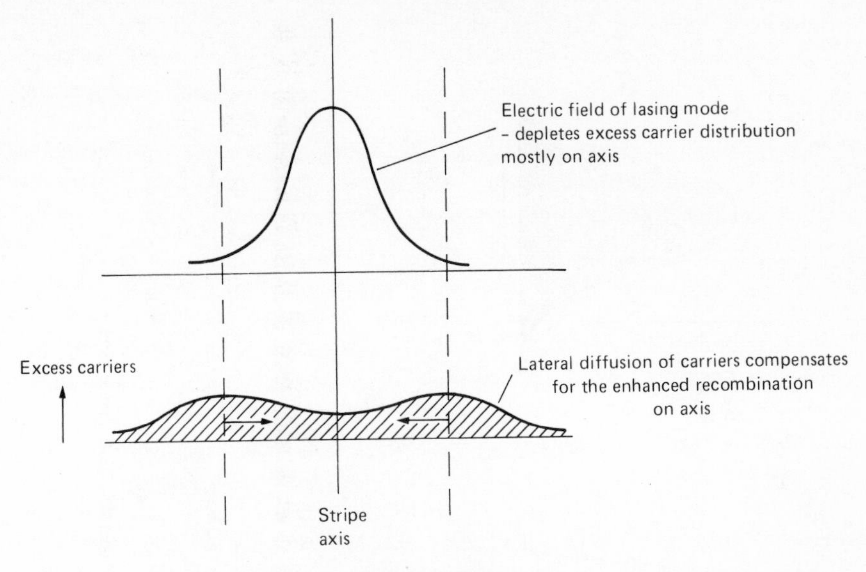

Figure 2.33.

and wide bandwidth the laser should operate stably in a single transverse mode, have a heavily doped *p*-type active layer for small τ_n, be short with high transmission facets for smallest τ_s, and have a narrow stripe which is pumped over a width greater than the optical field.

Laser drive circuits

Injection lasers can be driven directly by simply applying a current pulse, as in Figure 2.34a. A time delay which depends on the magnitude of the pulse current follows the rise of the current pulse from zero before the onset of lasing and large transient overshoots are observed which arise from the relaxation oscillations associated with each mode and the dynamically changing waveguide properties as the guide becomes established. This is not the best drive condition if a narrow spectral linewidth and a high signal-to-noise ratio is desired, but low-duty factors can be used to reduce power dissipation and effectively increase the laser life.

The current may be pulsed using a number of high-speed devices such as silicon bipolar transistors, GaAs FETs, Gunn-effect diodes, TRAPATT diodes and step-recovery diodes. For high data rates the turn-on delay can be reduced by increasing the amplitude of the current pulse or by maintaining the inversion near threshold between pulses by supplying a separate d.c. bias current (Figure 2.34b). The best spectral behaviour results if the laser is d.c. biased just above the threshold level, and pulsed from this pedestal with a current less than that required for the onset of higher-order transverse or longitudinal modes. This produces a pulse of radiation

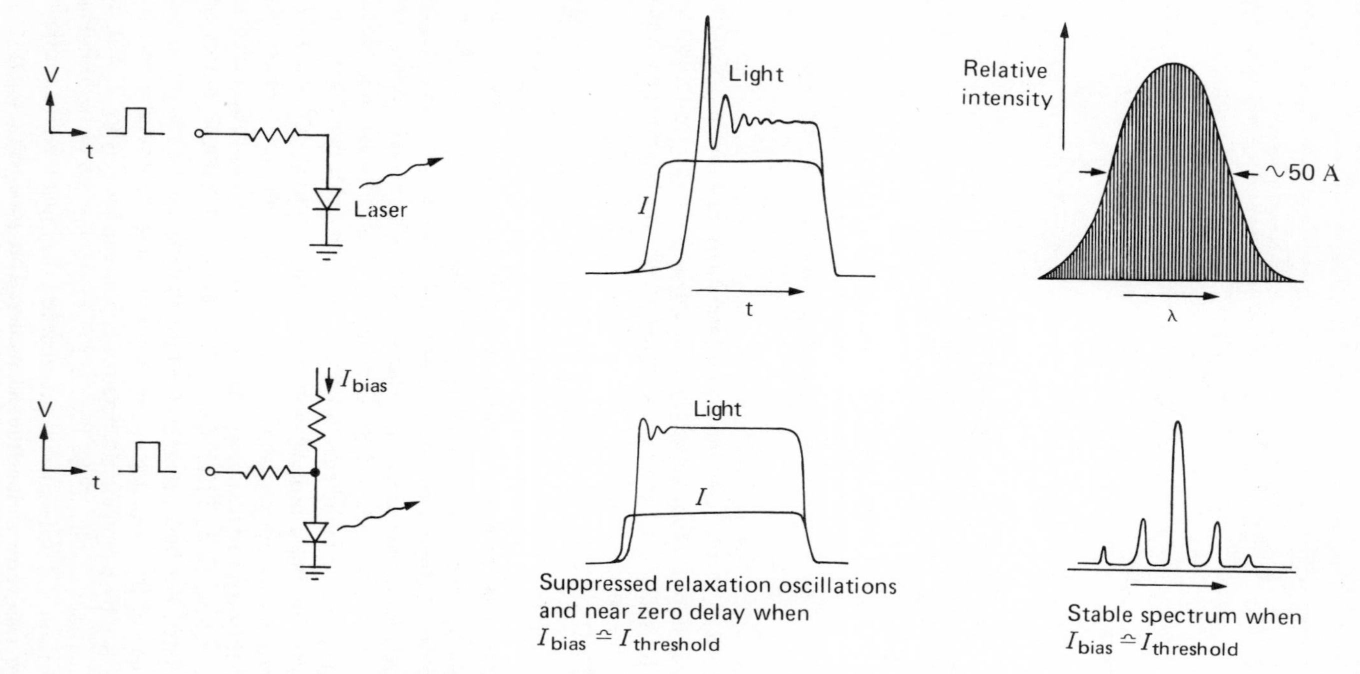

Figure 2.34. (a) Schematic of laser responses with the basic drive circuits

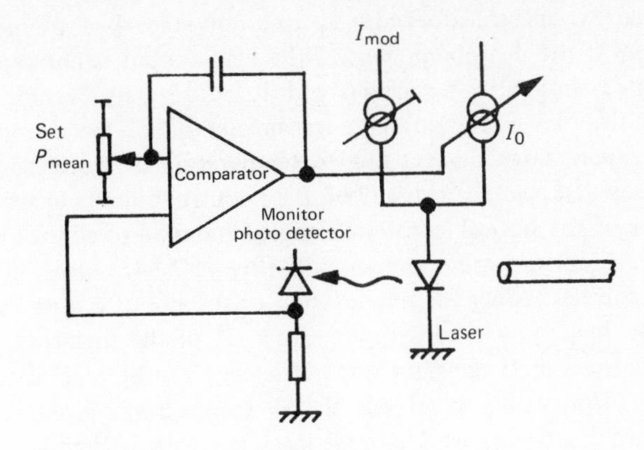

Figure 2.34. (b) Feedback circuit giving a stable mean output power[102]

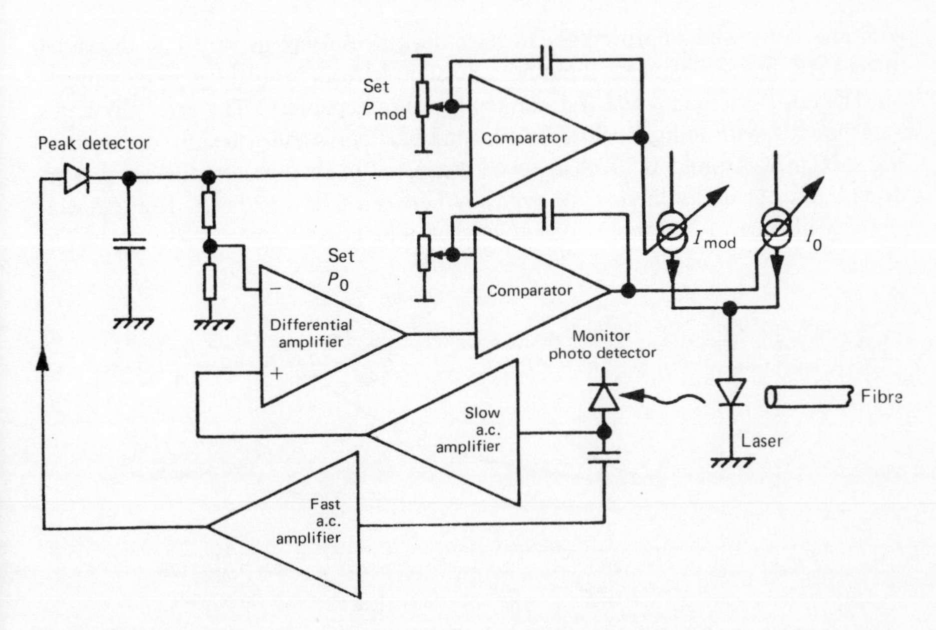

Figure 2.34. (c) Circuit which stabilizes both peak and minimum optical power levels for digital systems[102]

with constant narrow spectral linewidth up to a power level of around 10 mW in present-day lasers. If the d.c. component of the light output is not kept to a minimum, the detected shot noise associated with it will severely degrade the system performance so the bias conditions in a transmitter should be maintained using feedback. Device parameter changes due to temperature or degradation can then also be compensated. Epworth[102] described laser control circuits in which the different data levels of the optical signal are sampled and used to control the d.c. bias current and the RF current gain. Such circuits (Figure 2.34c) ensure constant output power and correct biasing for the duration of the life of a laser device at the expense of quite high power consumption because of the high-speed amplifiers and comparators needed. If constant slope efficiency can be maintained with life and temperature, Epworth pointed out that a feedback-stabilized RF electrical drive together with d.c. bias control with feedback is then sufficient.

Using the technique of d.c. prebiasing to just above threshold, Nawata and Takano[103] constructed a fibre-optic data link of 7 km operating at 800 Mbit/sec. This experiment demonstrated the very wide bandwidth capabilities of systems using laser transmitters.

2.4.6 Laser mounting and interfacing

With the continued improvement in laser design resulting in very low threshold current and high efficiencies, it has been possible to relax heat-sinking requirements considerably, yet maintain CW room-temperature capability. The early CW lasers were bonded with indium solder to diamond heat sinks and generally operated at 300–500 mA. Around 0.5 W had to be dissipated in these lasers whereas present stripe lasers operate with currents typically between 5 and 120 mA. The temperature-dependence of threshold current on heterojunction barrier height, shown in

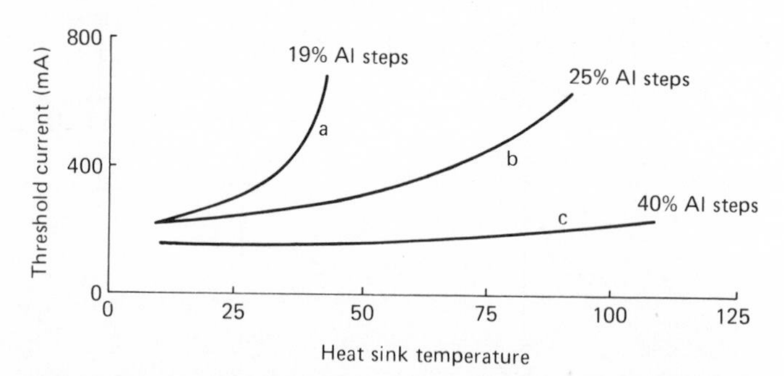

Figure 2.35. CW threshold current versus heat sink temperature for GaAlAs double heterostructure lasers with different heterojunction barrier heights. Curves (a) and (b) taken from Goodwin *et al.*[131] Curve (c) for a deep zinc diffused stripe laser of Goodfellow *et al*[132]

Figure 2.35, emphasizes the need to achieve consistently good thermal impedance when bonding the laser chips. Several analyses of the thermal impedance of multilayer structures have been made.

Joyce and Dixon[104] calculated thermal impedances assuming all heat is dissipated in the active region for the cases where the laser is bonded with indium,

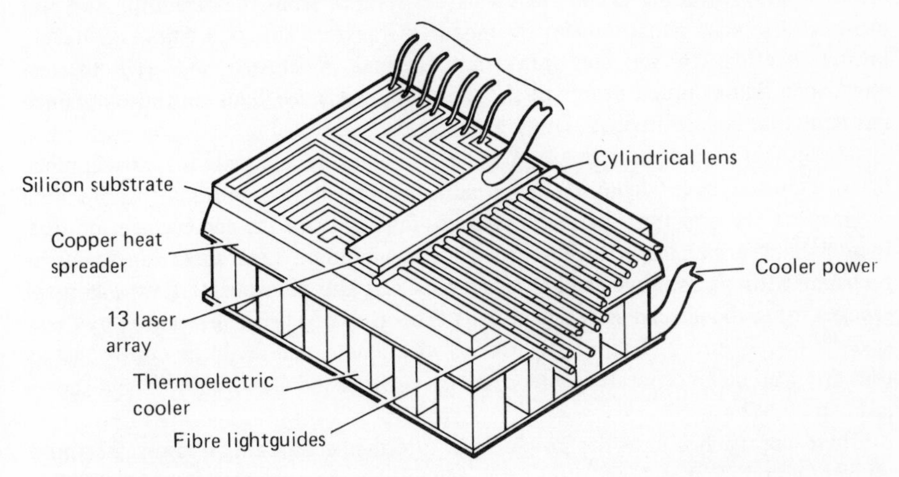

Figure 2.36. (a) Multichannel optical link package using a contoured silicon substrate for location of optical components[107]

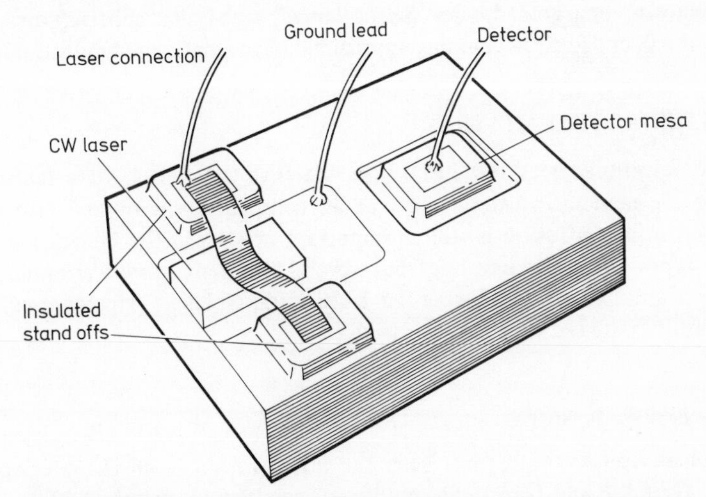

Figure 2.36. (b) Simple silicon submount for a stripe laser incorporating insulated stand-offs and a PIN detector to permit continuous monitoring of the output of the rear facet of the laser[108]

either layer up or down, on to copper or diamond heat sinks, and determined the advantages of having a thick gold pad electroplated on to the structure as a heat spreader. Kobayashi and Iwane[105] calculated the effects of voids, which are often included in the indium bonding layer, on the temperature uniformity of the active layer. Newman, Bond and Stefani[106] computed the thermal impedance of proton-isolated lasers with thick indium bonding layers on copper and silicon. They included heat-spreading effects in the layers remote from the heat sink, and also included the heat redistribution by radiative transfer. This was found to have a profound effect in reducing the total thermal impedance and gave thermal impedance figures much nearer to those actually measured than did those obtained assuming that heat is dissipated only in the active region.

Silicon and copper heat sinks can clearly be entirely adequate for most applications. This has been demonstrated practically but it is important that bonding surfaces are flat and that the solder contains few voids for the achievement of near-theoretical thermal impedance values. Silicon-submount heat sinks can easily be prepared with flat surfaces and the heat sink can also incorporate fibre-alignment grooves using the special selective etching properties of silicon as reported by Crow *et al.*[107] (see Figure 2.36a). Stand-offs for wire bonds, and feedback control detectors may also be incorporated as reported by Griffith and Goodfellow[108] and shown in Figure 2.36b.

These approaches ease the problems of interfacing fibres with lasers. For most efficient coupling, matching lenses are used to tailor the divergent emission distribution of stripe lasers to the narrow acceptance angle of most fibres. A short length of fibre is frequently used as a cylindrical lens, as in the grooved silicon submount system shown, but graded-index Selfoc lenses, high-index sphere lenses, tapered fibres and ball-ended fibre-coupling approaches have also been demonstrated.

2.5 MATERIAL PREPARATION

The first injection lasers and LEDs, like the first transistors, were fabricated by diffusing an element which behaves as an acceptor (or a donor) into a crystal substrate. In fact diffusion is still an important process in the fabrication of many current types of LED and laser but several different growth techniques have become established for producing the multi-epitaxial layers which are essential to the versatility required of fibre-optic sources.

2.5.1 Liquid-phase epitaxy

Liquid-phase epitaxy (LPE) has been the main workhorse in the development of GaAlAs, GaInAsP and GaAlAsSb multilayer structures. In Figure 2.37 we show the basic elements of the LPE technique. For example, to grow GaAlAs, gallium metal and aluminium metal are placed in a crucible with a sliding base and a source crystal of GaAs is placed on top and weighted. The crucible is then heated to temperature

T_0 (800 °C in the case of GaAlAs) and held for several hours at this temperature, during which time an amount of the source slice is dissolved into the melt and equilibrium is reached. When the temperature is slowly lowered, the liquid melt contains too much solute for equilibrium to be maintained and so a layer of crystal (GaAlAs) is deposited on to the source slice in order to restore the system back towards equilibrium. If before or during cooling the melt is moved over a substrate, a layer (of GaAlAs) is also deposited on to this and hence epitaxial growth by LPE occurs. Several melts containing different compositions are grown sequentially for double-heterojunction multilayer growth.

In principle, epitaxial growth will proceed with a very small drop in temperature corresponding with a small degree of supersaturation in the melt, but in practice, in order to ensure nucleation of growth of one layer on another the melt temperature is lowered by a step of 5–25°C and a large supersaturation is induced in the melts before growth.

LPE has produced some of the best epitaxial layer structures but has several problems; incomplete wipe-off of one melt before bringing the slice into contact with the next leads to the contamination of one melt with the material of another (melt mixing), and prevents re-use of old melts; between each layer the surface can be contaminated because it is exposed to the ambient of the apparatus; small thermal gradients and fluctuations can induce variations in layer thickness; the long melt equilibration times necessary can lead to oxidation of melt components due to oxygen residuals in the system atmosphere, and can lead to the inclusion of oxide particles in the epitaxial layer (this is especially a problem with melts containing aluminium); the need for a uniform temperature zone over the whole crucible length makes it difficult to grow multilayers on a large area of material. Some of these problems are avoided in the 'wipingless slider' approach.[109,110] In this method, successive melts are used to flush the previous melts away so that the problems associated with the separate wipe-off of each successive melt are alleviated. This goes some way towards tackling some of the problems of LPE. However, alternative epitaxial growth techniques afford growth on to large-area substrates and are therefore becoming more attractive for serious device production and are being actively developed, although most of the best devices produced to date have been produced in LPE material.

2.5.2 Vapour-phase epitaxy

The vapour-phase epitaxial technique (VPE) has three main variants but each is based on the principle that the constituent elements of the crystal are transported over the heated substrate, and hence deposited epitaxially. These three variants are:

(a) the trichloride system;
(b) the hydride system;
(c) the metallorganic system.

In the trichloride system, arsenic trichloride (in the case of GaAs epitaxy) is passed

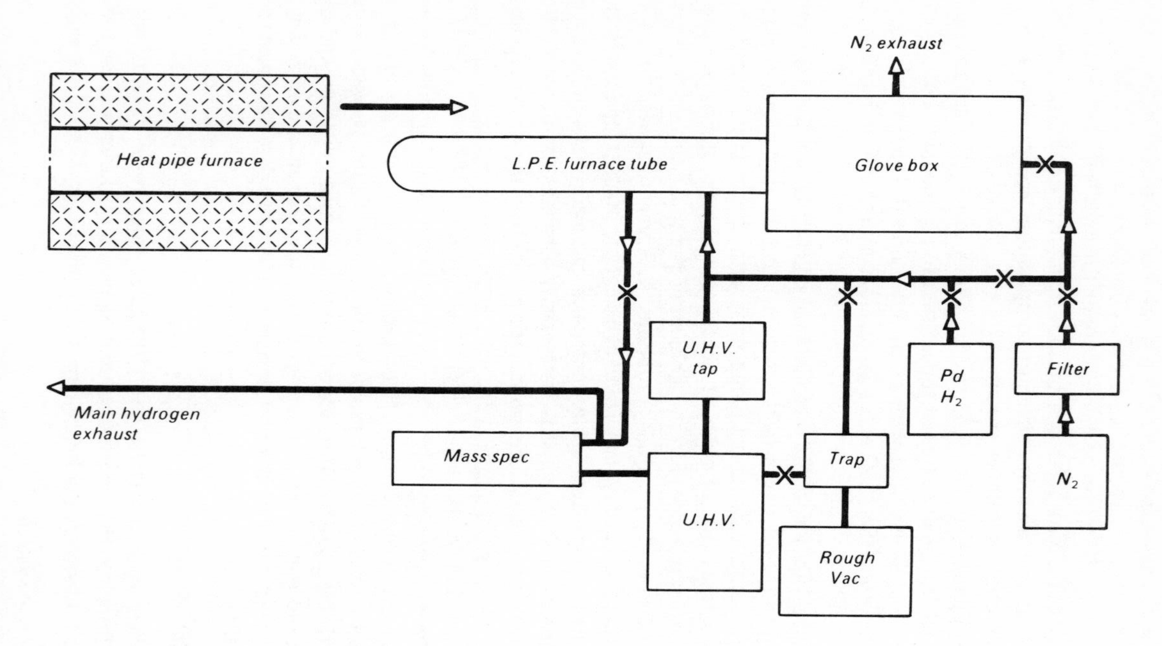

Liquid phase epitaxy system—schematic diagram

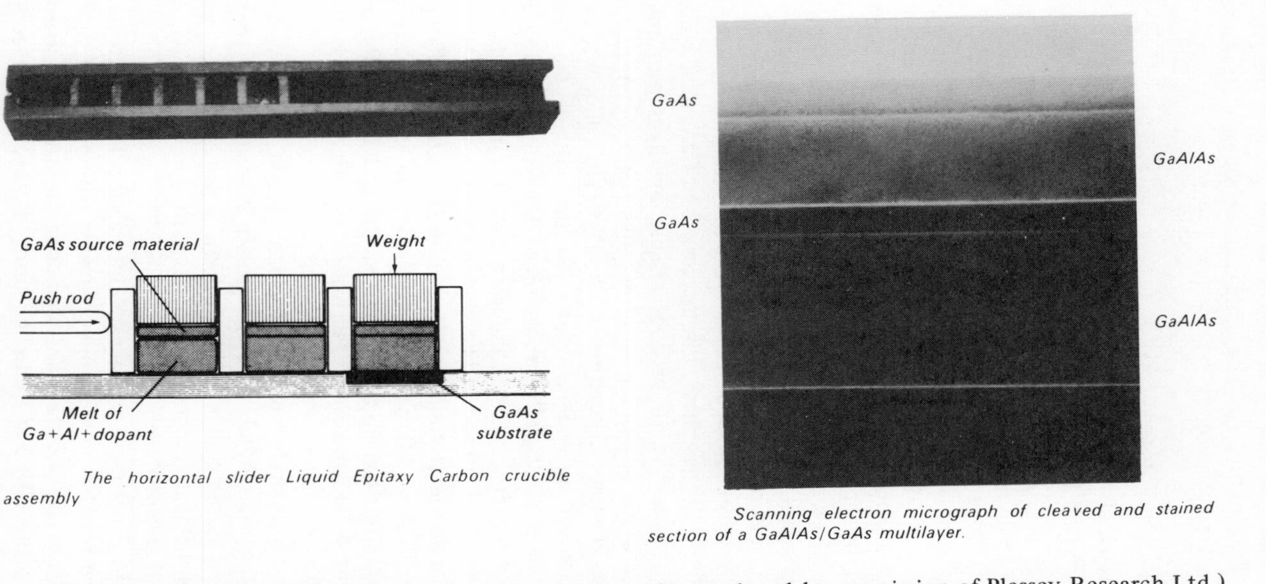

The horizontal slider Liquid Epitaxy Carbon crucible assembly

Scanning electron micrograph of cleaved and stained section of a GaAlAs/GaAs multilayer.

Figure 2.37. The basic liquid-phase epitaxy growth system. (Reproduced by permission of Plessey Research Ltd.)

over heated metallic gallium at temperature T_1 and the volatile gallium chloride and arsenic so formed are carried in pure hydrogen gas over the substrate at the somewhat lower temperature T_2. The epitaxial deposition of GaAs at the lower temperature follows the reaction[111]

$$4GaCl + 2H_2 + As_4 \rightleftharpoons 4GaAs + 4HCl$$

The relative fractions of As and Ga in the vapour phase are determined mainly by the temperature of the liquid $AsCl_3$ through which the hydrogen carrier gas is passed.

This process is employed extensively for the epitaxy of both GaAs and InP for microwave devices, where reproducible doping levels in the 10^{14}–10^{17} cm^{-3} range are required. Background doping levels are determined by silicon contamination due to dissociation of the silica reaction chamber commonly used, although this may be kept to an acceptably low level by adjustment of the relative mole fractions of the reactants in the vapour phase.

For the vapour-phase epitaxy of III–V alloys, precise control of the concentrations of the constituent elements in the vapour is required. This can be more readily achieved in the so-called 'hydride' process than in the trichloride method described above. Now the principal reactants are injected in metered amounts into the system directly in the gas phase. For the growth of GaAs these are arsine (gas), cracked at high temperature to produce arsenic and hydrogen, and hydrogen chloride (gas), passed over hot gallium to form gallium chloride. Again the substrate is held downstream at a low temperature to allow the epitaxial deposition of GaAs.

By the addition of an indium source to the system, the ternary alloy $Ga_xIn_{1-x}As$ has been successfully grown onto GaAs substrates by including a graded-composition buffer layer.[112] High-radiance LEDs fabricated in this material emitted efficiently at 1.06 μm and had bandwidths in excess of 200 MHz.

GaInAs *p–n* junction detectors fabricated in similar material exhibited quantum efficiencies up to 95% at 1.06 μm, but suffered from high leakage currents, a possible consequence of defects originating from the graded layer. Recently, Nuese *et al.*[59] have reported the growth and fabrication of GaInAsP/InP lasers using the hydride vapour-phase epitaxy growth technique.

The trichloride and hydride methods of vapour epitaxy are 'hot-walled' systems and are consequently not suitable for the growth of compounds containing the highly reactive element aluminium. This problem is avoided in the metallorganic system where the metallic compounds are transported cold in a hydrogen carrier gas as metal alkyls together with, say, arsine gas. This system is shown schematically in Figure 2.38. These gases are 'cracked' over a suitable substrate heated by an RF susceptor. The reaction is therefore contained within a small volume in the immediate vicinity of the substrate. A number of GaAlAs/GaAs structures have been prepared by the MOCVD technique using triethyl gallium and trimethyl Aluminium metal alkyls. These include low-threshold DH lasers operating CW at room temperature reported by Dupuis and Dapkus.[58] Excellent control of this

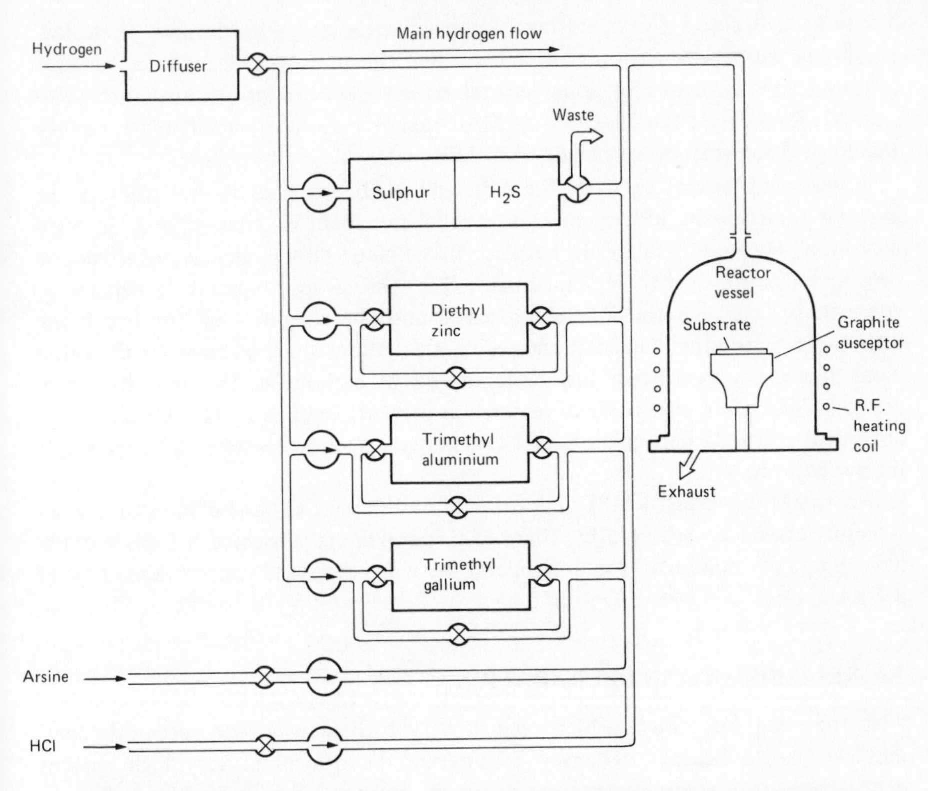

Figure 2.38. Schematic of the metallorganic vapour-phase epitaxy system

epitaxial process has been demonstrated by Dupuis and Dapkus[113] by the preparation of multilayer 'Bragg guided' lasers where sequential layers of GaAs and GaAlAs at 700 Å spacing were grown.

2.5.3 Molecular-beam epitaxy

In molecular-beam epitaxy (MBE), material is transferred directly as an ion beam from a heated source, known as a Knudsen cell, to a cooler single crystal substrate in an ultra-high vacuum (UHV) system, and conceptually is one of the simplest methods of epitaxy. The deposition of multicomponent semiconductors, such as the III–V alloys, is accomplished by using separate Knudsen cells for each element, with further cells for dopants, if desired. Alloy compositions and doping levels may be accurately controlled by means of the relative ion beam fluxes.

The close control achievable over alloy composition and depositions rate has allowed MBE growth of the thin multilayer structures for double-heterostructure GaAs/GaAlAs lasers.[114,115] Threshold current densities of 4000 A cm^{-2} were attained, resulting in CW operation in narrow stripe geometry devices. These low thresholds were achieved only after annealing the multilayer structures, perhaps indicating the existence of some crystal defect in the as-grown structure. Also GaAs/GaAlAs lasers fabricated from MBE material degraded much more rapidly under operation than devices made from LPE material.

A major difference between the LPE and MBE processes is that whereas the growing interface in LPE is in close equilibrium with its environment, in MBE conditions far from equilibrium exist, so that kinetic effects such as surface migration and desorption become important. To achieve good epitaxy, a balance or compromise between the kinetic processes must be sought – on the one hand, high surface mobility demands high substrate temperature, whereas on the other hand, low desorption needs low temperatures of deposition. The deposition processes in MBE are a major area of research at present, and this is aided by the *in situ* analytical methods pertaining to a UHV system, such as electron diffraction and mass spectrometry.

Recently the successful MBE deposition of multilayer GaInAs/InP lattice-matched structures has been achieved by Miller *et al.*[60] and lasers fabricated for this material have operated pulsed at room temperature with threshold current densities of 3.2 kA cm^{-2}.

2.6 RELIABILITY AND DEGRADATION

With the need for sources which emit at very high radiance for fibre-optic communications, it became necessary to operate the devices at very high current densities. Narrow-stripe lasers have threshold currents around 5 kA/cm^2, Burrus-type LEDs with emitting areas which match the fibre core operate around 7–15 kA/cm^2 and the smallest area devices which can be very efficiently coupled to fibres with microlenses operate at current densities in excess of 50 kA/cm^2. This contrasts strongly with the display LED devices, which operate typically at less than 1 kA/cm^2.

High current stresses were found to cause degradation problems in GaAs LEDs, when Biard *et al.*[67] showed that dark lines appeared in the emitting region which were aligned along major crystallographic directions. Lamorte *et al.*[116] showed that degradation was smaller in devices which had unbiased or passivated junction regions at the edges of the chip where they were exposed.

Early results on CW lasers were disappointing because of their rapid degradation but small-area GaAs homojunction and DH GaAlAs/GaAs LEDs operating at much greater current densities than the lasers, degraded very much more slowly. In fact, lives of tens of thousands of hours were achieved quite straightforwardly in GaAs high-radiance LEDs.[117]

Recently Kobayashi[7] reported results on a new GaAlAs double-heterostructure laser which had a *p–n* junction sited close to, but outside the active layer. This laser improved in performance during operation. Kobayashi proposed the explanation that the remote *p–n* junction extracted or gettered defects from the active layer, so reducing the absorption and increasing the spontaneous lifetime of the injected carriers. This development is presently in its infancy, but bodes well for the possibility for high-reliability, long-life lasers.

2.6.1 Degradation mechanisms

De Loach *et al.*[118] showed experimentally that the degradation rate in GaAs/GaAlAs lasers could be related directly to the bonding stresses and found that dark-line defects (DLDs) similar in character to those observed by Biard *et al.*[67] formed in the lasing stripe. Following this work, Petroff and Hartmann,[119] using transmission electron microscopy, revealed that one form of DLD was a dislocation network which was formed by the extension of a dislocation line threading the active layer. Other types of DLD form as a result of slip within the crystal lattice. Woolhouse[120] explained the formation of dislocation complexes by considering the diffusion of atoms along the dislocation line, i.e. pipe diffusion. This provides or removes the material needed for extension of the dislocation line which actually implies the growth or erosion of a crystal plane.

Improvements in substrates, crystal growth, device fabrication and selection have now largely eliminated the DLD problem from lasers and LEDs, and degradation is much slower and thought to be caused by other mechanisms. Hersee and Augustus[121] identified the long-term decay in output of GaAs LEDs (which gives predicted lives of 10^6 hours) with the diffusion of gold from the contact metallization reaching the active region. Yuasa *et al.*[122] identified residual degradation mechanisms in their lasers as due to photon-activated oxidation of the laser facets causing deterioration of reflectivity and also to interdiffisuion and alloying of the bonding metal solder causing increasing thermal impedance during operation. Other more fundamental degradation mechanisms have been invoked. These include the so-called Longini[123] and the Gold–Weisberg[124] phonon-kick mechanisms. The mechanism proposed by Longini concerns the movement of charged impurities in the depletion region. With zero bias the diffusion of charged impurities is balanced by the built-in field but under forward bias, diffusion can occur. If this mechanism applied and was responsible for degradation, the application of a reversed-bias field might be expected to reverse the effect. Burrus and Dawson[14] reported such a reversal in the degradation of LEDs which were pulsed into reverse bias periodically during normal forward-bias operation. Hwang[125] reported that the initial drop in output of lasers with GaAs active regions could be similarly restored by the application of a reverse bias. It appears that this mechanism may give rise to an initial degradation in output which saturates at a level dependent on the forward bias.

The Gold–Weisberg phonon-kick model concerns the formation of Frenkel defects in the lattice using the energy released from an electron–hole recombination process. The probability of such events depends on the binding energy of impurities in the lattice and on the magnitude of the energy released by a recombination event – which relates to the bandgap of the material. If this mechanism were dominant, it would perhaps predict better device lives for longer-wavelength (lower photon energy) emitters. Evidence against this is that both GaAlAs LEDs emitting around 0.82 μm and GaAs LEDs emitting at 0.9 μm have around 10^6-hour life extrapolations.

However, Kimmerling[126] argued that the phonon kick is an important factor in device degradation. He studied the electron–hole recombination enhanced annealing and motion of the point defects which had previously been introduced by bombarding the sample with 1-MeV electrons, and explained this in terms of processes like the phonon kick.

The degradation mechanisms we have just considered all depend on the device being forward-biased. However, it has been our experience that devices with non-ideal epoxy encapsulation can operate happily with almost zero degradation under normal drive conditions but a period of storage or temperature cycling can result in serious loss of output. This was traced to problems in the curing behaviour of the epoxy and shows that lifetesting must be very rigorously carried out if total reliability is to be assured.

2.6.2 Life tests and results

When device lifetimes extend beyond several thousand hours it is obviously necessary to devise an accelerated ageing procedure. In practice, devices are operated at several elevated temperatures and a failure rate or degradation rate is determined for each temperature. If a single degradation mechanism is involved, the Arrhenhius plot of degradation versus inverse temperature will be linear, allowing an activation energy to be determined. Using this activation energy, statistical data on the failure at high temperature can be used to predict mean times before failure (MTBF) for the intended operating temperature.

Complications arise if different degradation mechanisms dominate at the different selected operating temperatures so that there is an uncertainty in the activation energy, or if the test conditions differ significantly from the normal operating conditions. The latter complication applies when testing lasers. Laser thresholds increase with temperature so the elevated-temperature tests must be carried out either at increased current density if lasing is to be achieved or with the laser chip operated significantly below threshold, which is the condition usually adopted. However, long-term life tests with normal operating conditions are usually also carried out to give some assurance of the validity of the room-temperature life predictions derived from the accelerated ageing data.

Long-term 20 °C life tests on proton-isolated DH GaAlAs lasers have logged greater than 30 000 hours[127] as also have the 30 °C tests on zinc-diffused GaAs homojunction LEDs operating at 15 kA/cm^2.[128] Ikeda *et al.*[78] reported 12 000 hours of single-mode operation of a transverse-junction stripe GaAlAs laser.

Hsieh[20] has reported lives of 5000 and 7000 hours for CW GaInAsP/InP lasers which emit around 1.3 μm. This result is encouraging for this material in which CW laser operation was only first reported about 2 years earlier, and it is expected that improved contacts and materials preparation will lead to long life GaInAsP/InP devices.

We are not presently aware of any reports of long device lifetimes in any other DH materials but these may follow with further development. However, dark-line defects are expected to be a problem in materials which have defects and dislocations in the active region. This is most likely to occur where lattice mismatch between substrate and layers is accommodated using graded-lattice parameter layers. Such layers must necessarily contain misfit dislocations but these may be locked in the grading layers if the device layers are in compression. Nevertheless, we consider grading to be non-ideal for good epitaxy and best material quality, and therefore for best device reliability.

As yet, long device lives have only been reported for liquid-phase epitaxially grown junctions and for junctions formed by zinc diffusion into bulk crystal grown substrates. The GaAlAs multilayers prepared by molecular-beam epitaxy (MBE) were reported by Cho and Casey[57] to have excellent thickness and compositional control and to be of generally high quality, but device lifetimes were apparently very short. GaInAs/InP lasers prepared by MBE have recently been reported by Miller *et al.*[60] but no lifetime data has been given.

Lasers prepared using material grown by the metallorganic vapour epitaxy technique have been reported by Dupuis and Dapkus[58] but again long life data has not yet been presented.

It is obviously too early yet to make any conclusion about the reliability of LEDs and lasers prepared by MBE and the MOCVD growth techniques as it is also to conclude that GaInAsP/InP lasers will be long-lived, although we are optimistic that reliable devices can be made if sufficient control is applied to materials growth and device fabrication.

However, GaAlAs/GaAs LEDs and lasers have been operated continuously without failure for nearly four years and have extrapolated lives, at least in research environments, around 10^6 hours. These devices have sufficient reliability for most present day applications but hopefully this will be equalled and bettered in the newer sources.

ACKNOWLEDGEMENTS

We wish to acknowledge the help given us in the preparation of this chapter from several colleagues at the Allen Clark Research Centre, in particular A. Carter, G. Rees, S. Hersee and W. Stewart. We also thank M. Adams for some useful comments and references and finally the Plessey Company for assistance in preparation of the manuscript.

REFERENCES

1. K. C. Kao and G. A. Hockham, 'Dielectric-fibre surface waveguides for optical frequencies', *Proc. IEE,* **113** (7), 1151–1158, 1966.
2. H. Kroemer, 'A proposed class of hetrojunction injection lasers', *Proc. IEEE,* **51**, 1782, 1963.
3. Zh. I. Alferov and R. F. Kazarinov, Authors certificate 1032155/16–15 (USSR), 1963.
4. M. B. Panish, I. Hayashi and S. Sumski, 'A technique for the preparation of low threshold room temperature GaAs laser diode structures', *IEEE J. Quantum Electronics,* **5**, 210–212, 1969.
5. C. A. Burrus, J. Stone and A. G. Dentai, 'Room temperature 1.3 μm CW operation of a glass clad Nd: YAG single-crystal fibre laser end pumped with a single LED', *Electronics Lett.,* **12** (22), 600–601, 1976.
6. T. Kimura, S. Uehara, M. Saruwatari and J. Yamada, '800 MBit/s optical fibre transmission experiments at 1.05 μm', *Technical Digest of the Integrated Optics and Optical Fibre Communications Conference, Tokyo, Japan,* 489–491, 1977.
7. T. Kobayashi, 'Recombination-enhanced annealing effect of AlGaAs–GaAs remote junction heterostructure lasers', Paper presented at the IEEE International Semiconductor Laser Conference, San Francisco, 1978.
8. M. Horiguchi and H. Osanai, 'Spectral loss of low-OH-content optical fibres', *Electronics Lett.,* **12** (12), 310–312, 1976.
9. K. Nakagawa, T. Ito, K. Aida, K. Takemoto and K. Suto, '32 Mb/s optical fibre transmission experiment with 53 km long repeater spacing, 102–105, Supplement to Conf. Proceedings of the Fourth European Conference on Optical Communications, Geneva 1978.
10. S. Kobayashi, N. Shibata, S. Shibata and T. Izawa, 'Characteristics of optical fibres in infrared wavelength region', *Review of the Electrical Communications Laboratories,* **26** (3–4), 453–467, 1978.
11. D. A. Pinnow, A. L. Gentile, A. G. Standlee, A. J. Timpe and L. M. Bobrock, 'Polycrystalline fibre optical waveguides for infra red transmission', *Appl. Phys. Lett.,* **33** (1), 28–29, 1978.
12. L. G. Van Uitert and S. H. Wemple, '$ZnCl_2$ glass: A potential ultra low-loss optical fibre material', *Appl. Phys. Lett.,* **33** (1), 57–59, 1978.
13. G. Rees and W. Milne, 'Modulation efficiency in GaAs LED's', *Gallium Arsenide and Related Compounds,* 362–369, Edinburgh, 1976.
14. C. A. Burrus and R. W. Dawson, 'Small area high-current density GaAs electroluminescent diodes and a method of operation for improved degradation characteristics', *Appl. Phys. Lett.,* **17** (3), 97–99, 1970.
15. J. F. Womac and R. H. Rediker, 'The graded-gap $Al_xGa_{1-x}As$–GaAs heterojunction', *J. Appl. Phys.,* **43** (10), 4129–4133, 1972.
16. Zh. I. Alferov, V. M. Andreev, N. S. Zimogorova and D. N. Tret'yakov, *Sov. Phys. Semicond.,* **3**, 1373, 1970.
17. D. L. Rode, 'How much Al in the AlGaAs–GaAs laser', *J. Appl. Phys.,* **45**, 3887, 1974.
18. K. Sugiyama and H. Saito, 'AlGaAsSb DH heterojunction lasers', *Jap. J. Appl. Phys.,* **11**, 1057, 1972.
19. R. Nahory, M. A. Pollack and J. K. Abrokwah, 'Threshold characteristics and extended wavelength operation of $GaAs_{1-x}Sb_x/Al_yGa_{1-y}As_{1-x}Sb_x$ double heterostructure lasers', *J. Appl. Phys.,* **48** (9), 3988–3990, 1977.
20. J. J. Hsieh, J. A. Rossi and J. P. Donnelly, 'Room temperature CW operation

of GaInAsP/InP DH diode lasers emitting at 1.1 μm', *Appl. Phys. Lett.*, **28**, 709, 1976.
21. T. H. Glisson, J. R. Hauser, M. A. Littlejohn and C. K. Williams, 'Energy bandgap and lattice constant contours of III–V quaternary alloys', *J. Electronic Materials,* **7** (1), 1–17, 1978.
22. C. K. Williams, T. H. Glisson, J. R. Hauser and M. Littlejohn, 'Energy bandgap and lattice constant contours of III–V quaternary alloys of the form $A_xB_yC_zD$ or $AB_xC_yD_z$', *J. Electronic Materials,* **7** (5), 639–647, 1978.
23. R. A. Abram, R. W. Allen and R. C. Goodfellow, 'The coupling of light emitting diodes to optical fibres using sphere lenses', *J. Appl. Phys.*, **46** (8), 3468–3474, 1975.
24. Zh. I. Alferov, V. M. Andreev, B. V. Egoros and A. V. Syrbu, 'Heterojunction light-emitting Al–Ga–As diodes formed by negative profiling of the substrate', *Sov. Phys. Semicond.*, **11** (10), 1123–1127, 1977.
25. M. N. Zargar'yants, Yu. S. Mezin and S. I. Kolonenkova, 'Electroluminescent diode with a flat surface emitting continuously 25 W cm^{-2} Sr^{-1} at 300 °K', *Sov. Phys. Semicond.*, **4** (8), 1371–1372, 1971.
26. M. E. Ettenberg, K. Hudson and H. Lockwood, 'High radiance light emitting diodes', *IEEE J. Quantum Electron.*, **QE-9** (10), 987–991, 1973.
27. H. Kressel and M. E. Ettenberg, 'A new edge emitting (GaAl/As) heterojunction LED for fibre optical communications', *Proc. IEEE (Lett.),* **1975**, 1360–1361.
28. Y. Seki, 'Light extraction efficiency of the LED with guide layers', *Jpn J. Appl. Phys.*, **15** (2), 327–338, 1976.
29. M. Ettenberg, H. Kressel and J. Wittke, 'Very high radiance edge emitting LED', *IEEE J. of Quantum Electron.*, **QE-12** (6), 360–364, 1976.
30. J. P. Wittke, 'Spontaneous emission rate alteration by dielectric and other waveguiding structures', *RCA Review,* **36**, 655–660, 1975.
31. S. Horiuchi, K. Ikeda, T. Tanaka and W. Susaki, 'A new structure of GaAs–GaAlAs DH LED with a self aligned sphere lens for efficient coupling to low loss optical fibres', IEEE Specialist Conference on the Technology of LEDs, Japan, Sept. 1976.
32. S. Sugimoto, K. Minemura, K. Kobayashi, M. Seki, M. Shikada, A. Ueki, T. Yanase and T. Miki, 'High speed digital signal transmission experiments by optical wavelength division multiplexing', *Tech. Digest of the Integrated Optics and Opt. Fiber Comm. Internat. Conf.*, 497–500, Tokyo, Japan, 1977.
33. H. Ishio and T. Miki, 'A preliminary experiment on wavelength division multiplexing transmission using LED', *Tech. Digest of the Integrated Optics and Opt. Fiber Comm. Internat. Conf.*, 493–496, Tokyo, Japan, 1977.
34. Y. S. Liu and D. A. Smith, 'The frequency response of an amplitude-modulated GaAs luminescent diode', *Proc. IEEE,* **1975**, 542–544.
35. W. Harth, W. Huber and J. Heinen, 'Frequency response of GaAlAs light-emitting diodes', *IEEE Transactions on Electron Devices,* **1976**, 478–480.
36. H. C. Poon and H. K. Gummel, 'Modeling of emitter capacitance', *Proc. IEEE,* **1969**, 2182.
37. T. P. Lee, 'Effect of junction capacitance on the rise time of LEDs and the turn-on delay of injections lasers', *Bell Syst. Tech. J.*, **53**, 1975.
38. R. C. Goodfellow, A. C. Carter, I. Griffith and R. Bradley, 'GaInAsP/InP fast, high radiance 1.05–1.3 μm wavelength LEDs with efficient lens coupling to small N.A. silica optical fibres' *Transactions on Electron Devices*, August, 1979.

39. J. Heinen, W. Huber and W. Harth, 'Time and frequency response of different types of light emitting diodes', *Proceedings of the 2nd European Conference on Optical Fibre Communications, Paris,* 1976, 277.
40. T. P. Lee, C. A. Burrus, Jr and B. I. Miller, 'A stripe-geometry double-heterostructure amplified-spontaneous-emission (super-luminescent) diode', *IEEE J. Quantum Electron.,* **QE-9**, 820–828, 1973.
41. T. Ozeki and E. H. Hara, 'Measurement of non-linear distortion in light emitting diodes', *Electronics Lett.,* **12**, 78–80, 1976.
42. K. Asatani and T. Kimura, 'Analysis of LED non-linear distortions', *IEEE Transactions on Electron Devices,* **ED-25** (2), 199–207, 1978.
43. M. Abe, O. Hasagawa, H. Hamaguchi, I. Umebu, T. Yamaoka and H. Nakamura, 'Linearity of high efficiency GaAlAs LEDs', **OQE 1976** (105), 31–38 (in Japanese). See also *Tech. Digest of Integrated Optics and Optical Comm. Conference*, 109–112, Tokyo, Japan, 1977 (in English).
44. Y. Ueno and M. Kajitani, 'Colour TV transmission light emitting diode', *NEC Research and Development*, No. 35, 15–20, October 1974.
45. J. Strauss and O. J. Szentesi, 'Linearisation of optical transmitters by a quasi feed forward compensation technique', *Electronics Lett.,* **13** (6), 158, 159, 1977.
46. J. Strauss and O. I. Szentesi, 'Linear transmitters for fibre optic communications using high radiance LEDs', *Technical Digest 2nd European Conference on Optical Fibre Communications, Paris,* 1976, 209–212.
47. Asatani, K. and Kimura, T., 'Linearisation of LED non-linearity by pre-distortions', *IEEE Transactions on Electron Devices,* **ED-25** (2), 207–212, 1978.
48. R. C. Goodfellow and A. Mabbit, 'Wide bandwidth high-radiance gallium arsenide light emitting diodes for fibre optic communication', *Electronics Lett.,* **12** (2), 50–51, 1976.
49. A. C. Carter, R. C. Goodfellow and R. Davis, 'High speed GaAs and GaInAs high radiance light emitting diodes', Internat. Electron Devices Meeting, Washington 1977, 577–581.
50. A. Mabbit and R. C. Goodfellow, 'High radiance small area gallium indium arsenide 1.06 μm LEDs', *Electronics Lett.,* **11** (13), 274–275, 1975.
51. C. A. Burrus and B. I. Miller, 'Small area double heterostructure aluminium-gallium arsenide electroluminescent diode sources for optical fibre transmission lines', *Optics Commun.,* 1971, 307–399.
52. A. G. Dentai, T. P. Lee and C. A. Burrus, 'Small-area, high radiance CW InGaAsP LEDs emitting at 1.2 to 1.3 micron', *Electronics Lett.,* **13** (16), 484–485, 1977.
53. I. Umebu, O. Hasegawa and K. Akita, 'InGaAsP/InP DH LEDs for fibre optical communications', *Electronics Lett.,* **14** (16), 499–500, 1978.
54. D. R. Scifres, R. D. Burnham and W. Streifer, 'Single longitudinal mode operation of diode lasers', *Appl. Phys. Lett.,* **31** (2), 112–114, 1977.
55. M. G. A. Bernard and G. Duraffourg, 'Laser conditions in semiconductors', *Phys. Status Solidi,* **1**, 699–703, 1961.
56. M. B. Panish, I. Hayashi and S. Sumski, 'A technique for the preparation of low threshold room temperature GaAs laser diode structures', *IEEE J. Quantum Electrons,* **5**, 210–212, 1969.
57. A. Cho and H. Casey, Jr, 'GaAs–$Al_xGa_{1-x}As$ DH lasers prepared by molecular beam epitaxy', *Appl. Phys. Lett.,* **25**, 288, 1974.
58. R. D. Dupuis and P. D. Dapkus, 'Very low threshold $Ga_{1-x}Al_xAs$GaAs DH lasers grown by metalorganic chemical vapour deposition', *Appl. Phys. Lett.,* **32** (8), 473–475, 1978.

59. C. J. Nuese, G. H. Olsen, R. E. Emstrom and J. R. Appert, 'Vapour-grown CW lasers of InGaAsP/InP', Paper presented at the IEE International Semiconductor Laser Conference, San Francisco, 1978.
60, B. I. Miller, J. H. McFee, R. J. Martin and P. K. Tien, 'Room temperature operation of lattice-matched $InP/Ga_{0.47}In_{0.53}As/InP$ DH lasers grown by MBE', *Appl. Phys. Lett.*, **33** (1), 44–47, 1978.
61. R. E. Nahory, M. A. Pollack and J. K. Abrokwah, 'Threshold characteristics and extended wavelength operation of $GaAs_{1-x}Sb_x/Al_yGa_{1-y}As_{1-x}Sb_x$ double heterostructure lasers', *J. Appl. Phys.*, **48** (9), 3988–3993, 1977.
62. C. J. Neuse, G. H. Olsen and M. Ettenberg, 'Vapour grown CW room-temperature $GaAs/In_yGa_{1-y}P$ lasers', *Appl. Phys. Lett.*, **29** (1), 54–56, 1976.
63. H. C. Casey and M. B. Parish, 'Composition dependence of the $Ga_{1-x}Al_xAs$ direct and indirect energy gaps', *J. Appl. Phys.*, **40**, 4910, 1969.
64. H. C. Casey, Jr, M. B. Parish and J. L. Merz, 'Beam divergence of the emission from DH injection lasers', *J. Appl. Phys.*, **44** (12), 5470–5475, 1973.
65. R. E. Nahory and M. A. Pollack, 'Threshold dependence on active layer thickness in InGaAsP/InP DH lasers', *Electronics Lett.*, **14** (23), 727–729, 1978.
66. C. J. Hwang and J. Dyment, 'Dependence of threshold and electron lifetime on acceptor concentration in $GaAs–Ga_{1-x}Al_xAs$ lasers', *J. Appl. Phys.*, **44**, 3240–3244, 1973.
67. J. R. Biard, G. E. Pittman and J. F. Leezer, 'Degradation of quantum efficiency in gallium arsenide light emitters', *Gallium arsenide: 1966 Symposium Proceedings*, Reading, UK, 1966, 113.
68. K. Kobayashi, R. Lang, H. Yonezu, Y. Matsumoto, T. Shinohara, I. Sakuma, T. Suzuki and I. Hayashi, 'Unstable horizontal modes and their stabilisation with a new stripe structure', *IEEE J. Quantum Electron.*, **QE-13** (8), 659–661, 1977.
69. W. D. Schlosser, 'Gain induced modes in planar structures', *Bell Syst. Tech. J.*, **52** (6), 887–905, 1973.
70. Suematsu and Yamada, *Trans. IECE*, **11**, 1974 (in Japanese).
71. P. Kirkby, A. Goodwin, G. Thompson and P. Selway, 'Observations of self-focussing in stripe geometry SC lasers and the development of a comprehensive model of their operation', *IEEE J. Quantum Electron.*, **QE-13** (8), 705–719, 1977.
72. T. Paoli, 'Waveguiding in a stripe-geometry junction laser', *IEEE J. Quantum Electron.*, **QE-13** (8), 662–668, 1977.
73. J. C. Dyment, L. A. D'Asaro, J. C. North, B. I. Miller and J. E. Ripper, 'Proton-bombardment formation of stripe-geometry heterostructure lasers for 300 °K CW operation', *Proc. IEEE*, **60**, 726–728, 1972.
74. A. G. Steventon, P. J. Fiddyment and D. H. Newman, 'Low threshold current proton-isolated (GaAl)As DH lasers', *Opt. and Quant. Elect.*, **9**, 519–525, 1977.
75. R. L. Hartman, N. E. Schumaker and R. W. Dixon, 'Continuously operated (AlGa)As DH lasers with 70 °C lifetimes as long as two years', *Appl. Phys. Lett.*, **31** (1), 756–759, 1977.
76. H. Kumabe, T. Tanaka, H. Namizaki, M. Ishii and W. Susaki, 'High temperature single mode CW operation with a junction-up TJS laser', *Appl. Phys. Lett.*, **33** (1), 38–39, 1978.
77. D. D. Sell, H. C. Casey, Jr and K. W. Wecht, 'Concentration dependence of the refractive index for *n*- and *p*-type GaAs between 1.2 and 1.8 eV', *J. Appl. Phys.*, **45** (6), 2650–2657, 1974.
78. K. Ikeda, H. Kan, E. Oomura, K. Matsui, M. Ishii and W. Sasaki, 'TJS laser

maintaining a single mode behaviour even after 12,000 hours CW ageing', *Technical Digest of Internat. Conf. on Integ. Optics and Optical Fibre Comm., July 1977*, Tokyo.

79. H. Namizaki, 'Transverse junction-stripe lasers with a GaAs p–n homojunction', *IEEE J. Quantum Electron.*, **QE-11** (7), 427–431, 1975.
80. K. Aiki, M. Nakamura, T. Kuroda, J. Umeda, R. Ito, N. Chinone and M. Maeda, 'Transverse mode stabilised $Al_xGa_{1-x}As$ injection lasers with channelled-substrate-planar structure', *IEEE J. Quantum Electron.*, **QE-14** (2), 89–94, 1978.
81. N. Matsumoto, 'The bent-guide structure AlGaAs semiconductor laser', *IEEE J. Quantum Electron.*, **QE-13** (8), 560–564, 1977.
82. D. R. Scifres, W. Streifer and Burnham, R. D., 'Curved stripe GaAs:GaAlAs diode lasers and waveguides', *Appl. Phys. Lett.*, **32** (4), 231–234, 1978.
83. B. Frescura, C. Hwang, H. Luechinger and J. Ripper, 'Suppression of output non-linearities in DH lasers by use of misaligned mirrors', *Appl. Phys. Lett.*, **31**, 770, 1977.
84. D. Scrifres, W. Streifer and R. Burnham, 'GaAs/GaAlAs diode lasers with angled stripes', *IEEE J. Quantum Electron.*, **QE-14** (4), 223–227, 1978.
85. P. J. De Waard, 'Stripe geometry DH lasers with linear output/current characteristics', *Electronics Lett.*, **13**, 400, 1977.
86. T. Tsukada, 'GaAs–$Ga_{1-x}Al_xAs$ buried-heterostructure injection lasers', *J. Appl. Phys.*, **45** (11), 4906, 1974.
87. K. Saito, N. Shige, T. Kajimura, T. Tsukada and R. Ito, 'Buried-heterostructure lasers as light sources in fibre optics communications', *Technical Digests of the Integrated Optics and Optical Communications Conference*, Tokyo, Japan, 1977, 65–68.
88. N. Chinone, R. Ito and O. Nakada, 'Limitations of power outputs from continuously operating GaAs–$Ga_{1-x}Al_xAs$ double heterostructure lasers', *J. Appl. Phys.*, **47** (2), 785–786, 1976.
89. T. Ito, S. Machida, K. Nawata and T. Ikegami, 'Intensity fluctuations in each longitudinal mode of a multimode AlGaAs laser', *IEEE J. Quantum Electron.*, **QE-13** (8), 574–579, 1977.
90. A. Yariv and M. Nakamura, 'Periodic structures for integrated optics', *IEEE J. Quantum Electron.*, **QE-13** (4), 233–253, 1977.
91. T. Kuroda, S. Yamashita, M. Nakamura and J. Umeda, 'Channelled-substrate-planar structure distributed feedback semiconductor lasers', *Appl. Phys. Lett.*, **33** (2), 173–174, 1978.
92. K. Aiki, M. Nakamura and J. Umeda, 'Frequency multiplexing light source with monolithically integrated distributed feedback diode lasers', *Appl. Phys. Lett.*, **29** (8), 506–508, 1976. *IEEE J. Quantum Electron.*, **QE-13** (4), 220–223, 1977.
93. I. K. Reinhart, R. A. Logan and C. V. Shank, 'GaAs–$Al_xGa_{1-x}As$ injection lasers with distributed Bragg reflectors', *Appl. Phys. Lett.*, **27**, 45–48, 1975.
94. S. Wang, 'Proposal of periodic layered waveguide structures for distributed lasers', *J. Appl. Phys.*, **44** (2), 767–779, 1973.
95. J. B. Shellan, W. Ng, P. Yeh, A. Hariv and A. Cho, 'Transverse Bragg-reflector injection lasers', *Optics Letters*, **2** (5), 136–138, 1978.
96. H. Statz, C. L. Tang and J. M. Lavine, 'Spectral output of semiconductor lasers', *J. Appl. Phys.*, **35** (9), 2581–2585, 1964.
97. D. R. Scifres, R. D. Burnham and W. Streifer, 'Single longitudinal mode operation of diode lasers', *Appl. Phys. Lett.*, **31** (2), 112–114, 1977.
98. M. J. Adams, 'Rate equations and transient phenomena in semiconductor lasers', *Opto. Electron.*, **5**, 201–215, 1973.

99. K. Petermann, 'Theoretical analysis of spectral modulation behaviour of semiconductor injection lasers', *Opt. and Quant. Elect.*, **10**, 233–242, 1978.
100. J. Buus and M. Danielson, 'Carrier diffusion and higher order transversal modes in spectral dynamics of the semiconductor laser', *IEEE J. Quantum Electron.*, **QE-13**, 669–674, 1977.
101. W. Streifer, R. D. Burnham and D. R. Scifres, 'Dependence of longitudinal mode structure on injected carrier diffusion in diode lasers', *IEEE J. Quantum Electron.*, **QE-13**, 403–404, 1977.
102. R. E. Epworth, 'Subsystems for high speed optical links', *Proceedings of the 2nd European Conference on Optical Fibre Communications*, 377–382, Paris, 1976.
103. K. Nawata and K. Takano, '800 Mb/s optical-repeater experiment', *Electronics Lett.*, **12** (7), 178, 1976.
104. W. B. Joyce and R. W. Dixon, 'Thermal resistance of heterostructure lasers', *J. Appl. Phys.*, **46** (2), 855–862, 1975.
105. T. Kobayashi and G. Iwane, 'Three-dimensional thermal problems of DH SC lasers', *Jap. J. Appl. Phys.*, **16** (8), 1403–1408, 1977.
106. D. H. Newman, D. J. Bond and Jane Stefani, 'Thermal resistance models for proton-isolated double heterostructure lasers', *IEE Journal on Solid State Electron Devices,* 2 (2), 41–46, 1978.
107. J. D. Crow, J. S. Harper, L. D. Comerford, M. J. Brady and R. A. Laff, 'GaAs laser source package for multichannel optical links', Paper WB6-1, *Tech. Digest of Optical Fibre Transmission Meeting,* Feb. 1977, Williamsburg, USA.
108. I. Griffith and R. Goodfellow, 'Silicon submounts for injection lasers', 1977 (unpublished).
109. Z. I. Alferov, V. M. Andreyev, S. G. Konnikov, V. R. Larionov and G. N. Shelovanova, 'Liquid phase epitaxy of $Al_xGa_{1-x}As$–GaAs heterostructures', *Kristall und Technik,* **10**, 103–110, 1975.
110. K. Akita, Y. Nishitani, K. Nakajinia, A. Yamaguchi, T. Kusinoki, K. Kotani, H. Imai, M. Takusagawa and O. Ryuzan, 'An improved LPE growth method for (Ga, Al)As double heterostructures', *IEEE J. Quantum Electron.*, **QE-13** (8), 585–586, 1977.
111. J. R. Knight, D. Effer and P. R. Evans, *Solid State Electronics,* **8**, 178, 1965.
112. A. W. Mabbit, R. Goodfellow, W. Milne and J. Percival, 'Wide bandwidth small area high radiance LED sources for fibre optic systems', Conference on GaAs and related compounds, *Inst. Phys. Conf. Ser. No. 33a,* 1977, 354–361.
113. R. D. Dupuis and P. D. Dapkus, 'Room temperature operation of distributed-Bragg-confinement $Ga_{1-x}Al_xA_5$–GaAs lasers grown by metalorganic chemical vapour deposition', *Appl. Phys. Lett.*, **33** (1), 68–69, 1978.
114. A. Cho and H. Casey, Jr, 'GaAs–$Al_xGa_{1-x}As$ DH lasers prepared by molecular beam epitaxy', *Appl. Phys. Lett.*, **25**, 228, 1974.
115. H. C. Casey, Jr, A. Y. Cho and P. A. Barnes, 'Application of molecular beam epitaxial layers to heterostructure lasers', *IEEE J. Quantum Electron.*, **QE-11**, 467, 1975.
116. M. F. Lamorte, W. Agosto, G. Kupsky, H. W. Becke and N. Pennucci, 'Degradation phenomena of operating life in GaAs LEDs', *Gallium Arsenide, 1966 Symposium Proceedings,* Reading, UK, 1966, 118–125.
117. R. C. Goodfellow, 'High radiance small area GaAs lamps', Paper presented at the Specialist Conference on Electroluminescent Devices, held in Atlanta, USA, Nov. 1974.
118. B. W. Hakki, R. L. Hartman and L. A. D'Asaro, 'Degradation of C.W. GaAs

double-heterojunction lasers at 300 K', *Proc. IEEE,* **61** (7), 1042–1044, 1973.

119. P. Petroff and R. L. Hartman, 'Defect structure introduced during operation of heterojunction GaAs lasers', *Appl. Phys. Lett.,* **28** (8), 469–471, 1973.
120. G. R. Woolhouse, 'Degradation in injection lasers', *IEEE J. Quantum Electronics,* **QE-11** (7), 556–561, 1975.
121. S. Hersee and P. D. Augustus, 'GaAs LED failure by metallization diffusion', submitted to *Journal of Electronic Materials,* 1979.
122. T. Yuasa, M. Ogawa, K. Endo and H. Yonezu, 'Degradation of (AlGa)As DH lasers due to facet oxidation', *Appl. Phys. Lett.,* **32** (2), 119–121, 1978.
123. R. L. Longini, *Sol. St. Electron.,* **5**, 127, 1962.
124. R. D. Gold and L. R. Weisberg, *IRE Trans. Electron Devices,* **8**, 428, 1961. S. A. Steiner and R. L. Anderson, 'Degradation of GaAs injection devices', *Sol. St. Electronics,* **11**, 65–86, 1968.
125. C. J. Hwang, 'Initial degradation mode of long life lasers', *Appl. Phys. Lett.,* **30** (3), 167–169, 1977.
126. L. C. Kimmerling, 'Recombination enhanced defect reactions', Internat. Conf. on recombination in semiconductors, Southampton, Sept. 1978.
127. R. L. Hartman, N. E. Schumaker and R. W. Dixon, 'Continuously operated (Al, Ga)As double-heterostructure lasers with 70 °C lifetimeas as long as two years', *Appl. Phys. Lett.,* **3** (1), 756–759, 1977.
128. S. D. Hersee and R. C. Goodfellow, 'The reliability of high radiance LED fibre optic sources', *Proceedings of the 2nd European Conference on Optical Fibre Communication,* 213–216, Paris, 1976.
129. J. J. Hsieh and C. C. Shen, 'GaInAsP/InP double-heterostructure lasers for fibre optic communications', *Fibre and Integrated Optics,* **1** (4), 357–368, 1978.
130. K. Namizaki, H. Kan, Ishii and A. Ito, 'Current dependence of spontaneous carrier lifetimes in GaAs–GaAlAs double heterostructure lasers', *Appl. Phys. Lett.,* **24**, 486–489, 1974.
131. A. R. Goodwin, J. R. Peters, M. Pion, G. H. B. Thompson and J. E. A. Whiteway, 'Threshold-temperature characteristics of double heterostructure $Ga_xAl_{1-x}As$ lasers', *J. Appl. Phys.,* **7**, 3126–3131, 1975.
132. R. Goodfellow, I. Griffiths and R. Bradley, 'Deep and Shallow Zinc Diffused Stripe Lasers', Paper presented at the European Solid State Device Research Conference, Sussex, UK, 1977.
133. M. R. Matthews, D. R. Smith and L. J. Arnold, 'Fluctuations in the output of GaAs lasers', I.O.P. QE Conference, Oxford, 1975.
134. T. Rozzi, T. Itoh and L. Grun, 'Two-dimensional analysis of the GaAs double hetero stripe-geometry laser', *Journal of Radio Science,* **12**, 543, 1977.
135 J. C. Dyment, J. C. North and L. A. D'Asaro, 'Optical and electrical properties of proton bombarded p-type GaAs', *J. Appl. Phys.,* **44** (1), 207–213, 1973.

Optical Fibre Communications
Edited by M. J. Howes and D. V. Morgan

CHAPTER 3

Photodetectors for communication by optical fibres

T. PEARSALL

3.1 INTRODUCTION

The maximum distance between signal repeater stations in any telecommunication system is determined by the attenuation and distortion of the signal by the transmission medium. In the case where attenuation alone (through scattering or absorption) is the limiting factor, the performance of the communications system will be determined at the detector. Improving the detector characteristics thus permits the installation of fewer repeater stations and lowers both the capital investment and maintenance costs in the operation of an optical-fibre communication network. In a practical system, however, there are a number of repeater stations, required for the routing and switching of signals, which would not be eliminated even if sufficiently high detector performance permitted it. The general goal in photodetection for optical-fibre communication is to develop a detector sensitive enough to make unnecessary most remote repeater stations intermediate between switching stations. At this point, further improvement in detectors no longer affects the overall system performance.

In the following treatment, we will be most interested in detector characteristics which permit high bit rate (500 Mbit/sec) transmission at a 10^{-9} bit error rate over an optical-fibre link (15 km) with a moderate (100 μW) optical power coupled in from the source. It is almost always the case under such conditions that the system sensitivity is limited by the noise level generated in the preamplifier following the detector element. The degree to which the amplifier noise itself may be reduced depends on the required response time of the communication system and the electronic properties of the detector. It is therefore quite difficult to separate the detector performance from that of the following preamplifier. The effect of the preamplifier/detector noise can be somewhat alleviated by using avalanche gain. The increase in the detected signal-to-noise ratio which can be obtained in this

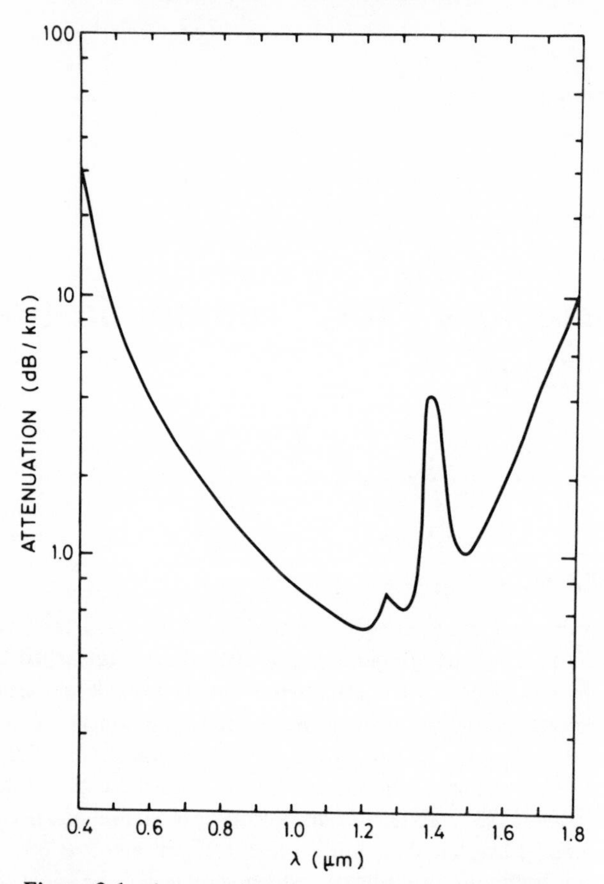

Figure 3.1. Attenuation as a function of optical wavelength in the spectral region 0.4 μm to 2.0 μm for a multimode fused silica fibre.[1] The spectral loss characteristic for a single-mode fibre shows uniformly greater attenuation. The loss in the 1.3 μm region of the spectrum is less than that at the GaAs laser wavelength. Over a fibre length of 15 km, the attenuation at 1.3,μm is about 3 orders of magnitude less. (Reproduced by permission of *IEE Electron, Lett.*)

manner depends on the noise properties of both the detector and source as well as the detailed behaviour of the avalanche gain.

Luminescent sources for the first-generation optical-fibre communication systems are being made from $Al_xGa_{1-x}As/GaAs$ compounds which emit light around 0.8 μm in the near infra-red.[1,2] Silicon detectors are a natural choice in

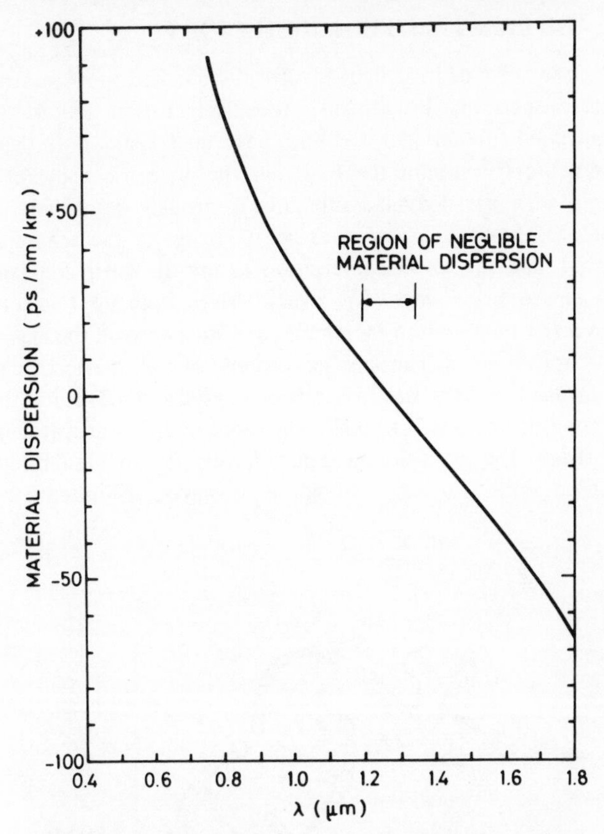

Figure 3.2. Material dispersion calculated for a phosphosilicate optical fibre.[2] Near 1.3 μm this dispersion is small. At the GaAs laser wavelength it is much larger. A laser pulse with a 10 Å optical line width broadens at the rate of 75 ps per km. Over a distance of 15 km, the laser pulse width alone limits the system bandwidth to 100 MHz. (Reproduced by permission of *IEE Electron. Lett.*)

these systems. Considerable advantages, however, can be gained in terms of diminished attenuation and improved bandwidth by transmitting the optical signals at 1.3 μm instead of 0.8 μm. Most of the discussion of specific detectors will be devoted to the properties of long-wavelength detectors, sensitive to 1.3 μm radiation, for comparison with silicon detectors whose performance is well known. The important point in this regard is the determination of operating characteristics of a detector at 1.3 μm of sufficiently high performance to take advantage of the optimized fibre transmission properties in this wavelength region.

3.2 VISIBLE AND INFRA-RED PHOTODETECTORS

The principal means of detecting light involve the conversion of photons either to heat or to electric currents. Broadband optical detectors in the visible (0.4 μm–0.7 μm) and infra-red (0.7 μm–30 μm) have been made using both these methods. Pyro-electric detectors[3,4] exploit the fact that the dielectric constant of a ferro-electric material such as tri-glycine sulphate, is strongly dependent on the temperature.[5] The absorption of photons causes a change in the temperature of the detector material, and the resulting variation in the dielectric constant is usually measured as a capacitance change. The signal derived from a pyro-electric detector depends only on the temperature increment, and thus only on the absorbed energy of the incident optical beam. The spectral response of a pyro-electric detector is, in principle and in practice, very flat over a broad spectral band. It has thus a special advantage in applications involving light emission at different wavelengths over a large spectral range. The pyro-electric effect is very fast, being limited by the dielectric relaxation time. The detector speed, however, is limited by the rate at

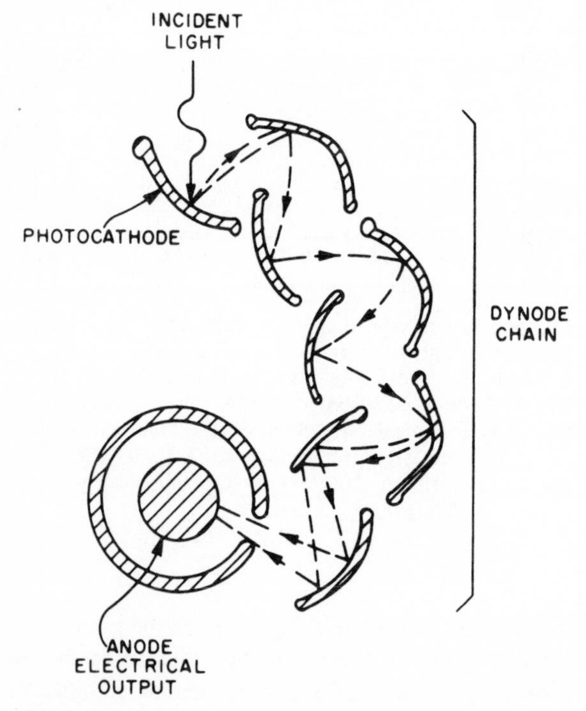

Figure 3.3. A schematic representation of a photomultiplier.[7] This device consists of a photocathode and an electron multiplier called a dynode chain. Photomultipliers are capable of very high gain ($G = 10^6$) and very low noise. (Reproduced by permission of North Holland Publishing Co.)

which the detector cools after excitation. Very high-speed response (r.t. ~ 1 ns) can be obtained, but this is accompanied by a corresponding loss in sensitivity. Pyroelectric detectors of this sort are suitable for detecting high-speed laser pulses.[6]

A photomultiplier consists of a photocathode and an electron multiplier packaged in a vacuum tube.[7] A schematic diagram of such a device is shown in Figure 3.3. Photons of energy greater than the photoelectric work function can eject electrons into free space where they are collected and amplified by the electron multiplier, usually a dynode chain. The photoemission involves two steps. First is the photoexcitation of an electron by the incident light; the second is getting the electron out of the cathode material. It is this second aspect that makes photoemission more difficult towards longer-wavelength ($\lambda = 1$–$2\ \mu$m) light. The work function of most materials is of the order of several volts. Electrons excited by photons of this energy can be ejected from the target cathode rather easily. As the energy of the light to be detected becomes less and less, the emission becomes more difficult. Two steps can be taken to overcome this problem. Instead of exciting the photoelectron directly into the vacuum, it is excited to an intermediate state such as from the valence to conduction band in a semiconductor (e.g. GaAs,[8] $InAs_xP_{1-x}$,[9] $Ga_xIn_{1-x}As$,[10] and $Ga_xIn_{1-x}As_yP_{1-y}$[11]). Next, it is necessary to fix the energy of the conduction band above that of the vacuum so that the photoelectrons can exit. This is accomplished by applying a thin film of Cs_2O and Cs to the photoemitting surface. This film forms a hetero-junction whose effect is to bend the band-structure of the semiconductor so that the conduction band is raised in energy as shown in Figure 3.4. Such a structure is called a negative electron affinity cathode. While this technique works extremely well for GaAs which is sensitive to photons with wavelengths shorter than 0.9 μm, it rapidly becomes less effective for longer wavelengths in the infra-red and ceases to work altogether beyond 1.2 μm, because there is no stable film which can bend the conduction band high enough in energy. The possibility of making photocathodes which operate at longer wavelengths exists by exploiting the Gunn effect: the transfer, by electron–phonon scattering, of electrons from the conduction band to a higher energy level in a strong electric field ($E > 3 \times 10^3$ V cm^{-1}), as shown in Figure 3.5. Photoelectrons are created in the conduction band of a material whose energy gap is too small for the standard Cs–Cs_2O coating to work. These electrons can gain energy from the electric field and then scatter to a stable energy level. If this level is more than 1 V above the valence band, then the normal cesiation of the surface will allow these electrons to be emitted. This acceleration and scattering in the electric field takes time, and some of the electrons will be lost by recombination, hence the efficiency of the cathode will be lower. But the low noise and high gain (10^6) of the dynode chain may make up for the loss in quantum efficiency. Utilization of this effect may extend the photocathode range to 2 μm. Promising results demonstrating this principle have been obtained.[12]

A photoconductor is a light-sensitive resistor. High-speed, high-sensitivity photoconductors such as we are concerned with are made from semiconductors. The

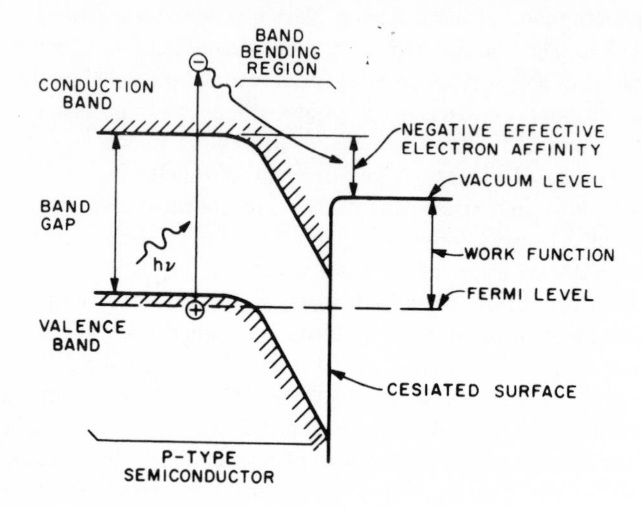

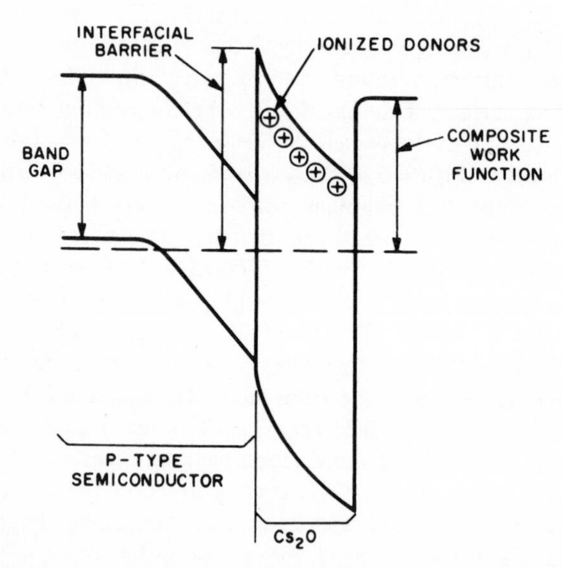

Figure 3.4. A simple energy-band diagram of a negative electron affinity photocathode.[7] A thin coating of caesium is used to bend the semiconductor photocathode conduction band above the vacuum level. Photoexcited electrons can then escape into free space. (Reproduced by permission of North Holland Publishing Co.)

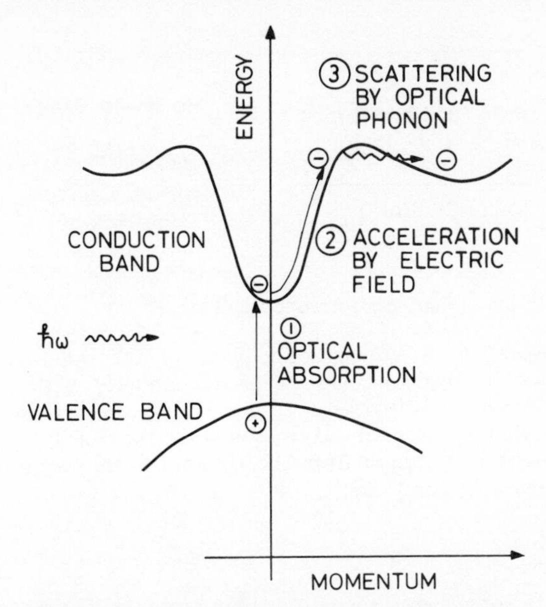

Figure 3.5. The Gunn effect is the scattering of energetic or hot electrons from one so-called 'valley' in energy–momentum space to another such valley. For this effect to be useful for photoemission, it is important that the two valley minima be quite different in energy. This is not usually a criterion for a good Gunn oscillator material

absorption of light creates an electron–hole pair which increases the conductivity during the lifetime of the pair. Although the excitation of an electron across the fundamental bandgap can be used to absorb the light, it is quite often the excitation of a carrier from an intentionally introduced impurity level in the forbidden gap to the conduction or valence band that is exploited in photoconductors.[13] By carefully choosing the impurity, one can obtain an absorption threshold energy corresponding to the wavelength of interest, as shown in Figure 3.6. The photoexcited carriers cross the photoconductor in a time T given by

$$T = L^2/\mu V \tag{3.1}$$

where μ is the carrier mobility, V is the applied voltage, and L is the sample dimension between the contacts. The ratio of the carrier lifetime τ to the transit time T is the number of carriers which cross the sample for each absorbed photon. Thus, if the transit time is much longer than the lifetime, the response of the photoconductor is severely attenuated. On the other hand, if the transit time is less than

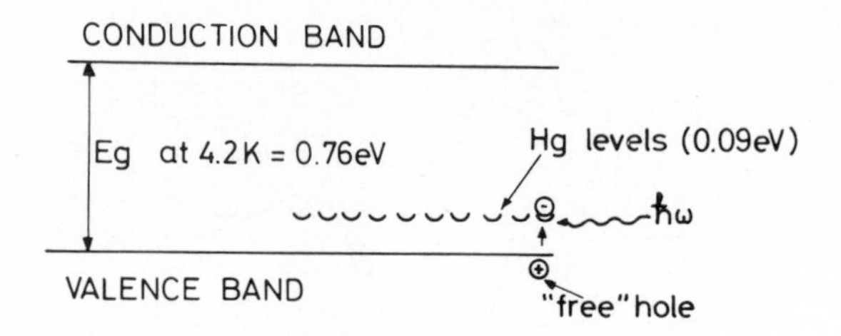

Figure 3.6. A schematic diagram of the impurity level in Hg-doped Ge – a photoconductor with a maximum sensitivity at 10 μm. The wavelength of maximum sensitivity corresponds to the energy of electron excitation from the Hg level to the valence band

the lifetime, the factor τ/T gives the photoconductive gain. The speed of response is fixed by the carrier lifetime, and thus the photoconductor obeys a well-defined gain–bandwidth product:

$$GB = \frac{\mu V}{2\pi L^2} = \frac{1}{2\pi T} \tag{3.2}$$

The upper limit on this product is given by the minimum possible transit time, i.e. the dielectric relaxation time. The practical applications of photoconductors have been principally restricted to detection in the far infra-red[14] where there is no other possibility for fast, sensitive detection. There is no fundamental reason why photoconductive detectors would not also make high-performance detectors for the shorter-wavelength regions of the optical spectrum of interest for optical-fibre communication.

A p–n junction is a metallurgical boundary between two semiconductors of different conductivity type. The difference in fermi level between the n and p materials causes a high-resistivity depletion region at the junction and a built-in electric field on the order of 10^4 V cm^{-1}. Photons absorbed in this region give rise to electron–hole pairs which are separated by the internal electric fields and collected at the edge of the depletion region. This mechanism is not fundamentally different from that in photoconductive detection. The major differences are that the internal electric field in a photodiode is several orders of magnitude greater than that habitually used in a photoconductor and that the photoexcited carriers are swept to regions where they are majority carriers. This last feature means that photodiodes are not governed by the gain–bandwidth product limitations for photoconductors given in Equation 3.2.

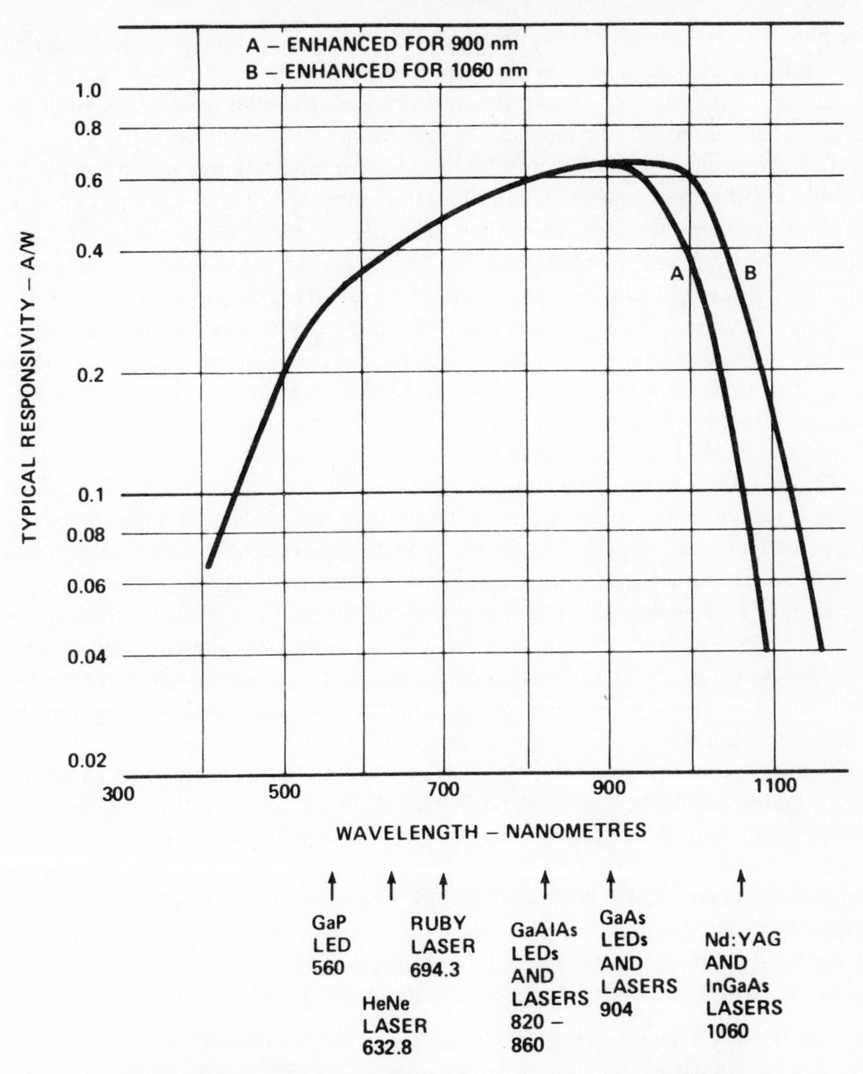

Figure 3.7. Spectral sensitivity of a silicon photodiode. (Reproduced by permission of Electro-Optics & Devices, RCA Corporation)

Internal gain in the photodiode signal is possible, and it is not based on a lifetime effect as is the case with photoconductors. The electric field in a photodiode biased near breakdown is of the order of 10^5–10^6 V cm^{-1}. Photoexcited carriers can gain several electron volts of energy in such a field in a distance of only a few hundred ångstroms. Such an energetic carrier can lose enough energy in a collision to promote an electron from the valence band to the conduction band, thus increasing the photoexcited current. This process, called impact ionization, and the associated gain, called avalanche gain, can be used to advantage in optical signal detection. Because the photodiode response time usually improves with reverse bias, both the gain and the bandwidth of a photodiode may be increased at the same time.

The choice of a photodetector for optical communications must be made by considering the total system design. To take advantage of the small size of glass fibres, it is desirable that the other elements in the system, the source and detector, be small too – of the order of a fibre diameter. While the photomultiplier has the highest performance in terms of sensitivity and speed of any of the detectors discussed above, its comparatively large physical size alone makes it a doubtful candidate for optical-fibre telecommunications in conventional applications. However, photomultipliers may be useful in highly specialized situations where their high gain ($G \sim 10^6$) and low noise may be preferred in links involving very few (10–100) fibres over long distances (100–200 km).

The minimum signal detectable by a pyro-electric detector in the 1.0–2.0 μm spectral region is about four orders of magnitude larger[3] than for a photodiode or photoconductor at 1 kHz. To operate at high speeds, the pyro-electric detector suffers an additional sensitivity loss related to the sample geometry[3] required to increases to cooling rate and to decrease the capacitance. The particular advantage of these devices – uniform spectral sensitivity over a broad optical wavelength range – is irrelevant when one considers that the information in an optical-fibre system will be transmitted at a fixed wavelength, well defined by the emission spectrum of the LED or laser source.

There is little fundamental difference between a photoconductor and a photodiode biased below the avalanche-gain region. It seems likely that the need for detectors for optical telecommunications will spur the development of new photoconductive detectors, especially for 1.3-μm photon detection.[15,16] However, under conditions which require both gain and high bandwidths, such as in high bit rate optical-fibre communications, the photodiode is superior to a photoconductor whose bandwidth often degrades with increasing gain.[17]

The special problem of optical-fibre communication requires detectors that are:

(a) small – the detector size should be similar to the fibre diameter;
(b) fast – modulation bandwidth on the order of 500 MHz; and
(c) sensitive – the capability of detecting 10^4 photons per bit, implying low noise and high efficiency at the optical wavelength of interest.

Only the photodiode satisfies all these major requirements.

3.3 DETECTOR CHARACTERISTICS FOR COMMUNICATION BY OPTICAL FIBRES

3.3.1 Spectral response

The fundamental aspect of photodiode performance is the spectral response. This response is usually presented as the quantum efficiency as a function of photon wavelength:

$$\eta_Q = \frac{\text{number of electrons collected}}{\text{number of incident photons}} \tag{3.3}$$

i.e. as the number of electrons collected at the contacts for each photon absorbed. The calculation of the quantum efficiency does not involve the energy of the photon. Sometimes the spectral response is given in terms of the absolute sensitivity, which is related to the quantum efficiency:

$$S = \frac{\text{electrons/sec}}{\text{incident photon power}} = \frac{q\eta_Q}{\hbar\omega} \quad \text{(A/W)} \tag{3.4}$$

The absolute sensitivity is a more useful expression because it gives the transfer characteristic of the photodiode: photocurrent per unit optical power. We note that it takes a greater optical power at 0.83 μm than at 1.3 μm to achieve the same

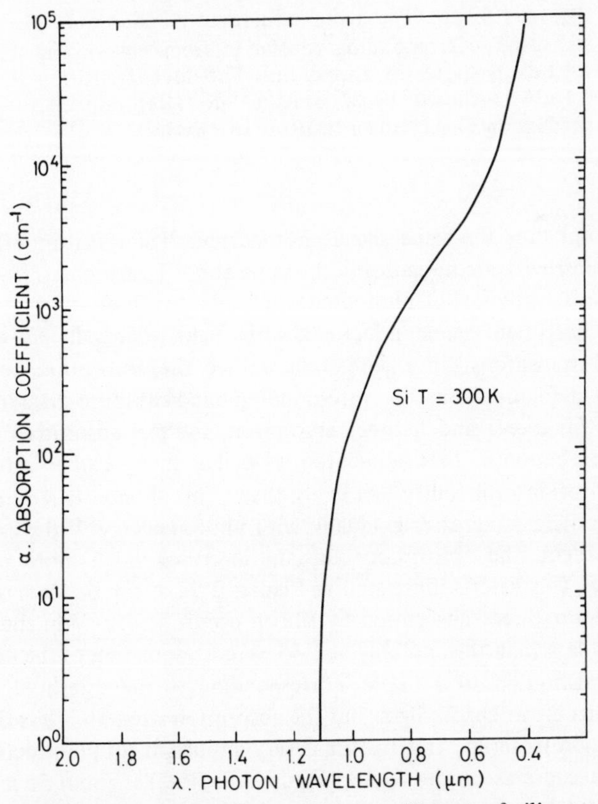

Figure 3.8. (a) The absorption coefficient of silicon at 300 K[18]

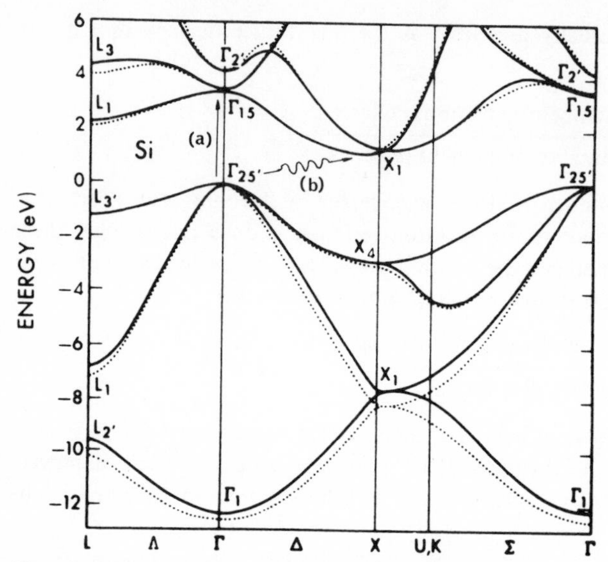

Figure 3.8. (b) The band structure of silicon.[19] The transition corresponding to the indirect absorption at 1.12 eV is shown by a wavy line. The direct transition at 4.1 eV is shown by a straight line. (Reproduced by permission of American Institute of Physics)

photon flux and thus the same signal photocurrent. The quantum efficiency and the absolute sensitivity are numerically the same at $\lambda = 1.24\ \mu m$.

The spectral sensitivities of photodiodes fall into two basic types depending on whether the constituent semiconductor absorbs light principally by direct or indirect optical transitions. In Figure 3.8a we see the absorption coefficient of silicon,[18] and in Figure 3.8b the corresponding band-structure diagram[19] showing the threshold for direct and indirect absorption. Indirect absorption requires the assistance of a phonon so that momentum as well as energy can be conserved. This makes the transition probability less likely than if no phonon is required, and the absorption coefficient increases gradually with photon energy. If the lowest energy transition is direct, the absorption constant increases much more rapidly with photon energy. This case is illustrated in Figures 3.9a, b for the case of GaAs.[19,20] The threshold for direct absorption by silicon occurs at 4.1 eV in the ultraviolet. At all energies less than this, notably at 1.45 eV corresponding to the energy of the GaAs laser emission, until 1.12 eV, corresponding to the threshold for indirect absorption, silicon can absorb light, but the absorption strength is less than that for a direct-transition material. This feature is very important for photodetectors made from silicon, because as can be seen in Figure 3.8a, it takes about 50 μm to absorb all the light at 0.83 μm, whereas with GaAs only 1 μm is required.

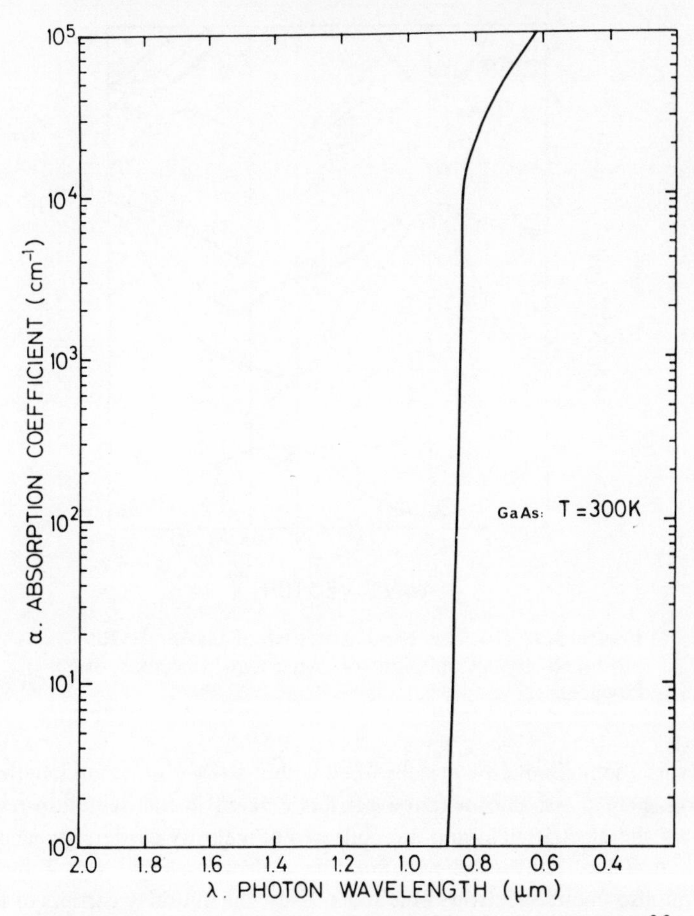

Figure 3.9. (a) The absorption coefficient of GaAs at 300 K[20]

Germanium is another material for which the lowest-energy optical absorption takes place by indirect optical transition. The absorption characteristic and band structure for germanium are given in Figure 3.10a, b.[18,19] In the absorption spectrum can be seen a strong increase in the absorption strength at 1.5 μm. Quick reference to the band structure shows that this wavelength corresponds to the direct transition in germanium. The photons of interest in optical communication all lie at shorter wavelengths than the direct edge in germanium, and so germanium will be strongly absorbing in this region. At 1.3 μm or 0.83 μm, the absorption in germanium is dominated by the direct transition, and the corresponding absorption length is on the order of 1 μm, similar to that of a direct-bandgap semiconductor.

Photon absorption in a photodiode is illustrated in Figure 3.11 where we show a simple energy level versus distance representation for a p–n junction. Photons,

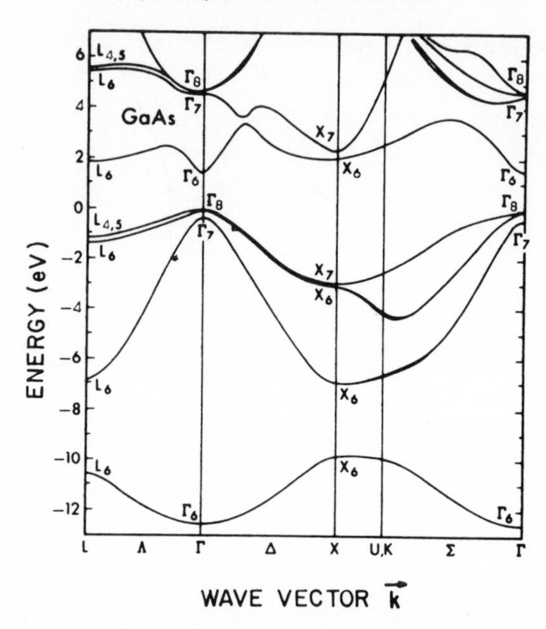

Figure 3.9. (b) The band structure of GaAs.[19] (Reproduced by permission of American Institute of Physics)

incident on the semiconductor are absorbed within a few absorption lengths, creating electron–hole pairs. Photoexcited carriers created in the depletion region are separated by the electric field and are collected as majority carriers on each side of the depletion region. Photons absorbed within a diffusion length (see Table 3.1) of the junction also create electron–hole pairs. Only the minority carriers of this pair can diffuse into the high-field depletion region. The majority carrier is automatically excluded by the electric field. For a photodiode of only moderate quality, the quantum efficiency of this process is nearly 100% for the photons absorbed in the semiconductor. A significant fraction of the photon flux, however, is reflected at the surface because of the discontinuity in the index of refraction between the air (usually), $n = 1$; and the semiconductor, $n = 3.5$. The reflection coefficient for normal incidence is:

$$R \frac{(n_1 - n_2)^2}{(n_1 + n_2)^2} = 30\% \tag{3.5}$$

To decrease the reflected power, the photodiode surface is coated with a material whose index of refraction is the geometric mean of the index of the semiconductor and air (usually)[2]:

$$n_{\text{coating}} = (n_1 n_2)^{1/2} \tag{3.6}$$

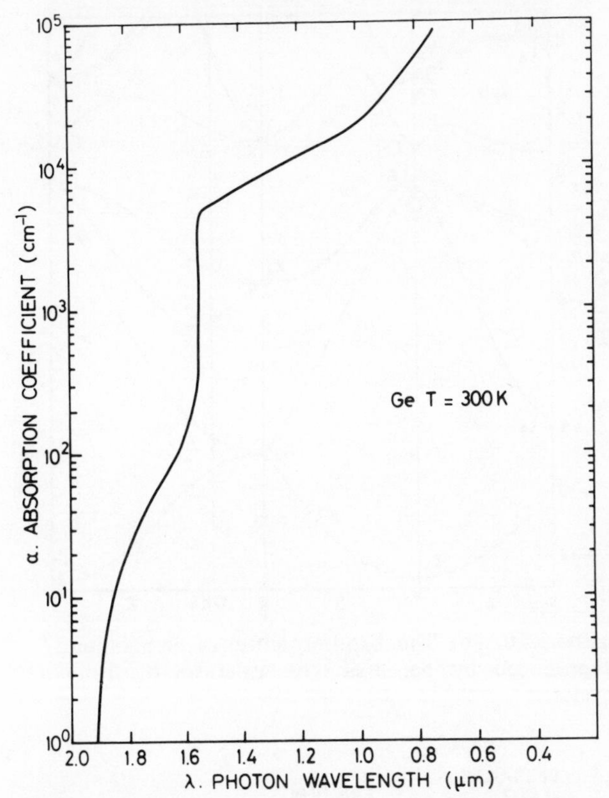

Figure 3.10. (a) The absorption coefficient of germanium at 300 K[18]

and whose thickness is one-quarter that of the wavelength of interest. Si_3N_4 with an index $n = 1.97$ works very nicely, and is easily deposited.[21] This coating does not decrease the reflectivity as much at other wavelengths so that the overall spectral response is more peaked around the wavelength of interest. In Figure 3.12 we show the spectral response of a direct-bandgap semiconductor photodiode with no anti-reflection coating. The sharp rise in the photoresponse at the band edge λ_g reproduces the absorption coefficient behaviour until a saturation point is reached corresponding to a quantum efficiency of 65–70% (compared with 90–100% for a coated detector). At higher photon energies, however, this response decreases. The loss in response can be attributed to the effects of surface recombination. The increase in the absorption coefficient above the band edge requires that a larger percentage of the photons are absorbed near the surface. The resulting minority carriers have a greater chance to be trapped in surface states where they subsequently recombine without contributing to the photocurrent. Surface recombination is a problem with any photodiode made from a semiconductor absorbing photons across

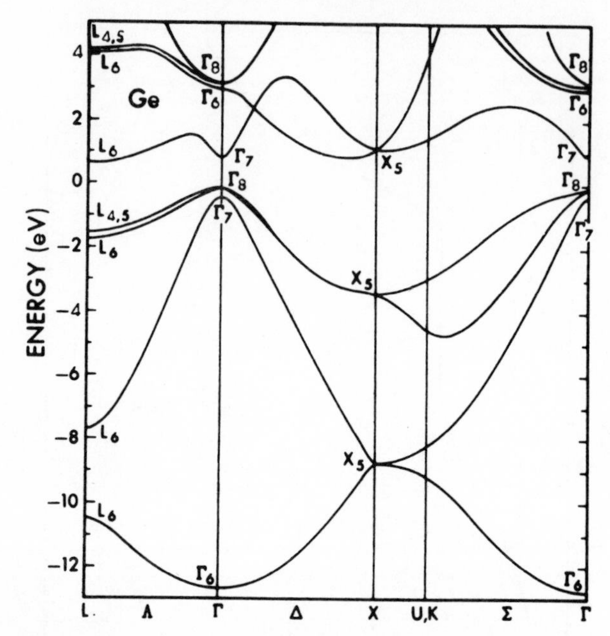

Figure 3.10. (b) The band structure of germanium.[19] (Reproduced by permission of American Institute of Physics)

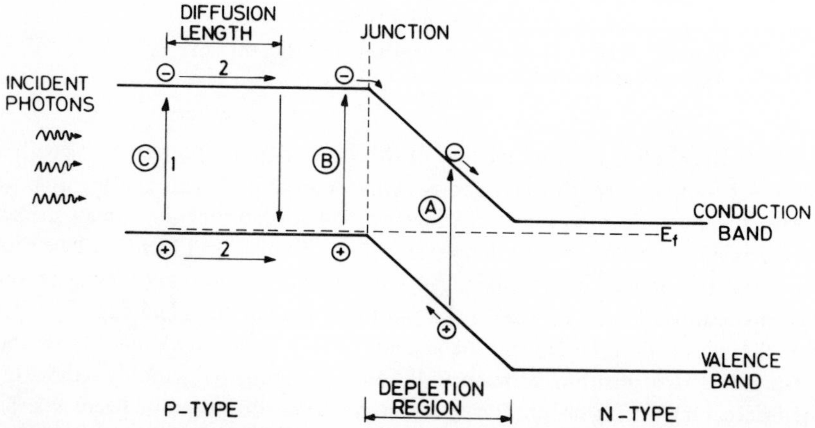

Figure 3.11. Photon absorption process in a *p–n* junction detector. Photons absorbed in the depletion region (A) created pairs of electrons and holes which are separated by the electric field. Photons absorbed within a diffusion length of the depletion region create pairs (B), but only the minority carriers diffuse to the junction. Photons absorbed more than a diffusion length from the depletion region (C) create minority carriers which recombine before reaching the depletion region

Table 3.1 Electronic properties of some semiconductor photodiode materials

	Bandgap (eV)		Mobility (cm^2-V^{-1} sec^{-1})		Intrinsic carrier concentration (cm^{-3})
	Indirect	Direct	μ_e	μ_h	n_i
GaAs	–	1.43	8000	300	10^7
Si	1.17	4.10	1500	600	1.6×10^{10}
$Ga_{0.47}In_{0.53}As$	–	0.75	10 000	200	5×10^{11}
Ge	0.66	0.90	4000	2000	2.4×10^{13}
InAs	–	0.33	33 000	4600	2×10^{16}

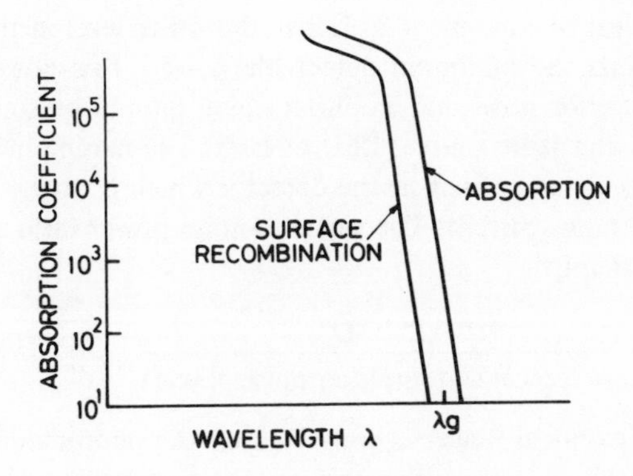

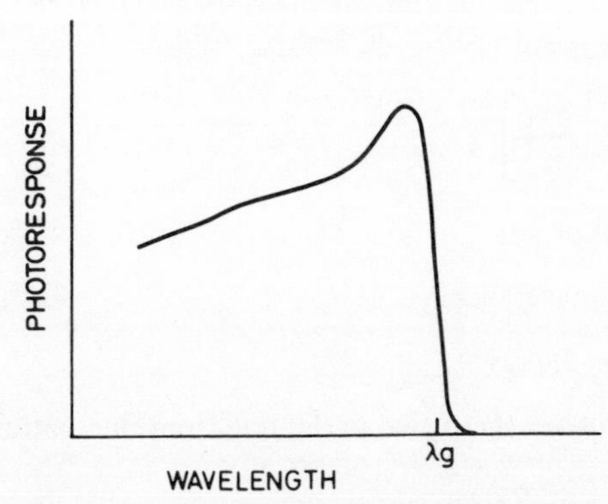

Figure 3.12. Spectral photoresponse typical of a direct-gap semiconductor photodiode showing how the energy dependence of surface recombination and absorption coefficient combine to make the maximum photosensitivity near the absorption edge λ_g

the *direct* gap. For this reason, it is always advantageous to match the absorption edge of such a photodiode to the emission wavelength of interest. The effects of surface recombination limit the maximum efficiency of Ge to 50% and of InAs to 30% at 1.3 μm. There is thus a factor of two to be gained in quantum efficiency by using an alloy semiconductor (whose maximum absorption can be designed to occur at 1.3 μm) rather than the commercially available alternatives Ge and InAs.

3.3.2 Noise and the minimum detectable signal

Detection in optical-fibre communication systems is concerned with very weak signals. The essential characteristic of the detector/preamplifier which describes the sensitivity of the detection system is the minimum detectable optical power. The minimum detectable power depends on two things only: the quantum efficiency and the noise. Since the quantum efficiency of photodiode detectors is usually quite near the ideal maximum, it is rather the noise level in the communication system which limits the minimum detectable power. This noise comes from two different sources: shot noise and amplifier noise. Shot-noise sources exist in both the detector and the light source, LED or laser. The minimum detectable optical power produces a signal current in the detector which is of the same order as the root-mean-square noise current. The signal-to-noise power ratio at the detector can be written down simply:

$$\frac{S}{N} = \frac{\text{(signal power)}}{\text{(shot-noise power)} + \text{(amplifier noise power)}} \tag{3.7}$$

Since the incident optical power is converted by the photodiode to a current, the signal power at the amplifier is proportional to the square of the optical power. For a full-depth sinusoidal modulation of the optical signal, the mean-square value of the detector photocurrent is:[17]

$$\langle i_s \rangle^2 = \frac{1}{2}\left[2P_{\text{opt}}\frac{q\eta_{\text{Q}}}{\hbar\omega}\right]^2 \tag{3.8}$$

where P_{opt} = optical power

η_{Q} = quantum efficiency

$\hbar\omega$ = photon energy

The shot-noise term consists of two parts: one from fluctuations in the signal,

$$\langle i_{\text{NS}} \rangle^2 = 2q\left[2P_{\text{opt}}\frac{q\eta_{\text{Q}}}{\hbar\omega}\right]B \tag{3.9}$$

and the other from the dark current in the detector,

$$\langle i_{\text{ND}} \rangle^2 = 2qi_{\text{D}}B \tag{3.10}$$

where B is the system bandwidth

and i_D is the detector dark current.

We note that in the case of communication by optical fibres, there is no contribution to the shot noise from fluctuations in the background light intensity because the field of view of the detector is extremely narrow, the angle being of the order of the fibre diameter divided by the fibre length.

The amplifier noise can be represented as the Johnson noise of the load resistor at an effective temperature T_{eff}.

$$\langle i_{NA}\rangle^2 = 4kT_{eff}/R_L \tag{3.11}$$

where k = Boltzmann's constant

R_L = load resistance.

The expression for the signal-to-noise ratio can now be written:[25]

$$\frac{S}{N} = \frac{2\left[P_{opt}\dfrac{q\eta_Q}{\hbar\omega}\right]^2}{\left[\left[2qi_D + 4q\left(P_{opt}\dfrac{q\eta_Q}{\hbar\omega}\right)\right] + \dfrac{4kT_{eff}}{R_L}\right]B} \tag{3.12}$$

The basic relation describing photodetector performance in optical communication systems is given by Equation 3.12. Of the three sources of noise in the denominator of Equation 3.12, the amplifier term is the most important in determining the system characteristics. A state-of-the-art FET amplifier has a noise figure of about 5 dB, which means that it is 5 dB (3 times) more noisy than a 50-Ω resistor at 300 K.[25] This noise figure translates into an effective temperature of ~1000 K. Considering Equation 3.11, one can calculate the amplifier noise-equivalent power as a function of bandwidth using the relations:

$$\text{NEP}_{amp} = 4kT_{eff}B^2(2\pi C) \tag{3.13}$$

and

$$B = (2\pi R_L C)^{-1} \tag{3.14}$$

That is, we pick the input load resistance of the amplifier to be as large as possible consistent with the bandwidth. For a detector with a 1 pF capacitance, the optimized input resistance at 100 MHz is 10^3 Ω, and the amplifier noise is 5×10^{-15} W. Using Equation 3.12 and setting $S/N = 1$ we obtain for the minimum detectable optical power:

$$P_{min} = \frac{\hbar\omega}{q\eta_Q} B[2\pi kT_{eff}C]^{1/2} \tag{3.15}$$

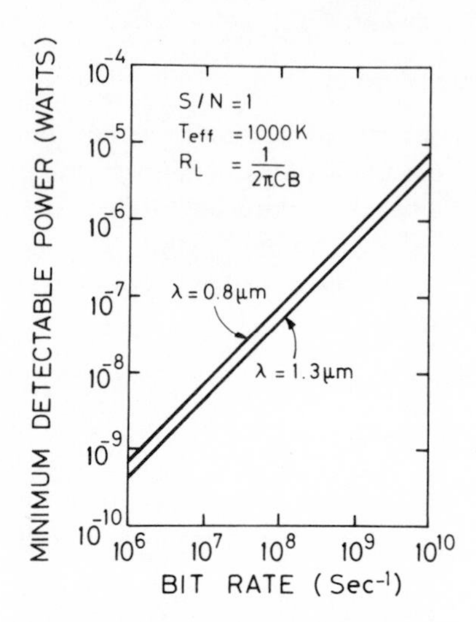

Figure 3.13. Minimum detectable optical signal as a function of bit rate. Detector capacitance is 1 pF and the detector load resistance is chosen using Equation 3.13

In the calculation of minimum detectable power, the signal and noise powers are taken as equal, hence the shot-noise contribution from the signal is always several orders of magnitude less. (In this limit the signal shot noise can be neglected. We will see, however, that the signal shot noise plays an important role when avalanche gain is concerned.) The minimum detectable signal as a function of bit rate (Equation 3.15) is plotted in Figure 3.13 assuming a detector capacitance of 1 pF for detection both at $\lambda = 0.8\ \mu m$ and $\lambda = 1.3\ \mu m$. The limit on the dark current is calculated from Equation 3.12 by equating the amplifier noise and the detector shot noise.

$$i_D = \frac{4\pi C k T_{eff} B}{q} \tag{3.16}$$

In Figure 3.14 we show the 'noise-equivalent dark current' as a function of bit rate. It can be seen at lower bit rates that the detector dark current must be less in order that the predominant noise contribution come from the amplifier. For a fibre detector area of $10^{-5}\ cm^2$, the dark current density of 10^6 Hz corresponds to $10^{-2}\ A\ cm^{-2}$. In Table 3.1 we list some characteristics of some state-of-the-art photodetectors. With the exception of InAs, whose dark-current characteristic is shown in Figure 3.15, the dark-current noise is always well below the limit imposed

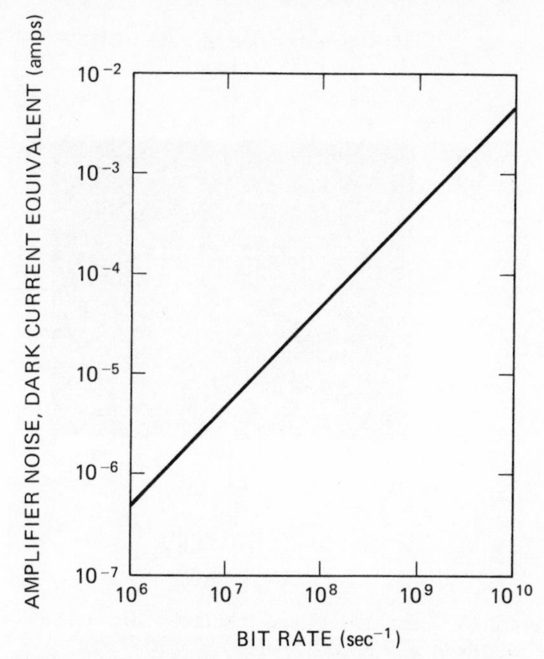

Figure 3.14. 'Noise-equivalent' dark current as a function of bit rate. This is the average current of the amplifier thermal noise with an effective noise temperature of 1000 K

by amplifier noise shown in Figure 3.14 for 1 MHz–1 GHz bit rates. For comparison with the case of InAs, we show in Figures 3.16–3.18 reverse characteristics of several other semiconductor diodes.

In the design of a communication system, the required signal-to-noise ratio at the receiver is determined by the specification of the desired bit error rate (b.e.r.). This standard for telecommunications is a 10^{-9} b.e.r. The relation between a specified b.e.r. and the required signal-to-noise ratio can be calculated from the statistics of signal detection. Two kinds of errors are possible. One is that a noise pulse will be mistaken for a signal pulse when in fact no signal occurs – a 'false alarm' error. The other is that no signal will be detected when in fact one is present – a 'threshold' error. The correct statistical distribution which describes the arrival rate of photons in an optical pulse is given by Poisson statistics.[24] The central-limit theorem of statistics, however, shows that for large numbers of photons per pulse, the correct Poisson distribution will be closely approximated by a Gaussian distribution. The number encountered in high bit rate optical communication is $\sim 10^4$ photons per bit, which is large enough to justify using Gaussian statistics to relate the signal-to-noise ratio and the error rate. In this approximation, the probability

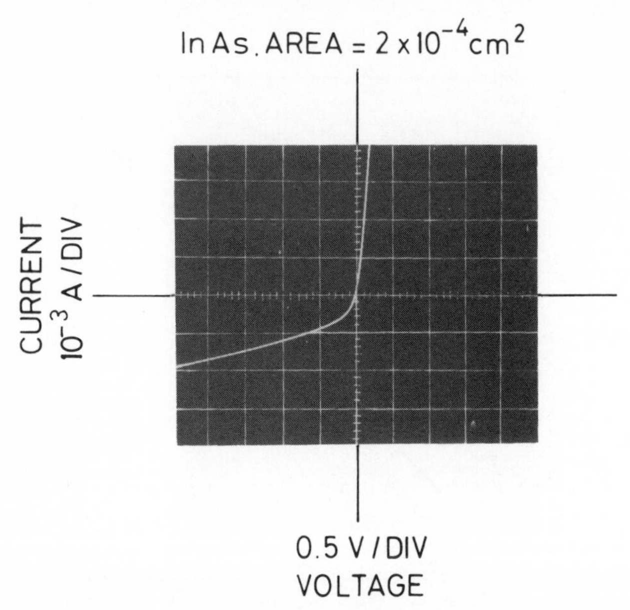

Figure 3.15. Current–voltage characteristic of an InAs photodiode at 300 K. Diode area = 2×10^{-4} cm^2

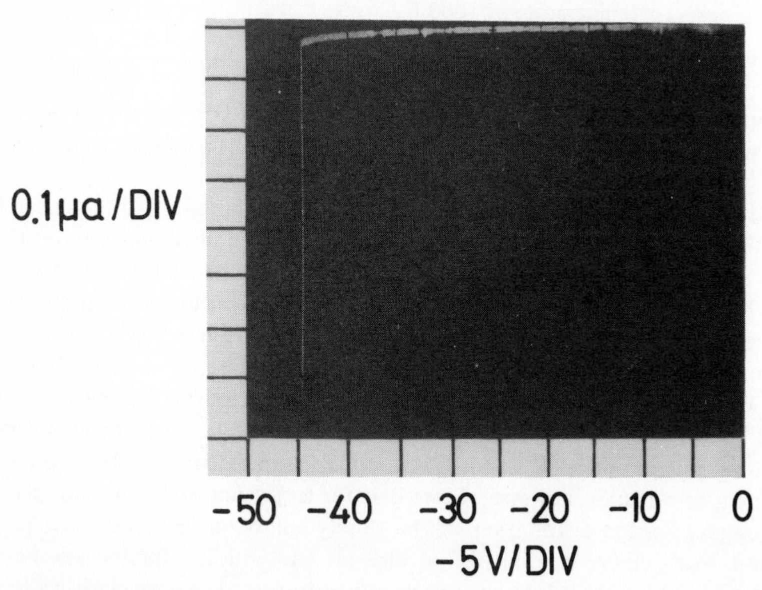

Figure 3.16. Reverse-current characteristic for a GaAs photodiode at 300 K. Diode area = 4×10^{-3} cm^2

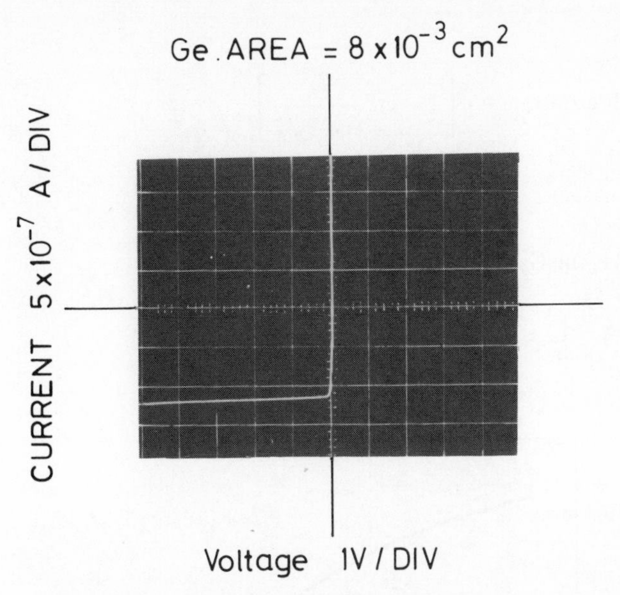

Figure 3.17. Reverse-current characteristic for a Ge photodetector at 300 K. Diode area = 8×10^{-3} cm^2

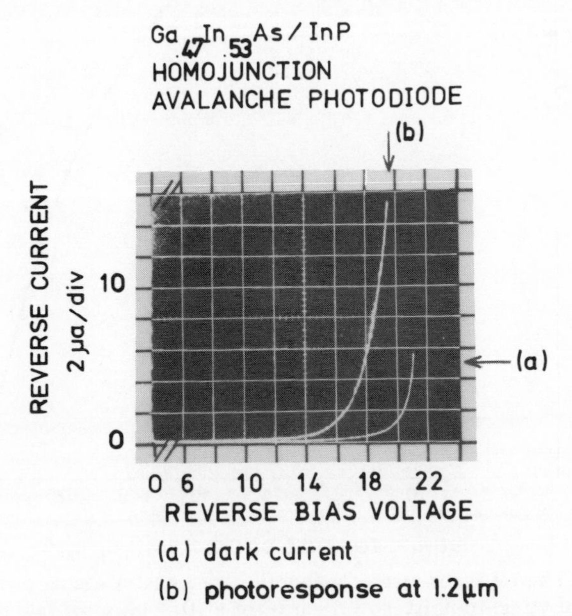

Figure 3.18. Reverse-current characteristic for a $Ga_{0.47}In_{0.53}As$ photodiode at 300 K. Diode area = 2.5×10^{-4} cm^{-2}

of an error is:

$$P_e = \text{bit error rate} = \frac{1}{2}\left[1 - \text{erf}\,\frac{i_s}{\langle i_N\rangle 2\sqrt{2}}\right] \tag{3.17}$$

$$= \frac{1}{2}\left[\text{erf}c\;\frac{i_s}{\langle i_N\rangle 2\sqrt{2}}\right] \tag{3.18}$$

where erf (x) is the Gaussian error function:

$$\text{erf}\,(x) = \frac{2}{\pi}\int_0^x e^{-z^2}\,dz \tag{3.19}$$

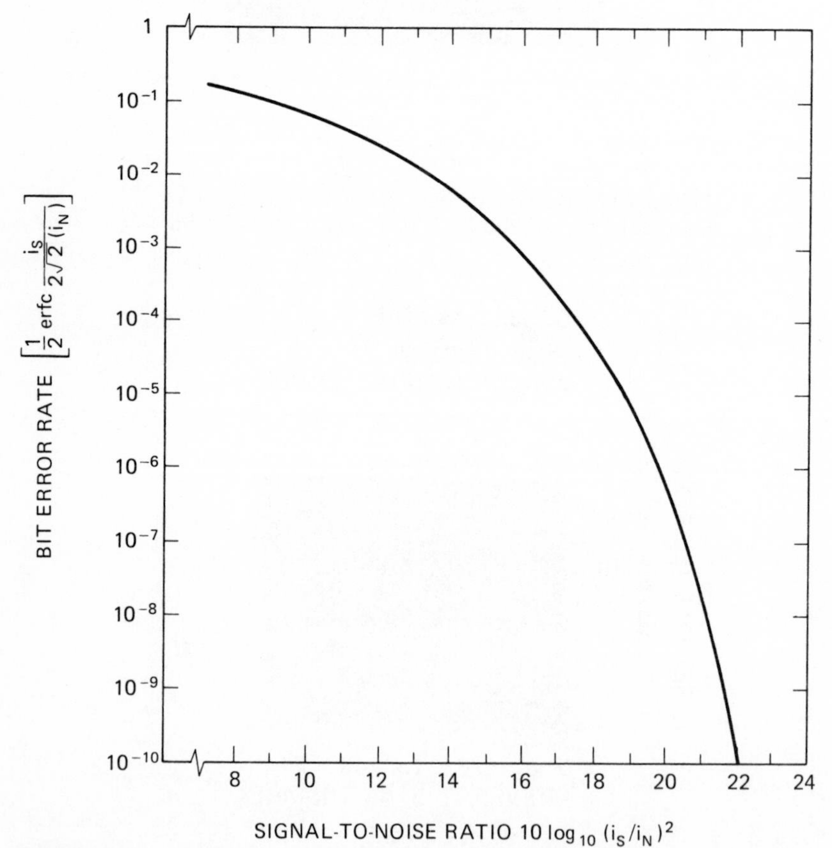

Figure 3.19. The bit error rate as a function of signal-to-signal power at the amplifier. The signal-to-noise power ratio for a 10^{-9} bit error rate is 23.8 dB. The optical signal-to-noise ratio is the square root of this figure, or 11.9 dB. (Reproduced by arrangement with Holt, Rinchart and Winston, New York, from *Introduction to Optical Electronics* 2/e by A. Yariv, © 1976)

Equation 3.17 is plotted in Figure 3.19. The required signal-to-noise power ratio is 21.5 dB for a 10^{-9} b.e.r. The signal-to-noise ratio of the optical power incident on the detector is the square root of this figure

$$\frac{S}{N_{\text{opt}}} = \frac{i_s}{\langle i_N \rangle} = 10.75 \text{ dB} = 11.9 \tag{3.20}$$

To achieve a 10^{-9} b.e.r., the optical power need be only 12 times larger than the minimum detectable power (Equation 3.15, Figure 3.13).

Using the results of Equation 3.20 we show in Figure 3.20 the optical power required at the receiver to achieve a 10^{-9} b.e.r. as a function of bit rate both at 0.8 μm and 1.3 μm. Using this information, we can calculate the maximum distance for transmission by optical fibres for signals sent at 0.8 μm and 1.3 μm. In these two examples, the power coupled into a monomode fibre is taken to be 100 μW, the attenuation at 0.8 μm to be 4 dB/km, and the attenuation at 1.3 μm to be 2 dB/km. The transmission of the fibre is assumed to be essentially distortion-dispersion free at both wavelengths, which is the case for a single transverse and longitudinal mode laser. For the detector we have assumed a 100% quantum efficiency at the wavelength of interest and a capacitance of 1 pF. The results are shown in Figure 3.21. For the parameters given above, the transmission distance is limited by the noise level in the receiver which is produced by the amplifier. The higher attenuation in the fibre at 0.8 μm alone accounts for a factor-of-two diminution of the transmission distance compared with the behaviour at 1.3 μm. In

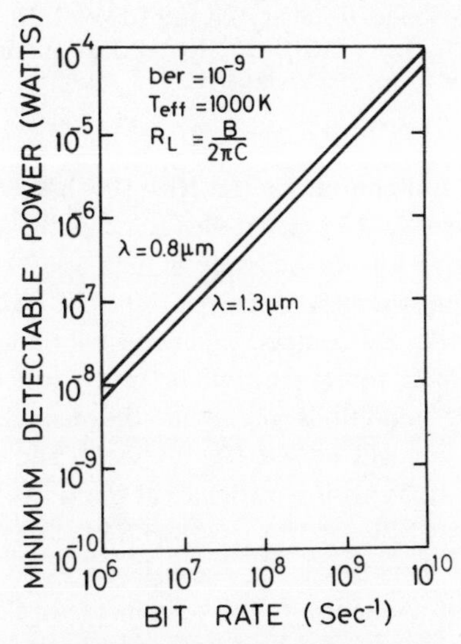

Figure 3.20. The minimum dectectable optical power at the receiver as a function of bit rate. Amplifier noise temperature is 1000 K

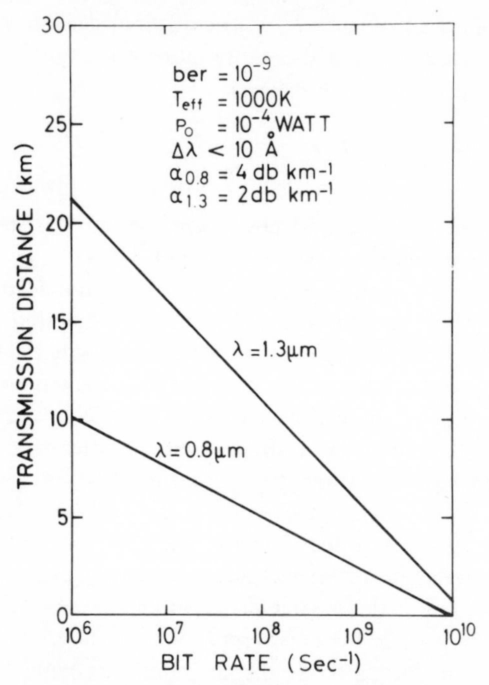

Figure 3.21. The maximum transmission distance of a 100-μW optical laser signal in a monomode fibre at 0.8 μm (α_L = 4 dB/km) and 1.3 μm (2 dB/km). Signal transmission is taken to be dispersion-free

the example above there is no bit rate in the 10^6–10^{10} bit sec^{-1} range for which the transmission distance exceeds 15 km. At 1.3 μm, on the other hand, the transmission distance exceeds 15 km for bit rates as high as 10^7 bits sec^{-1}. It is important to note that doubling the power coupled into the fibre, for example, would change this result very little. By contrast, reducing the fibre attenuation the values shown in Figure 3.2 (albeit for a multimode fibre) will improve the situation significantly. Under these conditions at 0.8 μm the maximum bit rate for which transmission exceeds 15 km will be 2×10^7 bit sec^{-1}, and at 1.3 μm, 2×10^9 bit sec^{-1}. Still, however, the transmission of signals at the GaAs laser wavelength under these optimistic conditions is limited to less than 5 km at 1 GHz. In order to transmit over longer repeater distances using GaAs lasers as sources, we need to use avalanche gain in the photodiode itself. This gain increases the signal power and the shot-noise power while, of course, the noise contribution from the amplifier is unchanged. The amount of avalanche gain available is that which raises the shot noise to the same level as the amplifier noise. It may be seen in Figure 3.21 that the required avalanche gain at 0.8 μm will be larger than that at 1.3 μm, and to achieve this gain that the shot noise power for a 0.8 μm detector such as silicon must be much less than that for a 1.3 μm detector.

3.3.3 Detector response time

The photodiode response time is limited by two factors: the transit time across the high field region, and the parameters of the measuring circuit (of which the photodetector is a part). The limitation imposed by the circuit is easily estimated by considering the equivalent circuit of a photodiode shown in Figure 3.22. The amplifier is characterized only by the load resistance R_L. This circuit behaves like a simple *RC* low-pass filter whose pass band is given by

$$B = 1/(2\pi R_t C_d), \tag{3.21}$$

where R_t is the sum of the load resistance and the diode series resistance and C_d is the diode capacitance including a contribution from the mounting. The transient response of real photodetectors does not follow strictly the decaying exponential response of an *RC* circuit, because some of the parameters may change with light frequency or intensity. This simple circuit model provides a useful guide to estimate the circuit limitations on the response time. For example, using Equation 3.14 we see that a photodetector with a 1 GHz bandwidth operating into a 50 Ω load must have a capacitance of $\leqslant$3 pF. For a diode of surface area 10^{-5} cm^{-2}, this corresponds to a depletion-layer width of about 1 μm for most semiconducting materials. The drift velocity in the depletion region is on the order of 10^6–10^7 cm/sec depending on the applied bias, hence the transit time is of the order of 10^{-11} sec. The response-time limitation from the circuit is 10^{-10} sec. The capacitance can be reduced further by making the diode area smaller, or by reducing the doping level in the depletion region. However, as the doping level is reduced, the depletion region becomes larger, hence the transit-time limitation becomes more important. For photodiodes of the dimension required for fibre-optic applications, the practical limit on response time is about 10^{-11} sec.

For a photodetector which absorbs light by indirect transition, such as silicon at

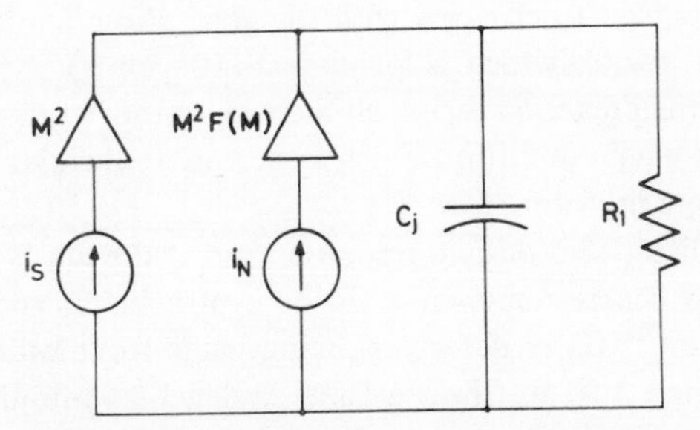

Figure 3.22. Equivalent circuit for a *p–n* junction photodetector

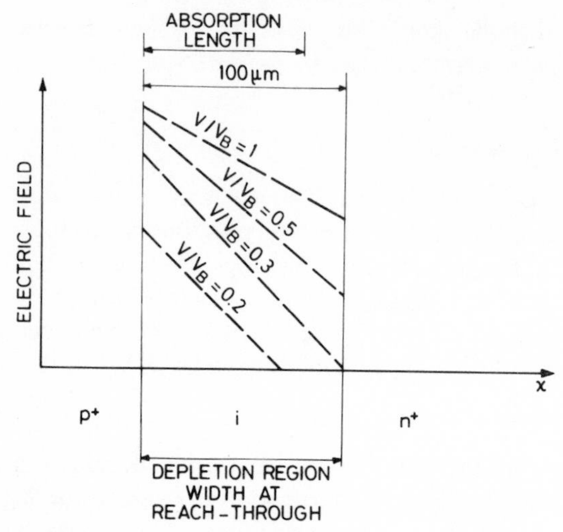

Figure 3.23. The 'reach-through' photodiode structure. This structure is used in silicon photodiodes where absorption is weak, to optimize with quantum efficiency and response time

0.8 μm, the situation is, however, different. Because of the weak absorption of silicon in the indirect regime, the absorption length is very long – ~50 μm at λ = 0.8 μm (see Figure 3.8a). The photoexcited carriers are distributed in a much greater thickness than would be the case for a material absorbing by the direct transition (such as germanium at 1.3 μm). The best way to collect these carriers as quickly as possible is to put the entire absorption region in an electric field whose magnitude is more or less uniform throughout this region. The response time in this case is always limited by the transit time of the carriers across this region (about 10^{-9} sec). The actual diode capacitance for a device with area: 10^{-5} cm^2 is a fraction of a picofarad, but the photodetector capacitance may be larger, being determined by the package. The basic device structure which incorporated these features is called a silicon 'reach-through' photodiode and is illustrated in Figure 3.23. The doping versus distance is designed to coincide with the absorption depth. The absorption takes place in the '*i*' region which is low-doped (10^{15} cm^{-3}) *n*-type material. The reverse bias extends the depletion region all the way to the heavily doped substrate. This is the 'reach-through' condition. Further increase in the bias raises the electric field and makes it more uniform.

State-of-the-art silicon photodiode response time at 0.8 μm is about 500 ps. In Figure 3.24 we show the response of a silicon avalanche photodiode to a pulsed GaAs laser at 0.83 μm.[25] GaAs detectors, being made from a direct-gap material, can be faster. In Figure 3.25 we show a GaAs avalanche photodiode with a 50 ps response close to the 10^{-11} s limit.[26] In Table 3.2 we list the measured response times of several photodiodes sensitive in the near infra-red.

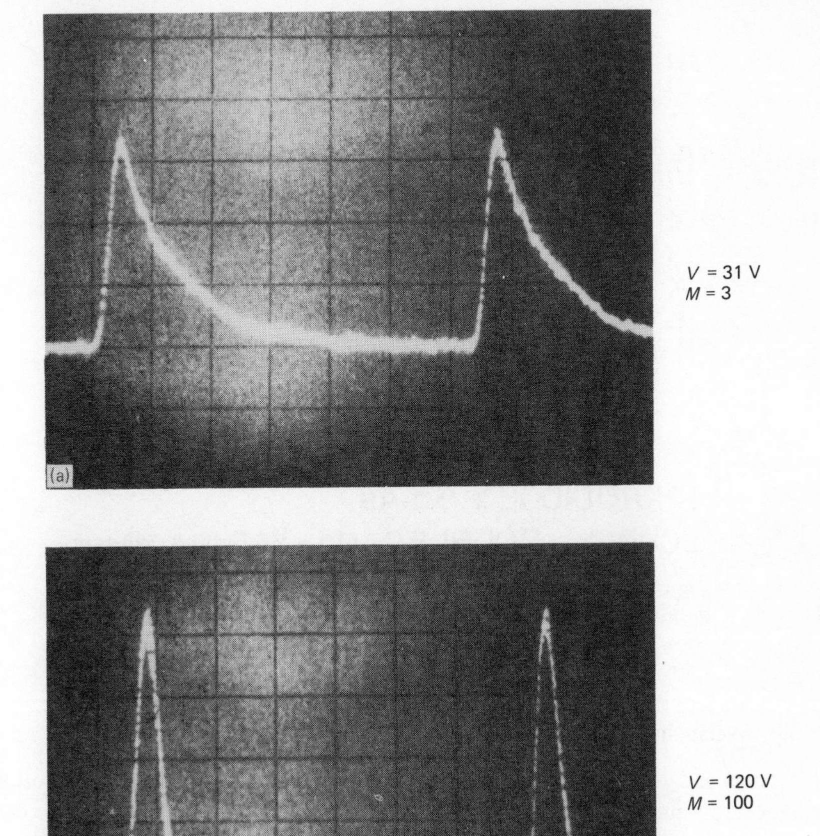

Figure 3.24. The response time for a state-of-the-art Si 'reach-through' avalanche photodiode at 0.83 μm. (Reproduced by permission of American Institute of Physics)

In summary, the response time of a photodiode with a sensitive area of 10^{-5} cm^2 and only moderate purity (10^{16} cm^{-3}) in the depletion region is shorter than 10^{-9} sec. The ultimate limit on response time is lower for materials absorbing via the direct transition than for those absorbing principally by indirect transition. This difference will be important only for optical-fibre communications with a bit rate in excess of 10^{10} sec^{-1}.

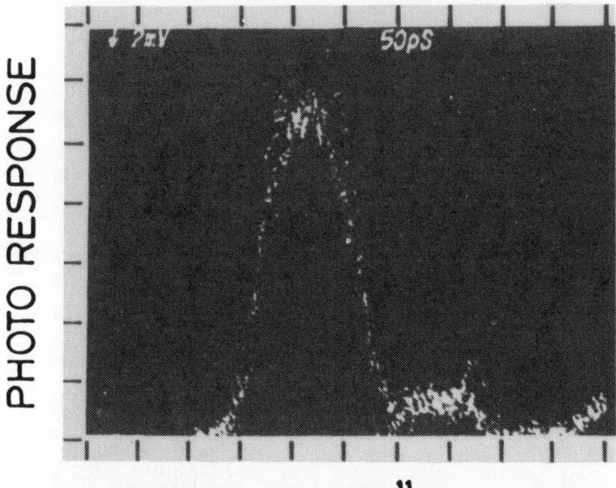

Figure 3.25. The pulse response of GaAs avalanche photodetector to a frequency-doubled Nd-YAG laser. The response time is about one order of magnitude less than that shown in Figure 3.24. (Reproduced by permission of Rockwell Int.)

3.3.4 Avalanche gain

Electrons and holes traversing the high-field region of a semiconductor photodiode gain kinetic energy. In a sufficiently high electric field they can acquire enough energy to create additional electron–hole pairs through an inelastic collision in which the energy lost is used to promote an electron from the valence band to the

Table 3.2 Semiconductor photodiode measured response times for detection in the near infra-red at room temperature

Photodiode material	Area (cm^2)	Optical wave-length (μm)	Measured response time (sec)	Reference
GaAs	3×10^{-5}	0.530	50×10^{-12}	Law *et al.*[26]
Si	6×10^{-4}	0.83	500×10^{-12}	Nishida *et al.*[25]
Si	1.8×10^{-4}	1.06	500×10^{-12}	Kanbe *et al.*[27]
$Ga_{0.83}In_{0.27}As$	1.2×10^{-4}	1.06	175×10^{-12}	Stillman *et al.*[28]
$Al_{0.16}Ga_{0.84}Sb$	2.8×10^{-5}	1.06	100×10^{-12}	Law *et al.*[29]
$Ga_{0.24}In_{0.76}As_{0.58}P_{0.42}$	1.8×10^{-4}	1.20	500×10^{-12}	Hurwitz and Hsieh[30]
$Ga_{0.47}In_{0.53}As$	5×10^{-4}	1.06	250×10^{-12}	Pearsall and Papuchon[31]
Ge	1×10^{-4}	1.15	120×10^{-12}	Melchior and Lynch[32]

conduction band. This collision event is called impact ionization, and the subsequent increase in junction current is called avalanche gain. Impact ionization is characterized by the ionization rates α for electrons, and β for holes. The ionization rate gives the number of secondary electrons created by a single initiating carrier per centimetre of travel through an electric field of magnitude ϵ. To initiate impact ionization, a carrier must have an energy at least as large as the bandgap. As a rule, impact ionization makes a significant contribution to the junction current for electric field greater than

$$\epsilon \gtrsim 2 \times 10^5 E_g \qquad (\text{V cm}^{-1}) \tag{3.22}$$

where E_g = Energy gap in eV.

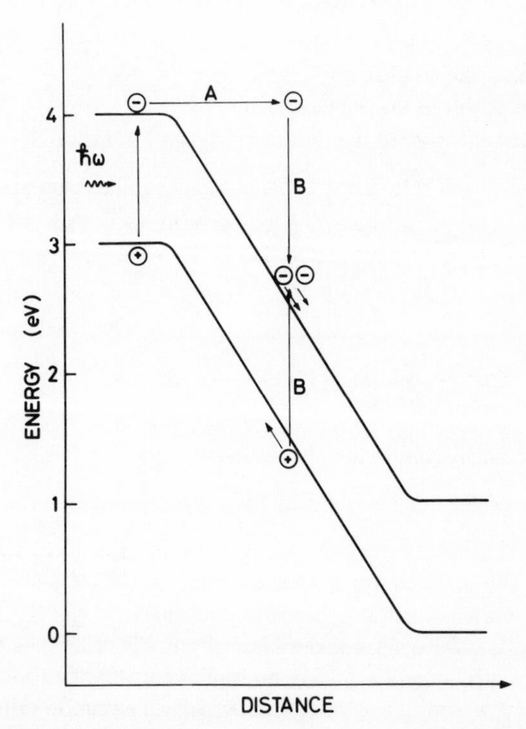

Figure 3.26. Avalanche gain in the photocurrent signal proceeds by impact ionization of electrons from the valence band to be the conduction band in the high-field depletion region of a reverse-biased semiconductor photodiode. The photon-excited carriers are accelerated by the electric field to energies of the order of energy gap above the band edge in order to initiate impact ionization

Fields of this magnitude are obtained in p–n junction diodes at reverse biases greater than $\frac{1}{2}V_b$, the breakdown voltage. Avalanche gain is useful in optical detection because it is internal to the photodiode itself (see Figure 3.26). Since the principal noise source is the preamplifier, avalanche gain provides a means of increasing the signal power without increasing the total noise level over a limited range of gain. The shot-noise current in the detector is multiplied along with the signal, and so the maximum useful avalanche gain is that which raises the shot-noise power to the same magnitude as the amplifier noise. If the shot-noise contribution from the laser or LED source is greater than the dark-current shot noise, the maximum avalanche gain will be limited by the noise in the photocurrent. The maximum useful avalanche gain is also determined by the rate at which the shot noise is multiplied as a function of the signal gain. In all avalanche diodes, the noise is multiplied faster than the signal. This feature exists because both holes and electrons contribute to impact ionization. The limits to avalanche gain are defined by noise:

(a) the noise level in the amplifier;
(b) the shot-noise power in the detector; and
(c) the excess noise intrinsic to the nature of impact ionization.

The signal-to-noise power ratio using avalanche gain can be calculated by modifying Equation 3.12:

$$\frac{S}{N} = \frac{2\left[P_{\text{opt}}\dfrac{q\eta_{\text{Q}}}{\hbar\omega}\right]^2 M^2}{\left\{\left[2qi_{\text{D}} + 4qP_{\text{opt}}\dfrac{q\eta_{\text{Q}}}{\hbar\omega}\right]M^2F(M) + \dfrac{4kT_{\text{eff}}}{R_{\text{L}}}\right\}B} \tag{3.23}$$

where M is the avalanche gain in the photocurrent, and

$F(M)$ is the excess noise introduced by avalanche given.

Because the signal photocurrent increases with M, the electrical signal power increases as M^2. The noise power increases faster, as $M^2F(M)$. The crucial point is that the amplifier noise contribution remains unchanged.

Avalanche gain is different from the kind of amplification in a photomultiplier. The dynode chain produces an orderly multiplication of electrons. The total gain can be expressed as a sum of a geometric series whose common ratio is the gain per stage, and whose number of terms is given by the number of stages in the chain. Gain by impact ionization has feedback as shown in Figure 3.27. At each stage where multiplication occurs two carriers are created – an electron and a hole. If the primary carrier is an electron as is the case in Figure 3.27, then the holes created by impact ionization are the 'feedback' carriers, because the holes are accelerated in the direction from which the primary carrier entered the junction. These holes may undergo ionizing collisions and generate new electrons which can be swept back

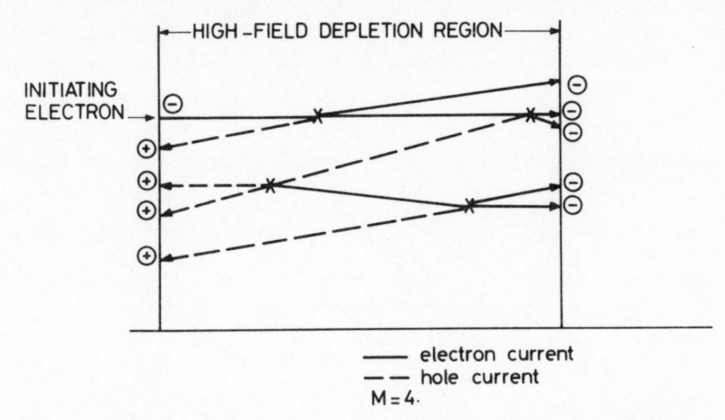

Figure 3.27. A schematic representation of carrier multiplication by avalanche gain in a semiconductor. The avalanche is initiated by an electron. The hole current represents the feedback. The strength of the feedback is given by (α/β). If $\alpha/\beta < 1$ the feedback is positive and the amplified signal is more noisy than if no gain were used

across the high-field region to ionize additional carriers. This process of amplification is called avalanche, to distinguish it from the electron multiplication in a photomultiplier. The avalanche gain is well represented by an exponential function of the bias voltage:[33]

$$M = \frac{1}{\left\{1 - \left(\frac{V}{V_B}\right)^a\right\}} \tag{3.24}$$

Avalanche gain is a much more critical function of bias voltage near breakdown than is the gain in photomultiplier.

The excess noise in avalanche gain is directly related to the amount of feedback and can be expressed in terms of the ionization rates α and β.[34] As the ratio α/β becomes more different from unity, only the carrier with the larger ionization rate contributes to impact ionization and the excess noise factor becomes smaller, tending toward its lower limit of 2. If the ionization rates are equal, the excess noise is at its maximum and F is at its upper limit of M. The detailed dependence of avalanche noise on the ionization rates has been calculated by McIntyre in the low-frequency limit (the carrier transit time is less than the reciprocal bit rate). The excess noise factor is expressed as[35]

$$F = M\left[1 - (1-k)\left(\frac{M-1}{M}\right)^2\right] \tag{3.25}$$

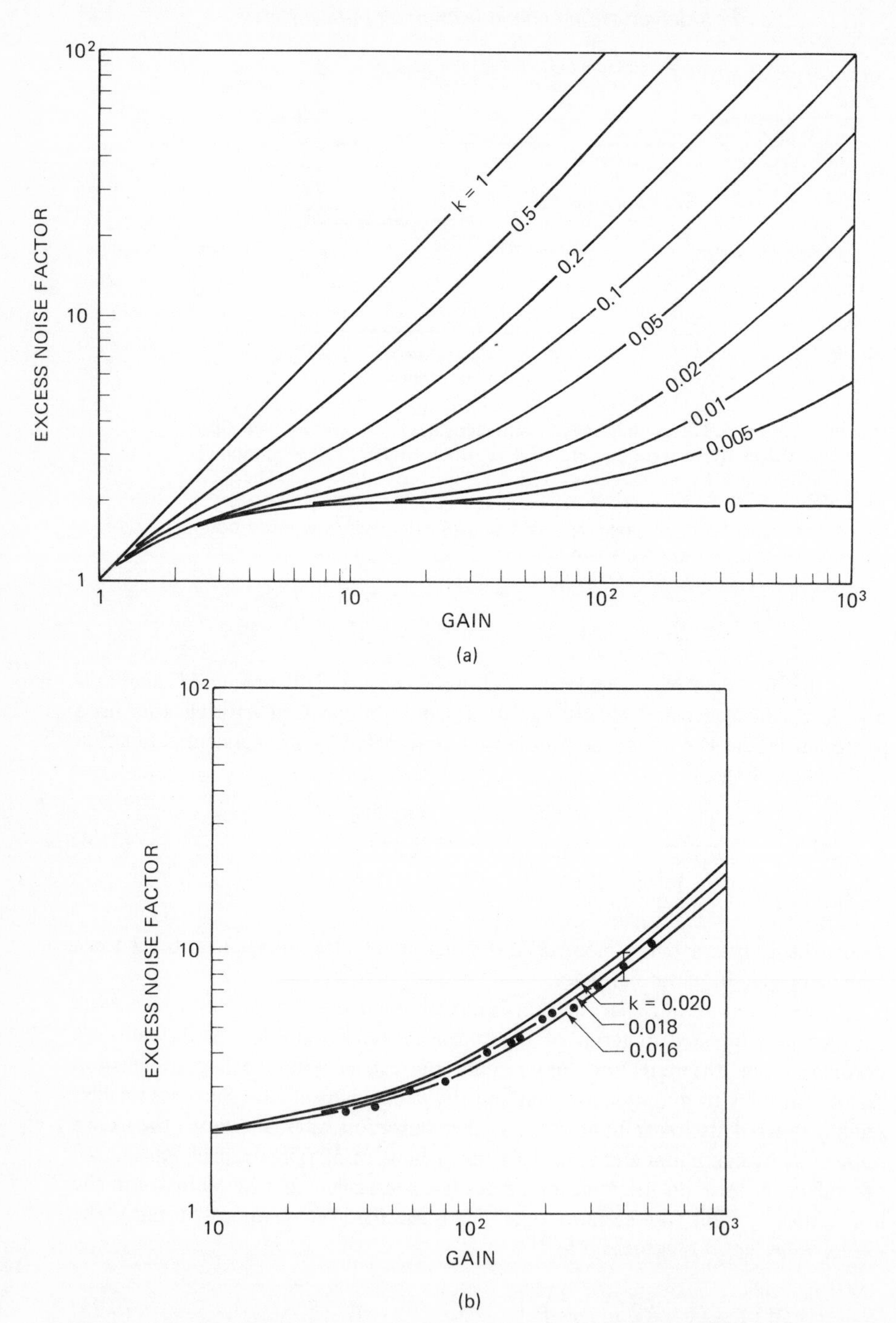

Figure 3.28. Excess noise factor $E(M)$ as calculated by McIntyre[35] (a) for general K, (b) experimental results for Si avalanche photodiodes. (Reproduced by permission of RCA Corporation)

for the case where electrons have the higher ionization rate, and

$$k = \frac{\beta}{\alpha} < 1$$

An equivalent expression exists for hole-initiated avalanche gain. In Figure 3.28 we show the result of McIntyre calculation for several values of the ionization-rate ratio. The noise behaviour in the high-frequency limit has been calculated by Gummel and Blue,[36] and at intermediate frequencies by Naqvi.[37] This result is shown in Figure 3.29 where the ratio of the actual avalanche noise to the shot noise multiplied by M^3 is shown for an avalanche photodiode in which electrons have the higher ionization rate. The work of Naqvi shows that as $\omega^2\tau^2M^2 > 1$, the excess noise generated by avalanche gain decreases. For practical avalanche photodiodes, this effect becomes important for frequencies above 1 GHz.

The essential feature of excess avalanche noise is that it is least when impact ionization is initiated by the carrier with the higher ionization rate. This means that the photons should be absorbed predominantly in a region where the minority carriers are those with the higher ionization rate. Using Equation 3.25 it can be seen

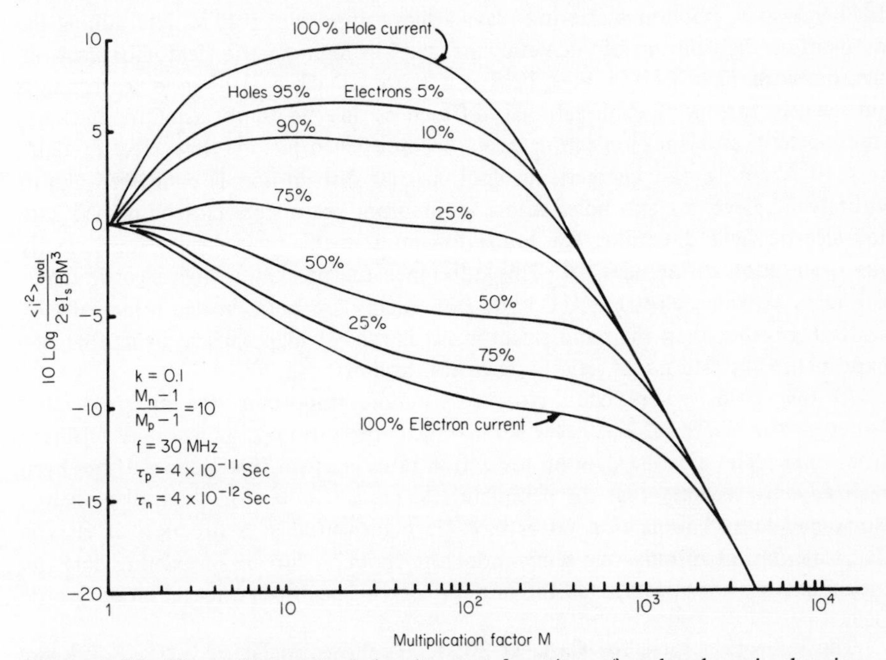

Figure 3.29. The noise spectral density as a function of avalanche gain showing the generalization of McIntyre's theory at high frequencies where ($\omega^2\tau^2M^2 >$ 1). This result, calculated by Naqvi,[37] shows that avalanche feedback becomes less-important at high frequencies so that the excess noise diminishes. (Reproduced by permission of Pergamon Press, *Sol. St. Electronics*)

that when the ionization rates are equal, the excess noise factor is equal to M so that the noise power increases as M^3. When the ionization rates differ by several orders of magnitude $F(M) \rightarrow 2$ and the noise power increases as M^2 just as the signal.

The relative magnitudes of the ionization rates can be understood to some degree by considering the physics of impact ionization. Charge carriers move in an electric field according to Newton's law:

$$\frac{\mathrm{d}}{\mathrm{d}t} \hbar \mathbf{k} = e\epsilon \qquad (3.26)$$

where **k** is the wave vector of the charge carrier.

The electronic bandstructure relates the carrier energy E to the crystal momentum $\hbar\mathbf{k}$, hence the carriers gain energy according to the restrictions imposed by the electronic bandstructure, and energy loss through phonon scattering (and eventually impact ionization).[38]

The electronic band structure is different for different crystal directions, and it is not the same for holes and electrons. The ionization rates, which are a measure, in part, of how easily carriers can gain energy in an electric field, are in general different for electrons and holes, and also orientation-dependent. As the electric field increases, phonon scattering plays a more important role in determining the momentum distribution of energetic carriers.[39] In high electric fields (for semiconductors with $Eg \approx 1$ eV, a high field is $\epsilon > 3 \times 10^5$ V cm^{-1}) phonon scattering is sufficiently strong that directional differences in the band structure becomes unimportant and the ionization rates become isotropic. At low electric fields ($\epsilon < 10^5$ V cm^{-1}), the energetic or 'hot' carrier distribution is composed almost entirely of electrons and holes whose momentum vectors are closely aligned with the electric field direction. The ionization rates in this limit display most clearly the orientation differences.[40,41] This difference means that at high electric fields, the rates at which electrons and holes gain energy are both limited principally by scattering rather than the band structure. It is thus at high electric fields that one expects the ionization rate ratio to be closest to unity.

At low fields where band structure is more important, the different band structure for electrons and holes act to make the ratio α/β increasingly different from unity. The existing data on ionization rates confirm this picture. It has been realized only recently that the parameters of impact ionization should be orientation-dependent.[42] Ionization ratios have been measured as a function of electric field orientation in only one semiconductor: GaAs.[40] For other existing data the measurements have been made for one (usually unspecified) orientation of electric field.

The ionization rates for GaAs at 300 K are shown in Figure 3.30a, 3.30b and 3.30c for the electric field oriented in the ⟨100⟩, ⟨110⟩ and ⟨111⟩ directions respectively. One sees clearly that the ⟨111⟩ direction is the optimum orientation for an avalanche photodiode. At a breakdown field of 3×10^5 V cm (corresponding to a doping in the depletion region = 10^{15} cm^3) the ionization rate ratio $\alpha/\beta > 10^2$. The

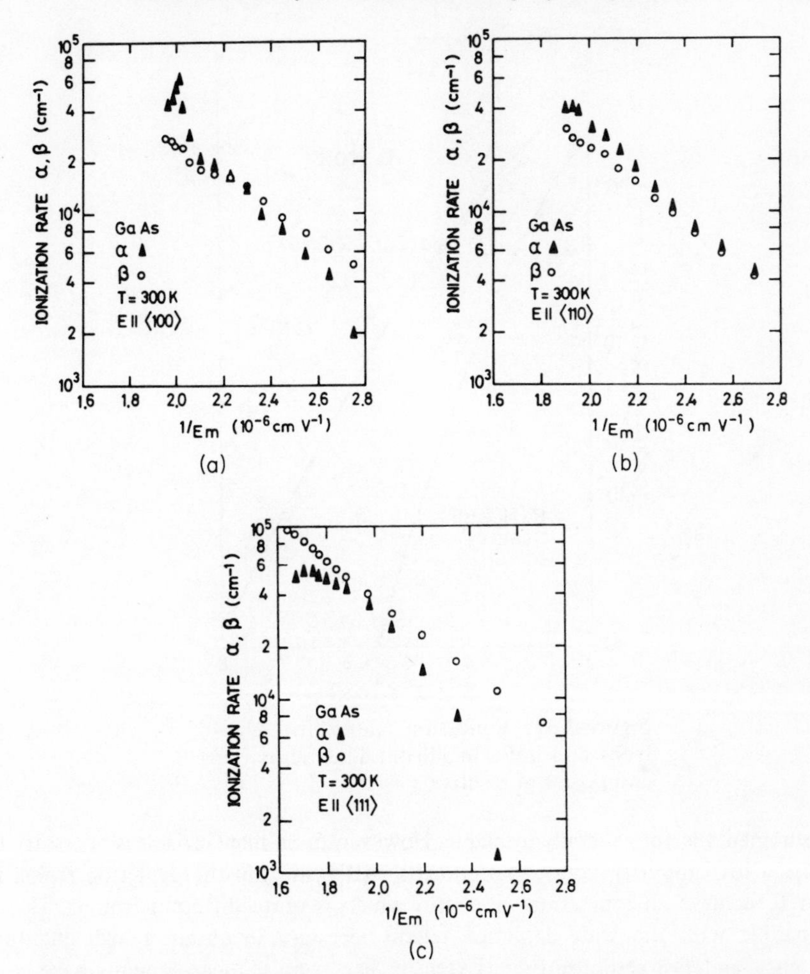

Figure 3.30. The ionization rates of GaAs versus reciprocal electric field. Electric field oriented: (a) parallel to the ⟨ 100 ⟩ direction; (b) parallel to the ⟨ 110 ⟩ direction; and (c) parallel to the ⟨ 111 ⟩ direction. Ionization rates are generally a function of electric field orientation and are different for electrons and holes. These curves show that the optimum noise properties in GaAs photodiodes are obtained for electric fields along the ⟨ 111 ⟩ direction

⟨100⟩ orientation is next best and the ⟨110⟩ is worst, with $\beta/\alpha \approx 1$. It should be carefully noted that the orientation differences in the ionization rate ratio may be very large, especially for low-field breakdown typical of avalanche photodiodes.

A composite curve of the measurement of ionization rates in silicon is shown in Figure 3.31.[43] Information on the orientation of the electric field during the various

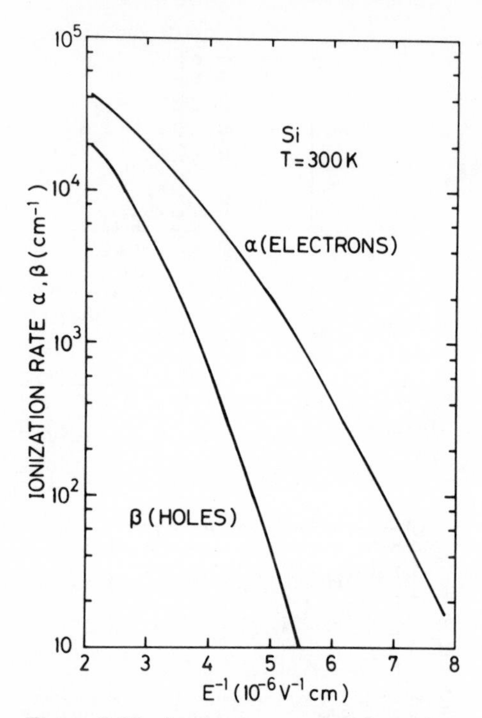

Figure 3.31. Ionization rates for electrons and holes in silicon. This figure is a composite of results

measurements is not generally available. However, in Si, like GaAs, it is necessary to choose a low impurity concentration (10^{14}–10^{15} cm^{-3}) in the depletion region in order to achieve an ionization rate ratio which is quite different from 1. This is compatible with the wide depletion region necessary to obtain a high quantum efficiency and fast response time. Extensive excess-noise measurements have been made for Si avalanche photodiodes, and the results show excellent agreement with theory.[44,45]

Several ionization-rate measurements for Ge exist and have shown that the ionization rates differ at most by a factor of 2.[46,33] No orientation information is available for these measurements. Since the ionization rates are nearly equal, the excess noise generated by Ge is near the theoretical maximum and such behaviour has been measured.[32] It would be worth while to study systematically the ionization rates in Ge to see if all major orientations of electric field show this extreme behaviour.

Ionization-rate ratios have been measured in a few III–V alloy semiconductors $Ga_{0.86}In_{0.14}As$,[47] $GaAs_{0.88}Sb_{0.12}$,[48] and $Ga_{0.47}In_{0.53}As$.[31] The measured ionization-rate ratios in all of these compounds is about five, which makes the excess-noise picture in these materials already much better than that for Ge. Measurements have not been made in low-doped alloy semiconductor photodiodes to see if a more favourable ionization-rate ratio exists at a depletion region doping on the order of 10^{15} cm^{-3}. However, the noise level measured in $Ga_{0.28}In_{0.72}As$ Schottky-barrier photodiodes indicates that the ionization-rate ratio β/α is greater than 10.[28]

A summary of ionization-rate ratio measurements is given in Table 3.3 for semiconductors of interest for optical communications, and for several impurity concentrations in the depletion region. The basic feature seen in all the measurements is that the ionization rates for electrons and holes become increasingly different as the carrier concentration in the lower-doped depletion region becomes smaller. The excess-noise factor subsequently becomes lower as the doping level is lowered provided that avalanche gain is initiated by the carrier with the higher ionization rate. Silicon, whose excess noise factor is small, is not unique among the semiconductors in Table 3.3. Measured ionization rates in GaAs indicate an excess-noise factor lower than that in Si for low doping in the high-field region. Similar behaviour may be shown by III–V alloy avalanche detectors at 1.3 μm; but sufficiently pure material has not been produced to verify this picture.

Table 3.3 Measured ionization-rate ratios for semiconductors whose bandgaps lie in the range 0.5–1.5 eV

Material	Bandgap at 300 K (eV)	Ionization rate ratio α/β Carrier concentration (cm^{-3}) $N_D = 10^{15}$	$N_D = 10^{16}$	Electric field orientation	Ref.
GaAs	1.43	0.12	1	⟨100⟩	(a)
		1	1	⟨110⟩	
		0.01	1	⟨111⟩	
Si	1.12	8.0	4.5	⟨111⟩	(b)
		60.0	4.5	Unknown	(c)
$Ga_{0.86}In_{0.14}As$	1.15	–	0.25	⟨111⟩	(d)
$GaAs_{0.88}Sb_{0.12}$	1.15	–	2.5	⟨100⟩	(e)
$Ga_{0.47}In_{0.53}As$	0.75	–	5.0	⟨111⟩	(f)
Ge	0.6	–	0.5	Unknown	(g)

(a) Ref. 41
(b) M. H. Woods, W. C. Johnson and M. A. Lampert, 'Use of a Schottky barrier to measure impact ionization coefficients in semiconductors', *Solid-State Electron.*, **16**, 381–394, 1973
(c) C. A. Lee, R. A. Logan, R. L. Batdorf, J. J. Kleimack and W. Weigman, 'Ionization rates of holes and electrons in silicon', *Phys. Rev.*, **134**, A761–A773, 1964
(d) Ref. 47
(e) Ref. 48
(f) Ref. 31
(g) Ref. 33

Using avalanche gain, we can increase the transmission distance in an optical-fibre communication system. The maximum useful avalanche gain can be calculated from Equation 3.22:

$$M^2(kM - k + 1) = \frac{2\,(\text{Ampnoise})}{\text{Shot noise}} \tag{3.27}$$

This expression can be solved for the multiplication provided the magnitude of the shot-noise term is known. It is composed of the detector dark current and the shot noise in the laser or LED source. The source shot noise (Equation 3.9) is proportional to the optical power. The absolute magnitude of the source shot noise at the detector will depend on the power coupled into the fibre, the attenuation, and the length of the fibre. In our case, the minimum detectable optical power determines the fibre length. Once this power is known (see Figure 3.20) the absolute source shot noise power can be calculated,[49] and this is shown in Figure 3.32.

The average signal shot-noise current can be compared to the dark current for several specific detectors. In Figure 3.33 we show the best measured dark-

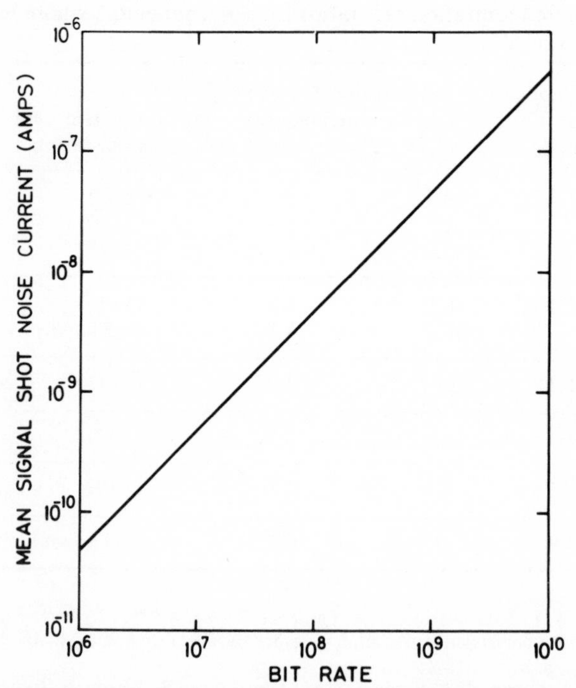

Figure 3.32. Average shot-noise current as a function of bandwidth for a semiconductor laser with an optical signal-to-noise ratio of 50 dB. The optical power is the minimum detectable signal for a 10^{-9} b.e.r. (see Figure 3.20)

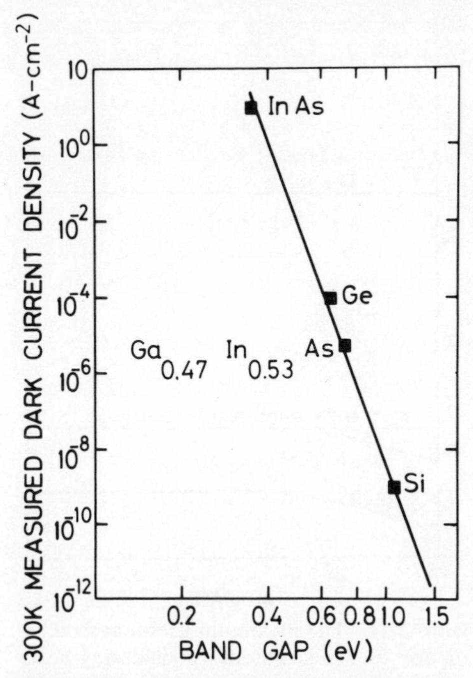

Figure 3.33. The measured dark-current density at $\frac{1}{2}$ V_B in semiconductor diodes as a function of bandgap from 0.3 eV to 1.1 eV. These points lie along a straight line on a log–log plot. The dark-current density in a 1.3 μm, III–V alloy photodetector may be estimated from this curve

current density in semiconductor diodes between 1.1 eV (Si) and 0.3 eV (InAs). The measured dark current is seen empirically to increase in an orderly way as the bandgap becomes smaller. It is possible to use this curve to estimate the dark-current density in a hypothetical III–V alloy photodiode at 1.3 μm (Eg = 0.9 eV). The actual dark-current level can be calculated using the device area 8×10^{-5} cm^2 corresponding to a 100-μm diameter photodiode. A comparison of the dark current to the shot current from the laser source shows that the maximum avalanche gain in Si will be limited by laser shot noise over the entire bandwidth range of interest for optical-fibre communications. For a 1.3 μm III–V avalanche photodiode, the gain will be limited by source noise also. For $Ga_{0.47}In_{0.53}As$ the gain will be limited by source noise above 10 MHz, and in Ge, the gain will be limited by photocurrent shot noise above 100 MHz. This result means that the advantage of the extremely low dark current in Si, and the moderately low dark current in a 1.3 μm, III–V alloy photodiode will be largely lost in optical communications applications.

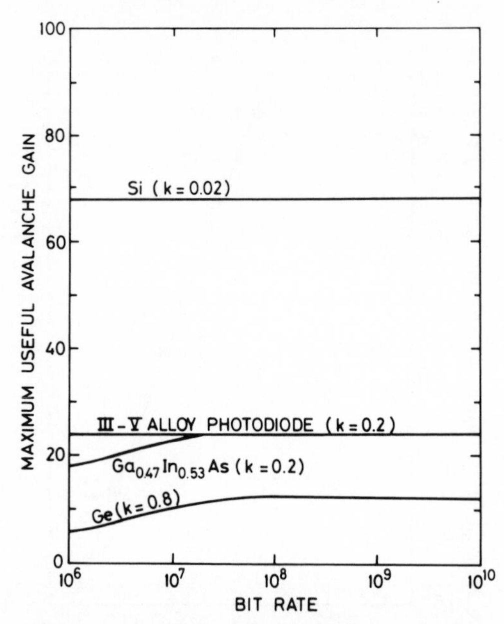

Figure 3.34. The maximum useful avalanche gain for four photodiodes considering both dark current and signal shot noise. The gain in Si and the III–V avalanche photodiodes is limited by signal shot noise over the most of bandwidth range $10^6 - 10^{10}$ Hz. The higher optimum gain of silicon avalanche photodiodes is the result of its low excess-noise factor

Table 3.4. Parameters used in the calculation of maximum transmission distance by optical fibres at 0.8 and 1.3 μm using avalanche gain

Optical wavelength, λ	Fibre attenuation, α_L	Quantum efficiency, η_Q	Amplifier noise temp., T_{eff}	Ionization rate ratio, α/β
0.8 μm:				
GaAs	–	–	–	0.02
Si	4 dB/km	90%	1000 K	50
1.3 μm:				
Ge	–	60%	–	1.25
$Ga_{0.47}In_{0.53}As$	2 dB/km	90%	1000 K	5.0
III–V alloy, $\lambda_g = 1.3\ \mu$m	–	90%	–	5.0

In the example which follows, we continue our earlier calculation and determine the maximum transmission distance in optical-fibre communication using avalanche gain in four specific semiconductor detectors: Si, Ge, $Ga_{0.47}In_{0.53}As$, and a hypothetical 1.3 μm, III–V alloy photodetector. We solve Equation 3.27 for the maximum avalanche gain. The results are shown in Figure 3.34. The relevant parameters are listed in Table 3.4.

The shot noise level in the photocurrent will be the limiting factor in long-distance telecommunication at 0.83 μm. This conclusion is independent of the fibre attenuation, although if the attenuation is low enough this limited repeater-free transmission distance may be long enough to satisfy system requirements. Using the results of Figure 3.34 we can calculate the maximum distance for optical-fibre transmission using avalanche gain. For transmission at 0.83 μm (Si), α_L is taken as -4 dB/km, and at 1.3 μm (Ge, $Ga_{0.47}In_{0.53}As$ and, III–V alloy detector) $\alpha_L = -2$ dB/km. We show the results of this calculation as a function of bit rate in Figures 3.35 and 3.36.

The considerable advantage of transmission at 1.3 μm is evident.[50] The poor

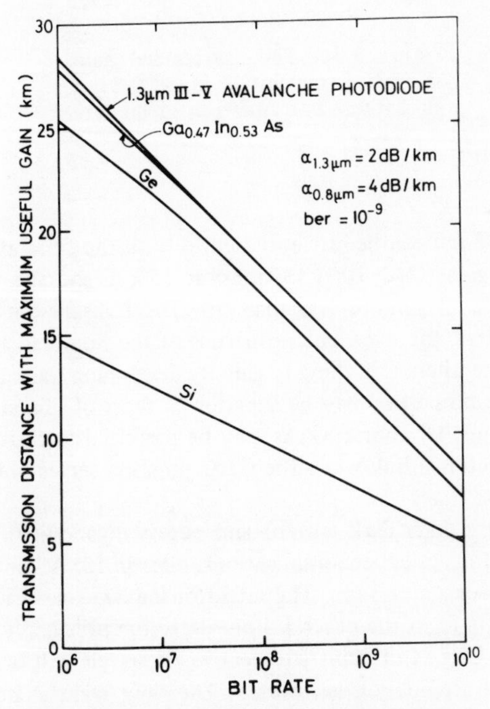

Figure 3.35. Maximum transmission distance as a function of bit rate for four photodiodes. The fibre attenuation α_L is 4 dB/km at 0.83 μm and 2 dB/km at 1.3 μm. The laser signal-to-shot-noise ratio is 50 dB and the amplifier noise temperature is 1000 K

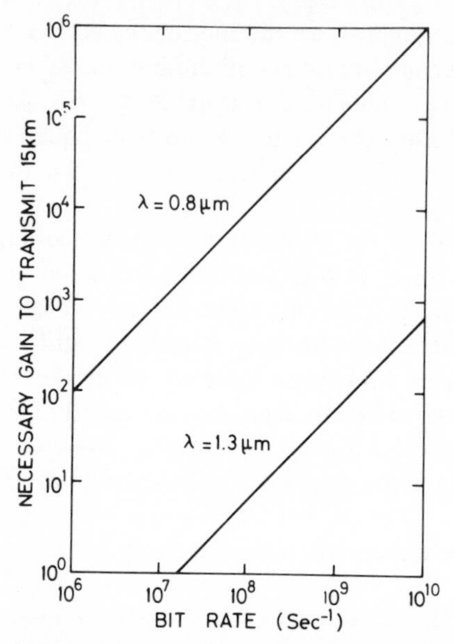

Figure 3.36. The avalanche gain required to transmit 15 km at 0.8 μm and 1.3 μm as a function of bit rate

performance at 0.83 μm can be attributed entirely to the fibre attenuation which requires very high gains ($M > 10^3$) to transmit 15 km, and the signal shot noise which limits the useful gain to less than 70. The transmission distance is thus largely independent of the excellent properties of the Si avalanche photodiode.[23] For this reason alone, there is nothing to gain by developing GaAs avalanche photodiodes, whose noise properties may be superior to those of silicon for optical-fibre detection applications. Of course, GaAs may be preferred to Si for other reasons, such as in a bi-directional link when the GaAs junction serves both as source and emitter.

Germanium, whose high dark current and excess noise might have seemed to make it unsuitable for optical communications, does in fact seem to be well suited for high bit-rate systems at 1.3 μm. The effect of the large dark current and excess noise differentiate Ge from the other 1.3 μm detectors principally at low bit rates. Transmission at 10 GHz is difficult irrespective of wavelength or detector, but is possible if the fibre attenuation is lowered. The only realistic hope for medium distance (15 km) high bit-rate telecommunications lies in transmission at 1.3 μm rather than 0.8 μm. The high performance possibilities of silicon avalanche photodiodes are largely vitiated by high fibre attenuation at 0.8 μm. At 1.3 μm, the Ge avalanche detector with its poorer noise characteristics seems nevertheless to be a good choice for a detector at 1.3 μm. Other 1.3 μm detectors – $Ga_{0.47}In_{0.53}As$, or a 1.3 μm, III–V alloy photodetector – are capable of detection over 20–25 km distance at low bit rates (10^6–10^7 Hz). The goal of 15 km transmission distance from 10^6–10^9 Hz is met nearly as well by Ge as by either of the other two 1.3-μm photodiodes.

3.4 1.3 μm DETECTORS

Germanium *p–n* junction photodiodes are at present the only commercially available uncooled photodetectors with sufficiently fast response to be suitable for optical communication at 1.3 μm at bit rates in excess of 1 Mbit/sec. As we have shown in the previous section, Ge avalanche photodiodes can be used in a 15 km link for bit rates up to 1 Gbit/sec with only a moderate avalanche gain required. Long-distance communication at higher rates will be limited by the dark current, and of course, the fibre attenuation. For most fibre-optic telecommunication links, Ge avalanche photodiodes appear to be a highly satisfactory choice. The spectral response of a Ge photodiode (commercially available from Judson infra-red), is shown in Figure 3.37. The typical Ge photodiode manufactured at the present time has a wide depletion region in order to maximize the sensitivity between 1.8 μm and the direct gap at 1.5 μm. A photodiode optimized for 1.3 μm does not need this sensitivity. Increasing the doping in the depletion region to the 10^{16} cm^{-3} level

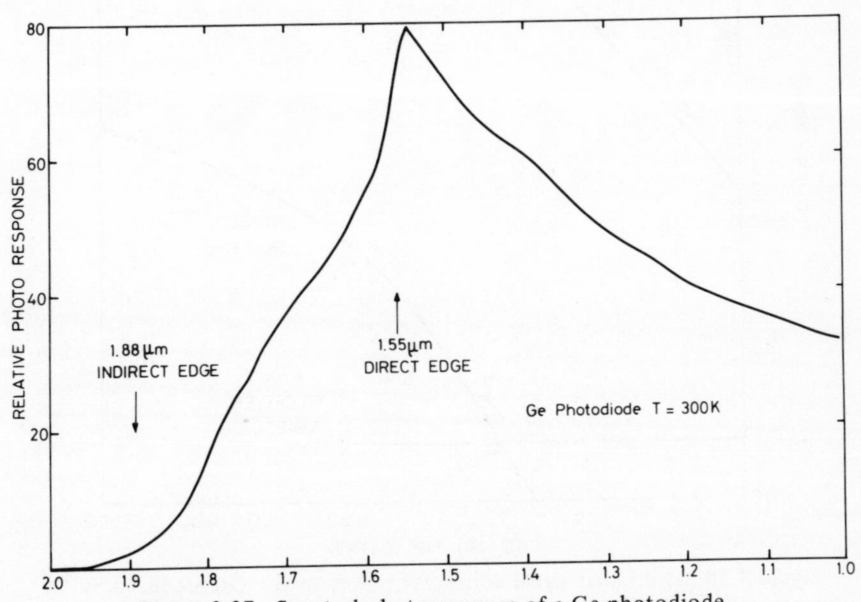

Figure 3.37. Spectral photoresponse of a Ge photodiode

will keep the depletion region to the order of 2 μm and have the benefit of reducing the dark current. Both the diffusion component[51]

$$J_{\text{diff}} = q\left[D_n \frac{n_p}{L_n} + D_p \frac{p_n}{L_p}\right] \qquad (\text{A cm}^{-2}) \tag{3.28}$$

and the generation component

$$J_{\text{gen}} = (qn_iW)/(2\tau') \qquad (\text{A cm}^{-2}) \tag{3.29}$$

$$\tau' = \tau_0 \cosh((E_i - E_f)/kT) \tag{3.30}$$

where n_i = intrinsic carrier concentration,

W = depletion layer width,

τ' = effective carrier lifetime, and

τ_0 = minority carrier lifetime

depend on the doping in the depletion region.[51] Increasing the doping reduces the

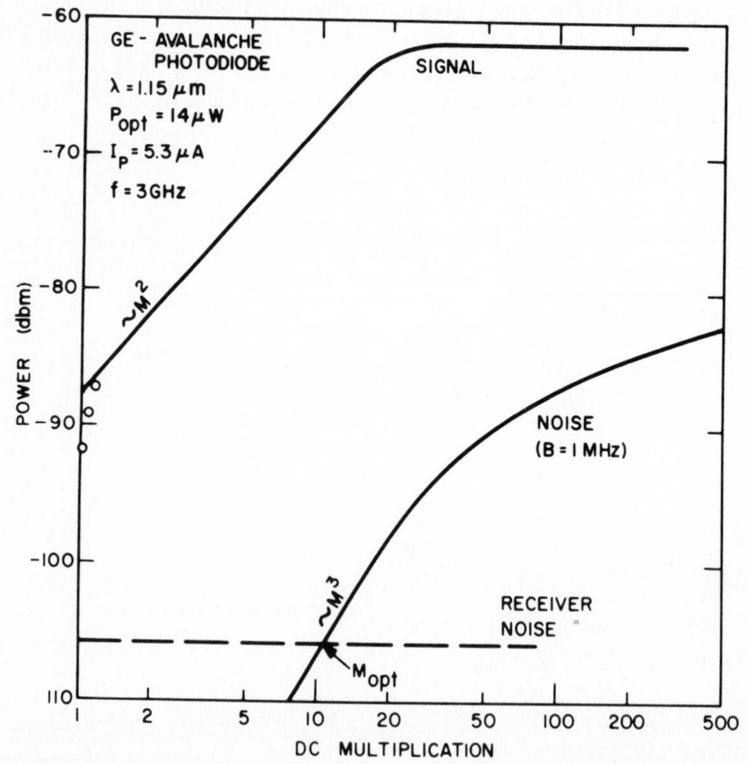

Figure 3.38. Multiplied signal and noise power in Ge. (Reproduced by permission of IEEE)

depletion layer width, W, and reduces the minority carrier densities at the edge of the depletion region.

Excess noise introduced by avalanche gain is high in Ge. The measured M^3 dependence of the noise power on the gain is shown in Figure 3.38.[52] Ionization-rate measurements in Ge are in conflict over which carrier has the higher rate,[53] and no information exists on the nature of the ionization-rate variation with electric field orientation. Thus, the ionization-rate data are incomplete. Since the measured noise dependence on gain is the theoretical maximum, any variation in the ionization rates with electric field orientation can be used to yield an avalanche photodiode with improved noise performance. The single feature which limits the Ge photodiodes at high bit rates is noise. For detection at 1.3 μm it seems that both noise sources in Ge (from dark current and avalanche gain) can be reduced. The quantum efficiency and speed of Ge at 1.3 μm are good.

In the calculation of useable avalanche gain in Section 3.3, we showed that the maximum gain in $Ga_{0.47}In_{0.54}As$ or a hypothetical III–V semiconductor alloy photodiode was limited by signal shot noise over most of the interesting frequency range for optical communications. Thus at high bit rates, there is no performance advantage to be gained in a detector with a bandgap optimized at 1.3 μm over $Ga_{0.47}In_{0.53}As$ except for a slightly higher (10%) quantum efficiency. Because there is greater leeway at 1.3 μm between the expected dark current shot noise and that which limits detection, it is expected that the yield in the manufacture of 1.3 μm alloy detectors with the required photodiode properties will be higher than that for the longer-wavelength materials.

Semiconductor alloys of III–V elements can be grown on III–V binary semiconductor substrates provided the lattice-constant difference between the substrate and the alloy is not too great ($\Delta a_0/a_0 < 5 \times 10^{-3}$). Although many alloy systems are possible,[54] there are two which are of particular interest for optical telecommunications at 1.3 μm: $Al_xGa_{1-x}Sb$, grown on GaSb; and $Ga_xIn_{1-x}As_yP_{1-y}$ grown on InP. Stable alloy compounds can be grown using both of these alloy systems over a considerable compositional range directly on the appropriate binary substrate. The bandgap range spanned by these two alloy systems is shown in Figure 3.39. $Al_xGa_{1-x}Sb$ is an indirect-gap semiconductor for all concentrations of AlSb greater than 20 ($\lambda_g = 1.5$ μm).[55] Lasers have been made using this system at wavelengths up to the direct–indirect transition.[56] Like $Al_xGa_{1-x}As$, there is little lattice-constant change as AlSb is alloyed with GaSb (see Table 3.5). By choosing a fourth element, it is possible to maintain the lattice constant of the substrate and alloy at the same value while adjusting the bandgap of the alloy.[57]

$Ga_xIn_{1-x}As_yP_{1-y}$ is a quaternary alloy which can be grown on InP with the same lattice constant and an absorption edge which can be varied between 0.9 μm (InP) to 1.65 μm ($Ga_{0.47}In_{0.53}As$). This quaternary alloy is a direct-gap semiconductor over the entire compositional range. The restriction that the lattice constant of the substrate be the same as that of the alloy requires that the arsenic atomic fraction be about two times the gallium atomic fraction, and one can write the

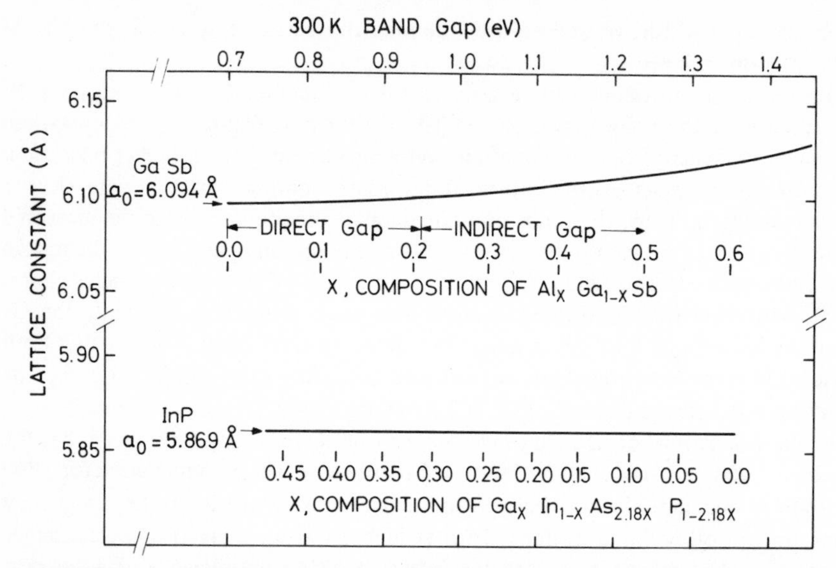

Figure 3.39. Bandgap and lattice constant as a function of alloy composition for $Al_xGa_{1-x}Sb$ and $Ga_xIn_{1-x}As_{2.18y}P_{1-2.18y}$

Table 3.5 Bandgap and lattice parameter dependence on composition for two imp important alloy semiconductors

	$Al_xGa_{1-x}Sb$	$Ga_xIn_{1-x}As_{2.18x}P_{1-2.18x}$
Bandbap variation with composition (eV) Direct Indirect	(Γ) $0.368x^2 + 1.129x + 0.700$ (X) $0.077x^2 + 0.492 + 1.020$ (L) $0.334x^2 + 0.746x + 0.799$	(Γ) $-0.80x^S_{Ga} + 1.35$ – (unavailable) – (unavailable)
Direct–indirect transition	$x = 0.22$	none
Composition at = 1.3 μm (0.95 eV)	$x = 0.21$	$x = 0.27$
Lattice constant of substrate (Å)	6.094 Å (GaSb)	5.868 Å (InP)
Lattice constant: Variation with composition (Å)	$0.0517x^S_{Al} + 6.094$	$0.38x^S_{As} - 0.83x^S_{Ga} + 5.868$

lattice-matched composition using only one parameter, x: $Ga_xIn_{1-x}As_{2x}P_{1-2x}$. At 1.3 μm, the lattice-matched composition is: $Ga_{0.27}In_{0.73}As_{0.60}P_{0.40}$. More precise details are given in Table 3.5.

$Al_xGa_{1-x}Sb$ has been used to make extremely fast avalanche photodetectors, sensitive to 1.4 μm. The response of a 100-μm diameter detector operating at a gain of 17 is shown in Figure 3.40. The diode is detecting 1.06 μm Nd–YAG laser emission. The measured rise-time is 65 ps. The measured dark-current density in this detector is large ($J \approx 10^{-2}$ A cm^{-2}), but it is apparently largely surface-leakage current which escapes multiplication by avalanche gain.[29]

$Al_xGa_{1-x}Sb$ grows normally p-type, very heavily doped. Producing a good p–n junction is difficult and must be done at low temperatures (400–500 °C) for growth by liquid-phase epitaxy. Low-doped n-layers are obtained by tellurium compensation of p-layers nominally doped in the 10^{17} cm^{-3} region. Uniform layers with a carrier concentration n^- in the 10^{15} cm^{-3} range have been obtained by this method.[58] The grown p–n junction photodiodes from this material show good quantum efficiency at 1.3 μm ($\eta_Q \approx 40\%$ for an uncoated detector).[59] The dark current (10^{-6} A for a 100-μm diameter diode) leads to a high shot-noise power which will nevertheless be smaller than the amplifier thermal noise for bit rates above 1 MHz. The excess-noise properties of this material are not very well known,

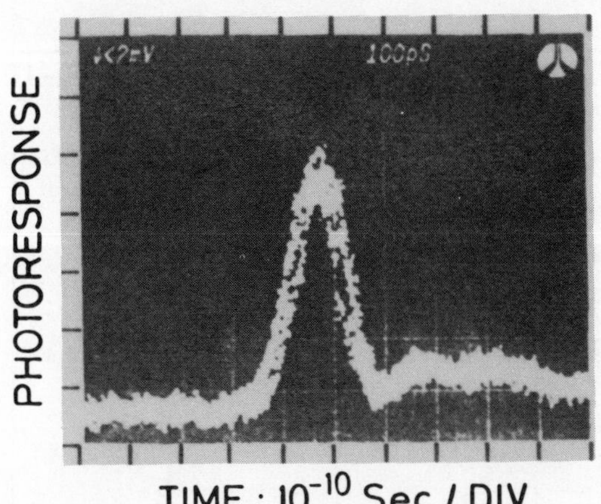

Figure 3.40. Transient photoresponse for an alloy photodiode made from $Al_xGa_{1-x}Sb$. The excitation is a mode-locked Nd-YAG laser at 1.06 μm. The measured rise-time is 100 ps. Diode area is 8×10^{-5} cm^2. (Reproduced by permission of Rockwell Int.)

and since the dark current shot noise is not known either, it is not possible to say how much of the gain shown in Figure 3.40 represents a useful increase in the signal-to-noise ratio. These detectors, however, in their early state of development do seem very promising for optical-fibre communication applications.

At the present time, no data exist on $Ga_xIn_{1-x}As_{2x}P_{1-2x}$ detectors with an absorption edge at 1.3 μm. This situation is expected to change quite soon.* However, results have been obtained on several other compositions of this quaternary system at shorter wavelengths[60,30] (and therefore not sensitive at 1.3 μm) and longer wavelengths.[31] In general, these results show that high quantum efficiency avalanche photodiodes can be made from $Ga_xIn_{1-x}As_{2x}P_{1-2x}$ over the entire compositional range. Considerable information on the growth of $Ga_xIn_{1-x}As_{2x}P_{1-2x}$ is now available because of its application as the active layer in long-lived CW room-temperature lasers.[61,62,63,64] This material is grown princi-

*see note added in proof.

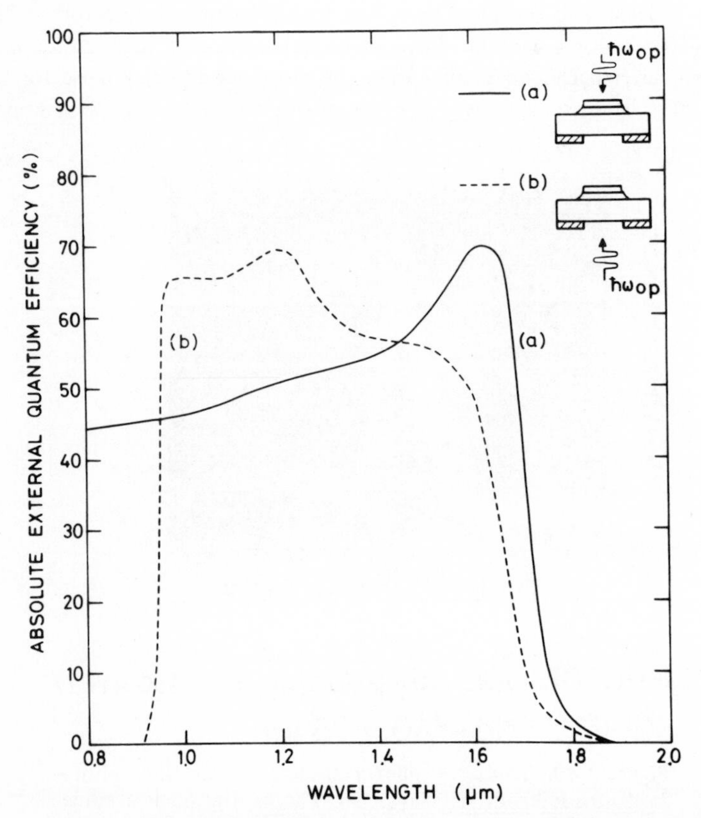

Fig. 3.41. Photodetection in $Ga_{0.47}In_{0.53}As$. Light is incident directly on the diode, trace (a) and through the substrate, trace (b).

pally by liquid-phase epitaxy, but it has also been produced by vapour-phase[65] and molecular-beam epitaxy.[66] While the quaternary can be grown both *n*- and *p*-type, *p*-doping is usually achieved using Zn which diffuses rapidly in both the quaternary and InP. It is therefore difficult to control the placement of the *p–n* junction which is consequently graded instead of abrupt.

At 1.2 μm, avalanche photodiodes with 1 ns response have been demonstrated using an n^+ on p^- structure. The measured dark current (10^{-8} A for a 200-μm diameter diode)[30] is higher than expected for a material with this energy gap (see Figure 3.33). However, it is still well below the amplifier noise limit, and the signal shot noise limit as well for most bit rates. No information on the excess noise is available for these devices.

At longer wavelengths, detectors have been made from $Ga_{0.47}In_{0.53}As$ with an absorption edge at 1.65 μm.[67] Photodiodes made from this material are also sensitive at 1.3 μm where the measured quantum efficiency for an uncoated detector is 40%. The substrate, being InP, is transparent at 1.3 μm and light can be coupled through the substrate or the detector surface. The spectral response in these two cases is shown in Figure 3.41. The higher quantum efficiency of the detector for

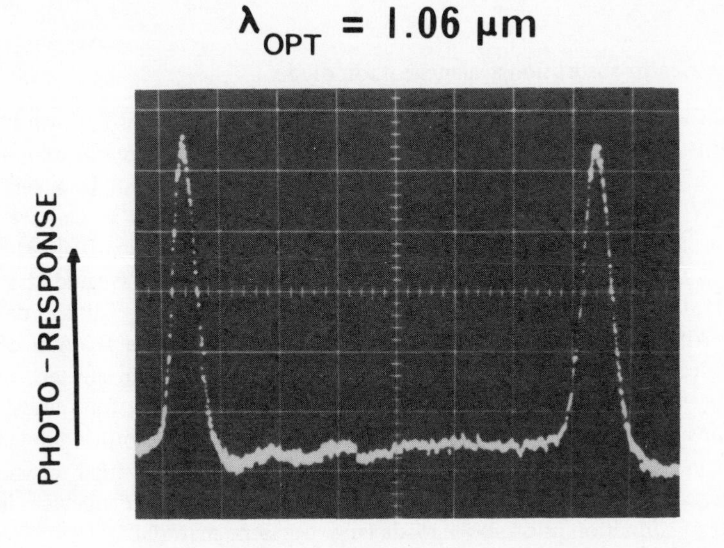

Figure 3.42. Transient response for a $Ga_{0.47}In_{0.53}As$ photodiode to a 1.06 μm Nd-YAG mode-locked laser. The rise-time is 250 ps. The diode area is 5×10^{-4} cm^2

light coupled through the substrate at photon energies above the band edge is the direct result of the reduction in surface recombination. Dark current of the order of 10^{-5} A cm^{-2} has been measured in these detectors, about the same as that seen in the quaternary diodes at 1.2 μm. $Ga_{0.47}In_{0.53}As$ detectors show avalanche gain. An ionization-rate ratio of 5 has been measured in this material for the electric field oriented along the ⟨111⟩ direction. Excess-noise measurements follow the McIntyre theory well.[34] Typical response times for a 250-μm diameter detector are in the low nanosecond range, although a faster response time of 250 ps has been measured and is shown in Figure 3.42.

3.5 ASSESSMENT OF DETECTORS FOR OPTICAL COMMUNICATIONS

In the preceeding sections we have presented both the principles of photodetection and the properties of some individual photodiodes. In addition to the discrete avalanche photodiodes that we have discussed, there are two other types of photodiodes that are of particular interest in an optical communication system: one is the waveguide, electro-absorption detector, and the other is the bi-directional detector/emitter.

3.5.1 The electro-absorption avalanche photodiode

Electro-absorption detectors have a sensitivity spectrum which shifts toward longer wavelengths at high applied bias. This class of *p–n* junction detector exploits the Franz–Keldysh shift in the absorption edge in an electric field. In a very pure material, $N_D \leqslant 10^{15}$ cm^{-3}, the spectral tuning can be large and in GaAs electro-absorption detectors, for example, sensitivity at 1.06 μm has been realized.[68] The electro-absorption detector can be used as a tap on an optical waveguide transmitting light at a wavelength beyond the normal absorption edge of the detector material. With no applied bias, the detector is transparent. The fraction of light detected depends on the bias, and the unabsorbed light is transmitted. Such a device may find applications in feedback stabilization of lasers for optical communications. The absorption characteristic of a GaAs electro-absorption detector is shown in Figure 3.43 (compare with Figure 3.9a). Note that at high reverse bias, avalanche gain can contribute to the response.[69] The speed of these devices is similar to that of a junction photodiode made from the same material.

3.5.2 The bi-directional detector-emitter

The basic feature of an optical-fibre system with a laser at one end and a detector at the other is its unidirectional nature. yet, any good emitter of light is also an absorber, and although absorption takes place at a wavelength shorter than that of emission, the results of the previous paragraph establish that proper design and biasing allows the tuning of the photodetector response to longer wavelengths.

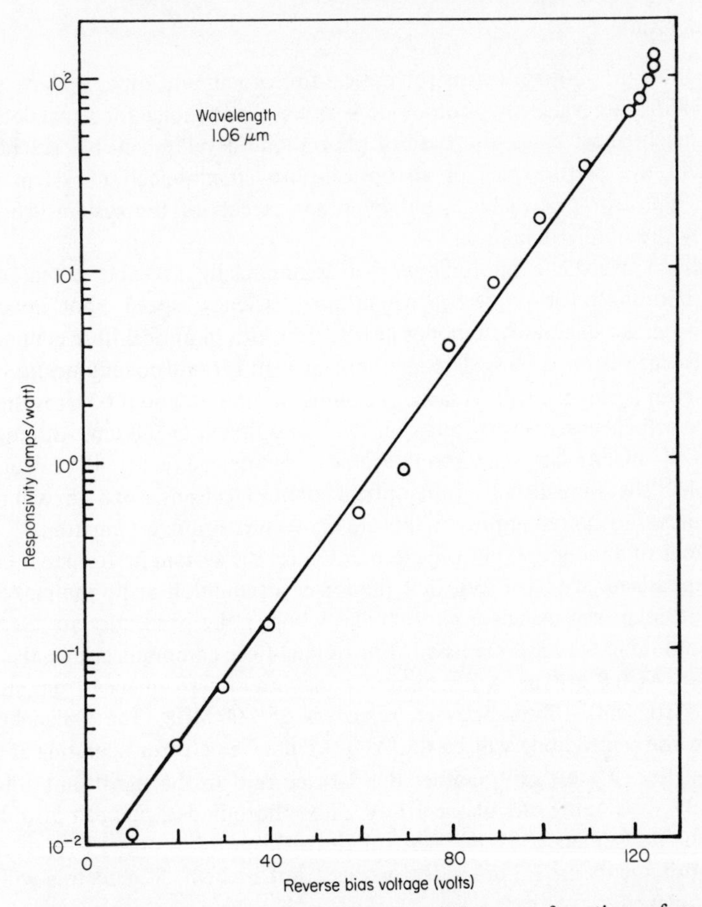

Figure 3.43. The absorption characteristic as a function of reverse bias for a GaAs electro-absorption detector at 1.06 μm, considerably beyond the band edge of GaAs $\lambda_g = 0.83$ μm. (Reproduced by permission of American Institute of Physics)

Under these conditions, it is possible to imagine a bi-directional optical communications link where the source and the detector are the same device. This device emits light under forward bias and under sufficient reverse bias acts as a detector. Initial experiments with such bi-directional devices have shown the principle to be valid, but the detectors suffer from low quantum efficiency and high capacity.[70] The detector performance will clearly be limited by device-geometry restrictions for long-lived emitters. The most logical configuration is a surface-emitting heterostructure LED with a 2 μm thick active region doped in the low 10^{16} cm^{-3} region. Such detector-emitters are best suited for high traffic, low data-rate links over short distances.

3.5.3 Conclusions

In this chapter on photodetectors for optical-fibre communication, we have shown that the solid-state avalanche photodiode is an excellent choice for signal detection applications. In most cases, the state of photodiode development has reached the point where the performance of an optical-fibre communication system is not limited by detector performance, but by other aspects of the system behaviour, particularly the fibre attenuation.

The silicon avalanche photodiode is a commercially available detector with excellent photodetector properties. Quantum efficiency, speed, shot noise, and excess noise make silicon photodiodes nearly ideal. But in optical-fibre communication at 0.8 μm, silicon is pushed to the limit in high bit-rate communication links. The maximum avalanche gain is signal shot-noise limited to about 65. It is the fibre attenuation which poses severe limits on the transmission at 0.8 μm. Although the 10^5 hour CW AlGaAs laser may soon become a commercial reality, the calculations here establish that long-distance fibre-optical communications at 0.8 μm will not be possible unless substantial improvements are made in the fibre attenuation.

The effect of a reduction in fibre attenuation on the system performance cannot be overemphasized. We have seen that the lower attenuation at 1.3 μm makes possible 15-km links even using Ge photodiodes whose noise properties are the worst of any photodiode we have discussed. For optical-fibre communications, the situation is clear. At 0.8 μm one will require the maximum performance from all components to transmit 10 km between repeaters at 100 MHz. The avalanche gain required in the photodiode will be 65. At 1.3 μm a Ge detector operating at a gain of 8 will suffice. At this gain, neither the dark current or the signal shot noise is a problem. By developing one of the III–V alloy photodiodes, one can introduce a large surplus performance margin, particularly with regard to noise. Thus while the quality requirements on 1.3-μm detectors need not be high, Si detectors will have to be hand-picked for use at 0.8 μm.

Ultimately, signal transmission at 1.3 μm will be limited by the same effects seen at 0.8 μm: fibre attenuation and shot noise. At 1.3 μm 15 km transmission at 16 GHz will be difficult. (At 0.8 μm, 15 km transmission is difficult above 16 MHz.) At these high rates, the performance differences between the various 1.3 μm detectors are small. Figure 3.36 shows clearly that the biggest differences are at very low bit rates, where the diode shot noise determines the gain and permits a larger avalanche gain for a 1.3-μm detector ($M = 24$) than for Ge photodiodes ($M = 6$). The advantage of a 1.3-μm alloy avalanche photodiode over Ge is that it can be used in long (25 km) repeaterless links at low bit rates (10^6–10^7 Hz). Whether or not this advantage can be exploited depends on the optical communications system design. If this design specifies for the majority of the system medium-distance (15 km) high bit rate (~100 MHz) links, it will be hard to distinguish between Ge and some other 1.3-μm detector on the basis of performance requirements, and Ge photodiodes, being commercially available, are likely to become the standard. This situation indicates the need for further work on Ge detectors. If there is a major need in fibre-optic networks for longer repeater-free links, then the alloy detectors: $Ga_{0.47}In_{0.53}As$ or a 1.3-μm alloy detector will be required instead of Ge. In either case, it is clear that the future for telecommunications by glass fibres lies at 1.3 μm. The choice of a detector at this wavelength can be made from several attractive alternatives according to the details of the communications system specifications.

REFERENCES

1. M. Horiguchi, 'Spectral losses of low-OH-content optical fibers', *Electron. Lett.*, **12**, 310–312, 1976.
2. D. N. Payne and W. A. Gambling, 'Zero material dispersion in optical fibers', *Electron. Lett.*, **11**, 176–178, 1975.
3. E. H. Putley, 'The pyro-electric detector', in *Semiconductors and Semi-Metals* (ed. R. K. Willardson and A. C. Beer), Vol. 5, 441–450, New York, Academic Press, 1970.
4. E. H. Putley, 'The pyro-electric detector – an update', in *Semiconductors and Semi-Metals* (ed. R. K. Willardson and A. C. Beer), Vol. 12, 441–450, New York, Academic Press, 1977.
5. M. E. Lines and A. M. Glass, *Principles and Applications of Ferroelectrics and Related Materials*, 559–578, Oxford, The Clarendon Press, 1977.
6. O. R. Wood, R. L. Abrams and T. J. Bridges, 'Mode-locking of a transversely excited atmosphere pressure CO_2 laser', *Appl. Phys. Lett.*, **17**, 376–378, 1970.
7. H. Melchior, 'Sensitive high-speed photodetectors for the demodulation of visible and near infra-red light', *J. Luminescence,* **7**, 390–414, 1973.
8. H. R. Zwicker, 'Photoemissive detectors', in *Topics in Applied Physics – Optical and Infra-red Detectors* (ed. R. J. Keyes), Vol. 19, 149–196, Berlin, Springer-Verlag, 1977.
9. H. Sonnenberg, 'InAsP–Cs_2O, a high-efficiency infra-red photocathode', *Appl. Phys. Lett.*, **16**, 245–246, 1970.
10. J. J. Uebbing and R. L. Bell, 'Improved photo-emitters using GaAs in InGaAs', *Proc. IEEE,* **56**, 1624–1625, 1968.
11. J. S. Escher, G. A. Antypas and J. Edgecumbe, 'High quantum efficiency photoemission from an InGaAsP photocathode', *Appl. Phys. Lett.*, **29**, 153–155, 1976.
12. J. Escher, P. E. Gregory, S. B. Hyder and R. Sankaran, 'Transferred-electron photoemission to 1.65 μm from InGaAs', *J. Appl. Phys.*, **49**, 2591–2592, 1978.
13. H. Levinstein, 'Extrinsic detectors', *Appl. Optics,* **4**, 639–647, 1965.
14. Paul W. Kruse, 'Indium antimionide photoconductive and photo electro magnetic detectors', in *Semiconductors and Semi-Metals,* Vol. 5, 15–83, New York, Academic Press, 1970.
15. Donald Long and Joseph L. Schmidt, 'Mercury-cadmium telluride and closely related alloys', in *Semiconductors and Semi-Metals* (ed. R. K. Willardson and A. C. Beer), Vol. 5, 175–255, New York, Academic Press, 1970.
16. G. Fiorito, G. Gasparrini and F. Svelto, 'Mercury cadmium telluride as a material for 1–1.3 μm room temperature photodiodes', *Infra-red Physics,* **18**, 59–61, 1978.
17. Hans Melchior, F. Arams and M. B. Fisher, 'Photodetectors for optical communications systems', *Proc. IEEE,* **58**, 1466–1486, 1970.

18. W. C. Dash and R. Newman, 'Intrinsic optical absorption in single-crystal germanium and silicon at 77 K and 300 K', *Phys. Rev.*, **99**, 1151–1155, 1955.
19. J. R. Chelikowsky and M. L. Cohen, 'Nonlocal pseudopotential calculations for the electronic structure of eleven diamond and zinc-blende semiconductors', *Phys. Rev.*, **14**, 556–582, 1976.
20. B. O. Seraphin and H. E. Bennett, 'Optical constants', in *Semiconductors and Semimetals* (ed. R. K. Willardson and A. C. Beer), Vol. 3, 449–543, New York, Academic Press, 1967.
21. J. P. Donnelly, *Ion Implantation in GaAs,* Symposium on GaAs and Related Compounds: St Louis, 1976, 166–190, Bristol, Institute of Physics, 1977.
22. L. K. Anderson and B. J. McMurty, 'High speed photodetectors', *Proc. IEEE,* **54**, 1335–1348, 1966.
23. J. R. Grierson and S. O'Hara, 'A theoretical assessment of low and high bandwidth gallium arsenide, germanium and silicon avalanche photodiodes for optical communications systems', *Solid-State Electron.*, **18**, 1003–1011, 1975.
24. *R.C.A. Electro-Optics Handbook,* 109–125, Harrison, R.C.A. Commercial Engineering, 1974.
25. K. Nishida, K. Ishii, K. Minemura and K. Taguchi, 'Double epitaxial silicon photodiodes for optical fibre communications', *Electron. Lett.*, **13**, 280–281, 1977.
26. H. D. Law, L. R. Tomasetta, K. Nakano and J. S. Harris, *Appl. Phys. Lett.* **35**, 180–182, 1979.
27. H. Kanbe, T. Mizushima, T. Kimura and K. Kajiyama, 'High-speed silicon avalanche photodiodes with built-in field', *J. Appl. Phys.*, **47**, 3749–3751, 1976.
28. G. E. Stillman, C. M. Wolfe, A. G. Foyt and W. T. Lindley, 'Schottky barrier $In_xGa_{1-x}As$ alloy avalanche photodiodes for 1.06 μm', *Appl. Phys. Lett.*, **24**, 8–10, 1974.
29. H. D. Law, L. R. Tomasetta, K. Nakano and J. S. Harris, '1.0–1.4 μm high speed avalanche photodiode', *Appl. Phys. Lett.*, **33**, 920–922, 1978.
30. C. E. Hurwitz and J. J. Hsieh, 'GaInAsP/InP avalanche photodiodes', *Appl. Phys. Lett.*, **32**, 487–489, 1978.
31. T. P. Pearsall and M. Papuchon, 'The $Ga_{0.47}In_{0.53}As$ homojunction photodiode – a new avalanche photo-detector in the near infrared between 1.0–1.6 μm', *Appl. Phys. Lett.*, **33**, 201, 1978.
32. H. Melchior and W. T. Lynch, 'Signal and noise response of high-speed germanium avalanche photodiodes', *IEEE Trans. Electron. Dev.*, **ED-13**, 829–838, 1966.
33. S. L. Miller, 'Avalanche breakdown of germanium', *Phys. Rev.*, **99**, 1234–1241, 1955.
34. P. P. Webb, R. J. McIntyre and J. Conradi, 'Properties of avalanche photodiodes', *RCA Review,* **35**, 235–277, 1974.
35. R. J. McIntyre, 'Distribution of gains in uniformly multiplying avalanche photodiodes', *Theory IEEE, Trans. Elect. Dev.*, **ED-19**, 703–713, 1977.
36. H. K. Gummel and J. L. Blue, *IEEE Trans. Elect. Dev.*, **ED-14**, 569–580, 1967.
37. I. M. Naqvi, 'Effects of time dependence of multiplication process on avalanche noise', *Solid-State Electron.*, **16**, 19–28, 1973.
38. G. A. Baraff, 'Distribution functions and ionization rates for hot electrons in semiconductors', *Phys. Rev.*, **128**, 2507–2517, 1962.

39. E. M. Conwell, *High Field Transport in Semiconductors,* Solid-State Physics Supplement 9, 105–160, New York, Academic Press, 1967.
40. T. P. Pearsall, R. E. Nahory and J. R. Chelikowsky, 'Orientation dependence of free-carrier impact ionization in semiconductors: GaAs', *Phys. Rev. Lett.*, **39**, 295–298, 1977.
41. T. P. Pearsall, F. Capasso, R. E. Nahory, M. A. Pollack and J. R. Chelikowsky, 'The band structure dependence of impact ionization by hot carriers in semiconductors: GaAs', *Solid-State Electron.*, **21**, 297–302, 1978.
42. C. L. Anderson and C. R. Crowell, 'Threshold energies for electron–hole pair production by impact ionization in semiconductors', *Phys. Rev.*, **B15**, (5), 2267–2272, 1972.
43. G. E. Stillmann and C. M. Wolfe, 'Avalanche photodiodes', in *Semiconductors and Semimetals,* Vol. 12, 291–393, Academic Press, New York, 1977.
44. T. Kaneda, H. Matsumoto, T. Sakurai and T. Yamaoka, 'Excess noise in silicon avalanche photodiodes', *J. Appl. Phys.*, **47**, 1605–1607, 1976.
45. Jan Conradi, 'The distribution of gains in uniformly multiplying avalanche photodiodes: experimental', *IEEE Trans. Elect. Dev.*, **ED-19**, 713–718, 1972.
46. R. A. Logan and S. M. Sze, 'Avalanche multiplication in Ge and GaAs p–n junctions', *Proc. Int. Conf. Phys. Semicond., Kyoto, 1966, Suppl. J. Phys. Japan,* **21**, 434–436, Tokyo, Phys. Soc. Japan, 1966.
47. T. P. Pearsall, R. E. Nahory and M. A. Pollack, 'Ionization rates for electrons and holes in $In_{0.14}Ga_{0.86}As$', *Appl. Phys. Lett.*, **27**, 329–331, 1975.
48. T. P. Pearsall, R. E. Nahory and M.A. Pollack, 'Ionization rates for electrons and holes in alloys of $GaAs_{1-x}Sb_x$', *Appl. Phys. Lett.*, **28**, 403–405, 1976.
49. T. Kobayashi, Y. Takanashi and Y. Furukawa, 'Reduction of quantum noise in very narrow planar stripe lasers', *Jap. J. Appl. Phys.*, **17**, 535–540, 1978.
50. Jan Conradi, Felix P. Kapron and John C. Dyment, 'Fiber optical transmission between 0.8 and 1.4 μm', *IEEE Trans. Elect. Dev.*, **ED-25**, 180–192, 1978.
51. A. S. Grove, *Physics and Technology of Semiconductor Devices*, John Wiley, New York, 1967, p. 176.
52. A. M. Burd, Y. A. Leichenko, B. N. Motenko and A. S. Popov, 'Temperature stabilization of a photoreceiver with an avalanche photodiode', *Instrument and Exp. Tech.*, **18**, 1224–1226, 1976; translation of *Prib. Tekh. Eksp* (USSR), **18**, 176–178, 1975.
53. J. Conradi, 'Temperature effects in silicon avalanche diodes', *Solid-State Electron.*, **17**, 99–106, 1974.
54. H. Kressel and J. K. Butler, *Semiconductor Lasers and Heterojunction LEDS*, New York, Academic Press, 1977, pp. 357–398.
55. S. M. Bedair, 'Composition dependence of the $Al_xGa_{1-x}Sb$ energy gaps', *J. Appl. Phys.*, **47**, 4145–4147, 1976.
56. L. M. Dolginov, L. V. Druzhinina, P. G. Eliseev, M. G. Milvidsk and B. N. Sverdlov, 'New uncooled injection heterolaser emitting in the 1.5–1.8 μm range', *Sov. J. Quant. Electron.*, **6**, 257, 1976.
57. G. A. Antypas, R. L. Moon, L. W. James, J. Edgecumbe and R. L. Bell, 'III–V Quaternary alloys', 1972 Symposium on GaAs and Related Compounds, Bristol, Institute of Physics, 1973, pp. 48–54.
58. S. J. Anderson, F. Scholl and J. S. Harris, 'AlGaSb alloys for 1.0–1.8 μm heterojunction devices', 1976 Symposium on GaAs and Related Compounds, Bristol, Institute of Physics, 1977, pp. 346–355.
59. H. H. Weider, A. R. Clawson and G. E. McWilliams, '$In_xGa_{1-x}As_yP_{1-y}$/InP heterojunction photodiodes', *Appl. Phys. Lett.*, **31**, 468–470, 1977.

61. J. J. Hsieh, 'Room-temperature operation of GaInAsP/InP double heterostructure diode lasers emitting at 1.1 μm', *Appl. Phys. Lett.*, **28**, 213–215, 1976.
62. K. Nakajima, T. Kusunoki, K. Akita and T. Kotani, 'Phase diagram of the In–Ga–As–P quaternary system and L.P.E. growth conditions for lattice-matching on InP substrates', *J. Electrochem. Soc.*, **125**, 123–127, 1978.
63. T. P. Pearsall, R. Bisaro, R. Ansel and P. Merenda, 'The growth of $Ga_xIn_{1-x}As$ on (100) InP by liquid phase epitaxy', *Appl. Phys. Lett.*, **32**, 497–499, 1978.
64. M. A. Pollack, R. E. Nahory, J. C. DeWinter and A. A. Ballman, 'Liquid phase epitaxial $In_{1-x}Ga_xAs_yP_{1-y}$ lattice-matched to ⟨100⟩ InP over the complete wavelength range 0.92 μm–1.65 μm', *Appl. Phys. Lett.*, **33**, 364, 1978.
65. C. J. Neuse, R. E. Enstrom and J. R. Appert, '1.7 μm heterojunction lasers and photodiodes of $In_{0.53}Ga_{0.47}As$/InP', *IEEE Trans. Elect. Dev.*, **ED-24**, 1213, 1977.
66. B. I. Miller and J. M. McFee, 'Room-temperature lasers from $Ga_{0.47}In_{0.53}As$/InP by M.B.E.', *IEEE Trans. Elect. Dev.*, **ED-25**, 699, 1978.
67. T. P. Pearsall and R. W. Hopson, Jr, 'Growth and characterization of lattice matched epitaxial films of $Ga_xIn_{1-x}As$/InP by liquid phase epitaxy', *J. Electron. Mat.*, **7**, 133–146, 1978.
68. G. E. Stillman, C. M. Wolfe, C. O. Bozler and J. A. Rossi, 'Electroabsorption in GaAs and its applications to waveguide detectors and modulators', *Appl. Phys. Lett.*, **28**, 544–546, 1976.
69. K. H. Nichols, W. S. C. Chang, C. M. Wolfe and G. E. Stillman, 'GaAs waveguide detectors for 1.06 μm', *Appl. Phys. Lett.*, **31**, 631–633, 1977.
70. J. J. Geddes, P. E. Peterson, D. Chen and K. P. Koeneman, 'Semiconductor junction transceiver for fiber optic communication', *Electron. Lett.*, **14**, 214–216, 1978.

Note added in proof: Since this manuscript was written, significant progress has been made in $Ga_xIn_{1-x}As_{2.18x}P_{1-2.18x}$ photodiodes. A representative list of publications follows.

71. T. Ito, T. Kaneda, K. Nakijama, Y. Toyoma, T. Yamoaka and T. Kotani, 'Impact Ionisation Ratio in $In_{0.73}Ga_{0.27}As_{0.57}P_{0.43}$', *Electron Lett.* **14**, 418–419, 1978.
72. G. H. Oslen and H. Kressel, 'Vapour Grown 1.3 μm InGaAsP/InP Avalanche Photodiodes', *Electron Lett.* **15**, 41–142, 1979.
73. H. D. Law, L. R. Tomasetta, and K. Nakano, 'Ion-Implanted InGaAsP Avalanche Photodiodes', *Appl. Phys. Lett.*, **33**, 920–922, 1978.
74. R. E. Yeats and S. H. Chiao 'Long Wavelength InGaAsP Avalanche Photodiodes', *Appl. Phys. Lett*, **34**, 581–583, 1979.
75. Y. Takanashi and Y. Horikoshi 'InGaAsP/InP Avalanche Photodiode', *Jap. Jour. Appl. Phys.* **17**, 2065–2066, 1978.

Optical Fibre Communications
Edited by M. J. Howes and D. V. Morgan

CHAPTER 4

Optical waveguide components

D. B. OSTROWSKY

4.1 INTRODUCTION

Integrated optics is a rapidly developing field that could have an important impact on future optical communication systems. Its main objective is the development of a family of optical and electro-optical elements in a thin-film planar form that will allow the integration of a large number of devices on a single substrate. This should result in the same advantages, *vis-à-vis* interconnected individual components, that were accrued when electronics underwent integration such as lowered unit cost, higher packing density, increased reliability, etc.

The basic idea underlying integrated optics was proposed by Anderson in 1965.[1] He pointed out that since the microfabrication technology developed for electronic circuits permitted the realization of structures having sections of the order of a micrometre, it was possible to use these techniques to fabricate monomode optical devices in semiconductor or dielectric materials. It was, however, only after the appearance of the September 1969 edition of the *Bell System Technical Journal* that the field began to undergo a rapid expansion. In an introductory article[2] the long-range possibilities were outlined and the form of various components was suggested. In Figure 4.1 we show some of these suggestions and state whether the elements were realized.

A fundamental point to note is that the majority of these elements are based on essentially *monomode* optical waveguides. For this reason, most of integrated optics is incompatible with the *multimode* fibre systems currently being installed or projected for installation in the near future. The main impact of integrated optics on optical communication technology will, therefore, be reserved for a time at which monomode fibre systems begin to be installed. Such systems, in conjunction with integrated optical devices, will permit optical communication to attain its ultimate limits.

In this article we shall concentrate on the aspects of integrated optics which the author believes to be directly related to optical communication systems. The

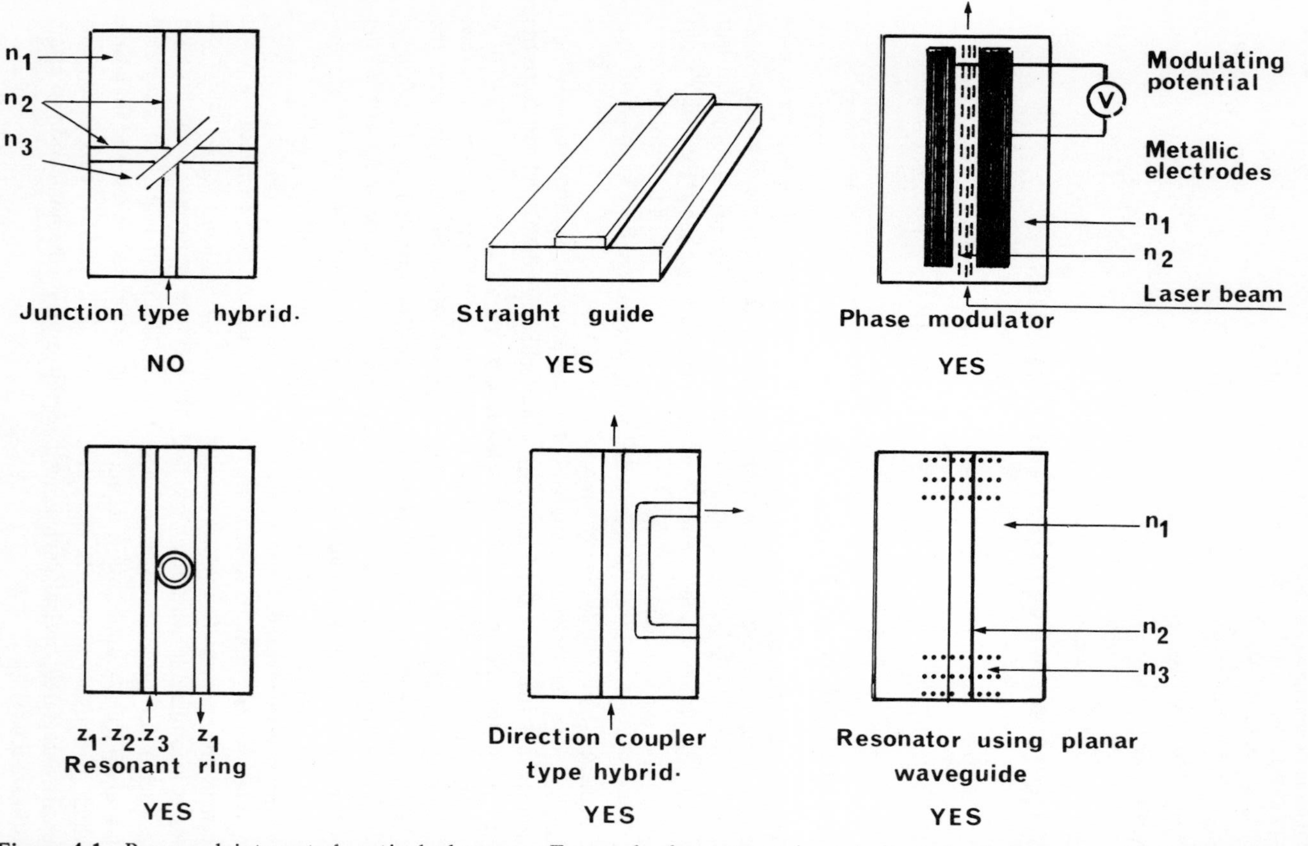

Figure 4.1. Proposed integrated optical elements. For each element we have indicated whether or not a prototype version has been realized

development is device-oriented and readers interested in basics of planar waveguide theory are referred to the appropriate article.[3] We shall try to outline what has already been accomplished in integrated optics with regard to the essential components for optical communication systems and suggest what developments can be expected in the near future. The specific elements covered will be modulators and switches, filters, sources and detectors as well as techniques for coupling such components to monomode fibres. We shall also mention several cases of integrated opto-electronics in which various essentially electrical components (detectors, Gunn oscillators) have been integrated with injection lasers. While it is not clear that such devices are, strictly speaking, integrated optical components, they could have an important impact in the near future since they are compatible with multimode fibres.

The basic elements will be presented in the order given above since we believe this represents the order of importance of their development to date. We begin, therefore, with a discussion of the integrated optical modulators and switches.

4.2 MODULATORS AND SWITCHES

Integrated optics has certainly had its greatest impact on optical modulators and switches. In the case of modulators it has led to a reduction of several orders of magnitude of required drive power. In the case of switches, the impact is even more dramatic: it has spawned a family of switches that have no bulk counterparts.

A variety of effects (electro-optic, acousto-optic, magneto-optic, etc.) have been employed, in a number of configurations (channel waveguides, planar guides, etc.) to realize these modulating and switching devices. In this article we shall concentrate on the channel waveguide electro-optic devices which appear to be the most likely to find extended use in optical communication systems.

We begin by discussing the channel waveguide electro-optic phase modulator, whose function is the basis for most of the components discussed. These include amplitude modulators, switches and hybrid opto-electronic components such as two- and four-port bistable devices.

4.2.1 Channel waveguide electro-optic phase modulator

A number of crystals exhibit the linear electro-optic (Pockels) effect which consists of a change of refractive index directly proportional to an applied electric field given by

$$\Delta n = -\tfrac{1}{2} n_0^3 r E \tag{4.1}$$

where n_0 is the original index of refraction, r is the electro-optic coefficient and E is the applied electric field. Some representative values are given in Table 4.1 for a number of materials.

Evidently, light propagating a distance l through the crystal will accumulate

Table 4.1

Material	Index of refraction	Electro-optic coefficient ($\times 10^{-12}$ m/V)
GaAs	3.34	$r_{41} = 1.6$
$LiNbO_3$	$n_0 = 2.295$	$r_{33} = 30.8$
	$n_e = 2.203$	$r_{13} = 8.6$
$LiTaO_3$	$n_0 = 2.175$	$r_{33} = 30.3$
	$n_e = 2.180$	$r_{13} = 5.7$
Quartz	$n_0 = 1.54$	$r_{41} = 0.2$
	$n_e = 1.55$	$r_{63} = 0.93$
KDP	$n_0 = 1.51$	$r_{41} = 8.6$
	$n_e = 1.57$	$r_{63} = 10.6$

a phase change given by

$$\Delta\phi = k_0 \Delta n l \tag{4.2}$$

where $k_0 = 2\pi/\lambda$ and λ is the vacuum wavelength.

To use this effect efficiently one adopts a configuration in which the electric field is applied transversely to the direction of light propagation (Figure 4.2). In this case

$$\Delta\phi = -\frac{k_0}{2} n_0^3 r V \frac{l}{d} \tag{4.3}$$

and one immediately sees that in order to reduce the drive voltage V, it is advantageous to construct modulators that have the largest possible ratio of l/d.

In a bulk modulator, one in which light is not guided but rather focused to pass through the crystal, l and d are not independent. The transverse dimension d is diffraction limited to a value of $\sqrt{2l\lambda/n}$. In a waveguide modulator the transverse dimension will be on the order of λ and d_{guide}/d_{bulk} will therefore be proportional

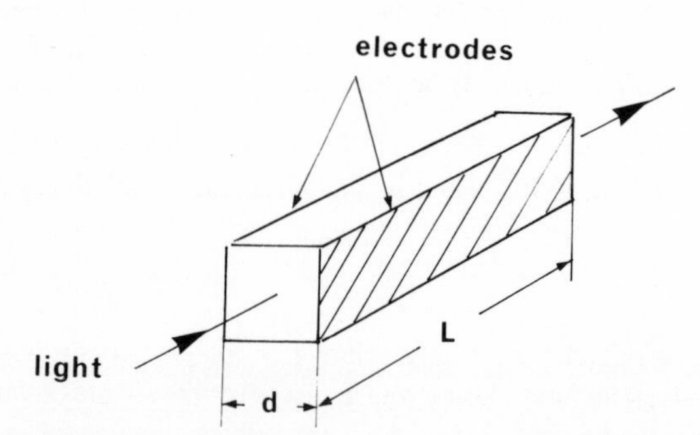

Figure 4.2. A transverse electric field electro-optic phase modulator

to $\sqrt{\lambda/l}$, which for $\lambda = 1\ \mu m$ and $l = 1$ cm yields a drive voltage on the order of 100 times smaller for the waveguide case.

A commonly used figure of merit for modulators is the specific energy, $(P/\Delta f)_z$, defined as drive power per unit bandwidth for $\Delta\phi = 2$ rad. One can show quite generally[4] that in this case

$$\frac{(P/\Delta f)_{z\ \text{guide}}}{(P/\Delta f)_{z\ \text{bulk}}} \propto \frac{\lambda}{l}$$

and will typically have values of approximately 10^{-4}.

Let us consider a realistic example. By polishing a crystal one can arrive at crystals having transverse dimensions on the order of 25 μm and lengths of the order of 1 cm, or an l/d ratio of about 40. Using a $LiTaO_3$ crystal ($n = 2.18$, $r = 30 \times 10^{-12}$ m/V) one arrives at voltages on the order of 50 V for $\Delta\phi = \pi$. Integrated optics allows one to do considerably better. One can fabricate, by diffusion of Nb into the $LiTaO_3$, a microguide (Figure 4.3) of Nb : $LiTaO_3$ of the order of two by 2 μm and 2 cm long, which has an l/d ratio of 1000 and a voltage, V_π, of the order of 1 V.

The best phase modulator of this type realized to date[5] allowed modulation with a $(P/\Delta f)_z = 6.8\ \mu V/MHz$ which is a three to four order of magnitude improvement over the bulk modulators. This clearly demonstrates the basic, essentially geometric, advantage of the integrated-optics configuration. In the following sections we shall see a number of ways in which integrated optical structures, more complex than simple channel waveguides, can be used to transform phase modulation to amplitude modulation.

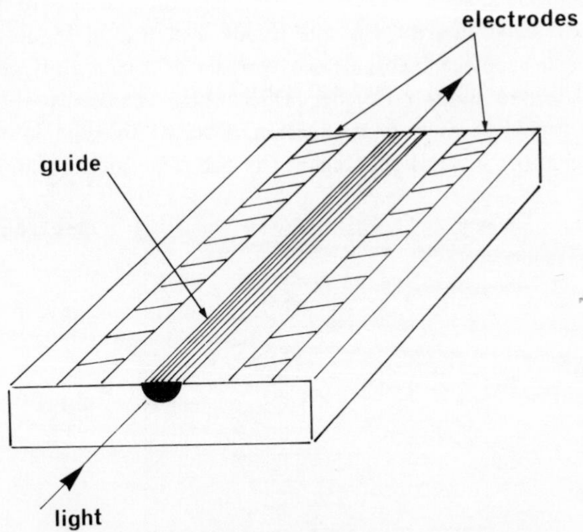

Figure 4.3. A strip waveguide phase modulator

4.2.2 Amplitude modulators

Waveguide modulators

One of the simplest ways of using phase modulation to create amplitude modulation, which is usually desired in optical communication systems, is shown in Figure 4.4. In such a configuration the symmetric separation and recombination regions play the role of beam-splitters in a Mach–Zender interferometer. Half the entering beam goes into each of the two branches. If the beams in the two branches are in phase at the recombination section, they will constructively interfere to excite the output guide. If, however, one has applied a voltage V_π across one branch, the two beams will be out of phase, interfere destructively at the output guide entrance, and radiate into the substrate.

The best device of this type,[6] was fabricated, by Ti diffusion in a Y-cut plate of $LiNbO_3$ which permitted using the r_{33} electro-optic coefficient for TM polarized waves. For a wavelength of $\lambda = 0.5145$ μm an extinction coefficient of 93% was attained for a command voltage of 1.6 V on a 50 Ω line with rise and fall times of the order of 700 ps. This demonstrates that the integrated optical technology is already capable of providng gigahertz modulation capability with low drive powers.

Planar modulators integrated with lasers

In general the planar modulator structure, in which the light is guided in the film but not laterally bounded, will not perform as well as the channel waveguide modulator. This is simply due to the fact that one again runs into the diffraction limit in the transverse direction and one cannot increase the modulator length independently. Nevertheless, such modulators have been integrated with injection lasers and could be of some interest for multimode systems. In Figure 4.5 we show schematically such a device.[7] This device consists of a large optical cavity (LOC) laser, using distributed Bragg reflectors (DBR: these are discussed in Chapter 2), followed by a forward-biased p–n junction. Current injected into the junction modulates the gain (or loss) of this region (by the same mechanism that gives rise

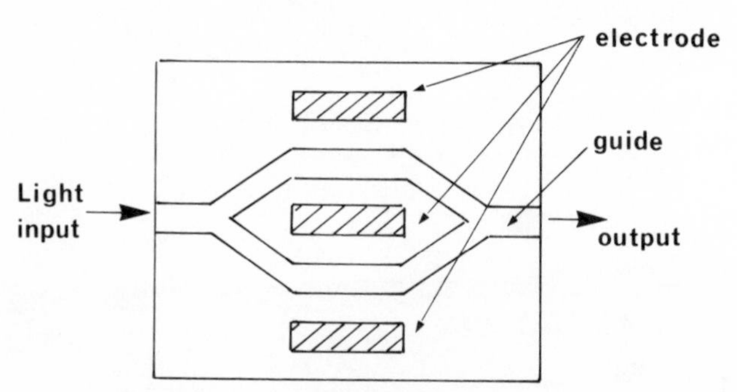

Figure 4.4. An interferometric modulator

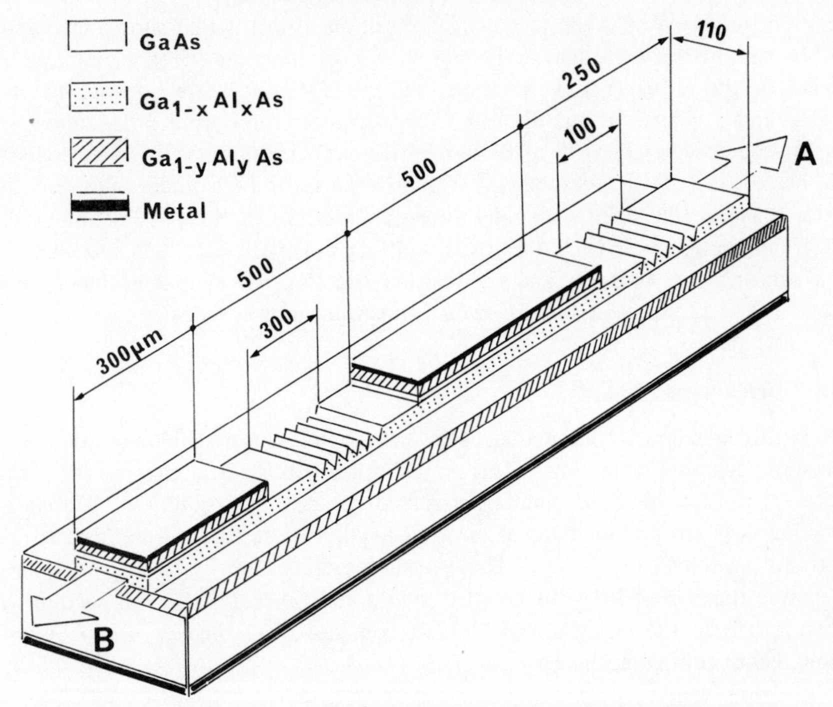

Figure 4.5. An integrated DBR laser-injection modulator structure

to laser action) and hence leads to a modulation of the light emitted by the laser. With this configuration, extinction ratios of the order of 10 have been reported for modulator current densities on the order 6 kA/cm^2, which corresponded to currents on the order of 1–2 A. It is clear that this represents a prelimary experiment and an improvement of several orders of magnitude in modulator current will be needed to render such a device operationally attractive.

Another type of planar modulator which could be useful if integrated with

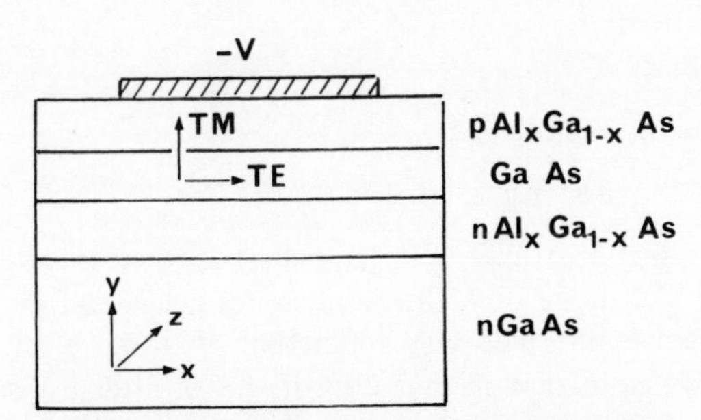

Figure 4.6. An integrable laser-polarization modulator structure

lasers emitting beyond 1 μm (to avoid absorption losses) is that shown in Figure 4.6. In this modulator[8] light is guided in a *p–n* junction. If this direction is parallel to the ⟨110⟩ crystal direction, TE and TM modes are coupled to one another and a change of the applied voltage changes the output polarization. In this configuration a $(P/\Delta f)_z$ of 0.15 mW/MHz was measured with an extinction ratio better than 20 dB. As expected, the drive power is 1–2 orders of magnitude greater than for the $LiNbO_3$ channel-waveguide modulator. It is possible, however, that if integrated with GaAlAsP lasers emitting beyond 1 μm, such modulators, by separating the emission and modulation function, to eliminate ringing and hysteresis effects, could find use in optical communication systems.

4.2.3 Guided wave switches

It is in this domain that integrated optics has made its most striking contribution to communications technology. This contribution consists of a class of switching devices that have no bulk counterparts, and offer the possibility of bilaterally switching data streams encoded at rates far beyond those that can be handled by electronic switching technology. These optical switches are based on controlling the power transferred between coupled guides and we shall begin this section by outlining briefly the essentials of this phenomenon, and then go on to discuss various device configurations.

Coupled-wave power transfer

If two dielectric waveguides are placed sufficiently close to one another their evanescent waves will overlap, thereby coupling the fields propagating in the guides (directional coupler configuration).

Consider the configuration shown in Figure 4.7. If we initially excite guide I the field E_I can be written as

$$E_I(z) = \cos(z\sqrt{K^2+\delta^2}) + j\delta\,\frac{\sin(z\sqrt{K^2+\delta^2})}{\sqrt{K^2+\delta^2}} \tag{4.4}$$

and that in guide II as:

$$E_{II}(z) = -jK\,\frac{\sin(z\sqrt{K^2+\delta^2})}{\sqrt{K^2+\delta^2}} \tag{4.5}$$

where K is the coupling coefficient, $\delta = (\beta_I - \beta_{II})/2$, where β represents the propagation constant of the mode of each (monomode) guide, and $j = \sqrt{-1}$. In Figure 4.7 we show the power in guide I as a function of distance for the cases where $\delta = 0$ and $\delta \neq 0$. We note that if, and only if, $\delta = 0$, after a distance called the coupling length L_c, all the power that was originally in guide I will be transfered to guide II. We also note that if we destroy the resonance ($\delta \neq 0$), less energy will

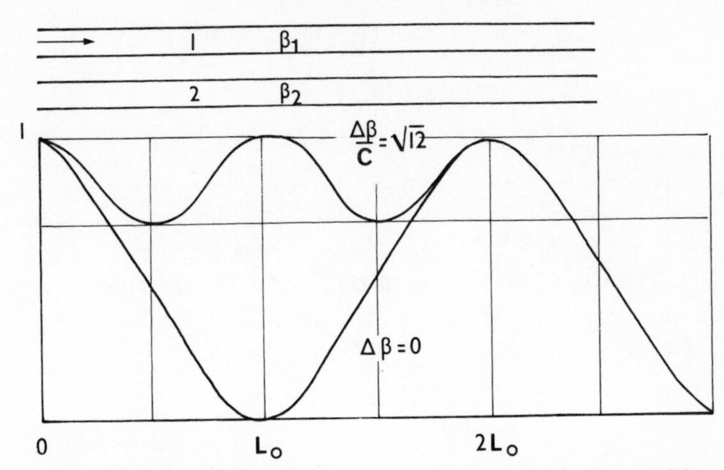

Figure 4.7. Directional coupler configuration and the resulting power transfer

be exchanged but it will be exchanged more 'quickly'. We have shown in Figure 4.7 the case where, by correctly choosing δ we find at L_c all the energy in guide I, the one that was originally excited. We shall, following Kogelnik and Schmidt,[9] call the state in which at L_c all the energy is in II the *crossed* state (denoted by ⊗), and the state in which at L_c all the energy is in I the *parallel* state (denoted by ⊜). It should now be evident that if we can construct two resonant guides which interact over the length L_c, and if we can control their resonance (i.e. change $\delta = 0$ to $\delta \neq 0$), we can switch light entering one guide to the other guide at the output. In the following two sections we discuss the configurations actually being used to accomplish this.

The COBRA configuration The most widely used electrode configuration for such devices is that suggested, and demonstrated, by Papuchon and coworkers.[10] The electrodes are placed directly above the two guides (Figure 4.8). With an electric field applied, the opposed vertical field components will, if the crystal axes have been correctly chosen, induce opposed changes of the index of refraction in the guides, thereby creating the desired δ. While such devices have been realized in a number of materials ($LiNbO_3$, $LiTaO_3$, GaAs, etc.) the best devices reported to date have been those realized in Ti-doped lithium niobate, which we denote as Ti:$LiNbO_3$. By using a crystal with the face normal to the C-axis (Figure 4.8), TM modes will have a Δn proportional to the largest electro-optic coefficient in this material, the r_{33} coefficient. The inherent flaw of such devices, however, lies in the fact that if one does not fabricate them with perfectly resonant guides having an interaction length of *exactly* L_c, one cannot switch *all* of the light from the originally excited guide to the other, to attain the crossed state. For example, in the original COBRA device a transfer of only 60% of the energy was reported.

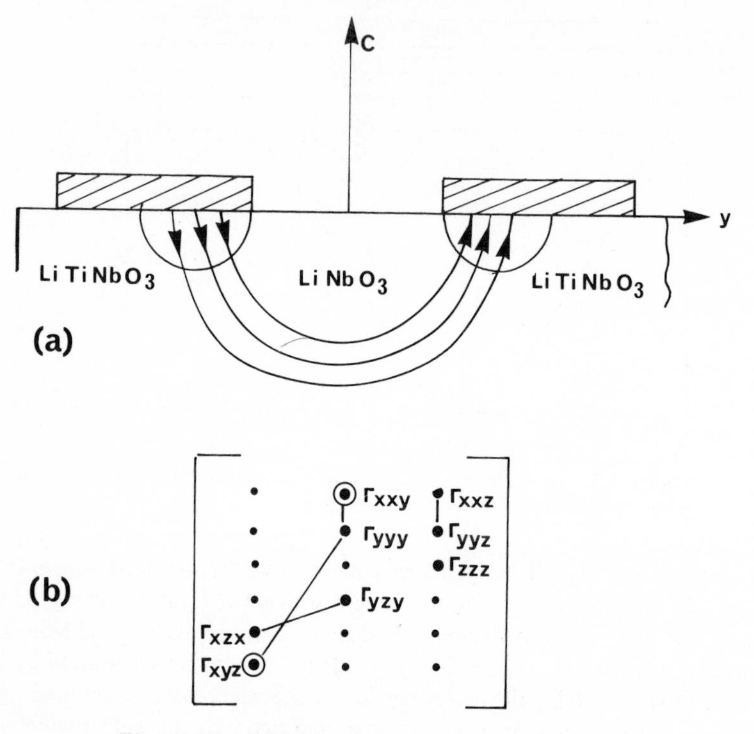

Figure 4.8. COBRA electrode configuration

In the next section the technique developed by Kogelnik and Schmidt[9] to overcome this defect is discussed.

Stepped $\Delta\beta$ coupler The stepped $\Delta\beta$ configuration is shown in Figure 4.9. The electrodes are each split into two sections and voltages of opposite polarity are applied so that the induced Δn in each section has an opposite sign. Using a matrix analysis, Kogelnik and Schmidt have shown that if one fabricates a device having an actual length lying between 1 and 3 interaction lengths, it is *always* possible to

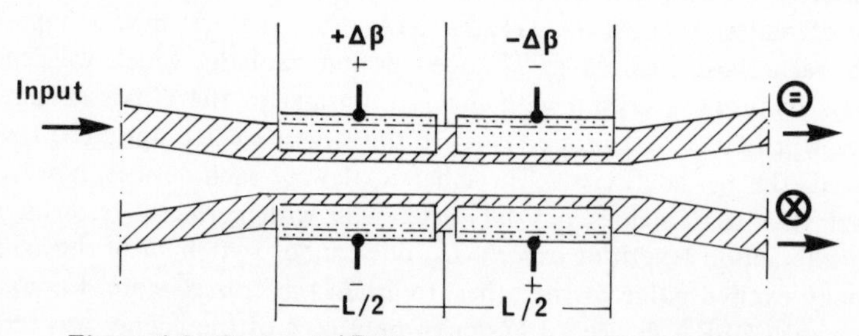

Figure 4.9. Stepped ΔB directional coupler configuration

attain, by applying different values of the (reversed) voltage V, both the crossed and the parallel state. In particular, the crossed state will be attained whenever

$$\frac{K^2}{K^2+\left(\frac{\Delta\beta}{2}\right)^2}\sin^2\frac{1}{2}\sqrt{K^2+\left(\frac{\Delta\beta}{2}\right)^2}=\frac{1}{2} \tag{4.6}$$

and the parallel state whenever

$$\left(\frac{L}{l}\right)^2+\left(\frac{2L\delta}{\pi}\right)^2=(4\gamma)^2 \tag{4.7}$$

where γ is an integer.

In addition to providing a large tolerance on coupling length, the stepped $\Delta\beta$ configuration also permits compensating for non-identical guides, and, using differing reversed voltages, for a lack of longitudinal homogeneity.

Individual switches have been produced in Ti: $LiNbO_3$ having a cross-talk below -30 dB and switching speeds of less than 2 ns with applied voltages of 9 and 5 V.[6] While multisection versions of the switch[11] have permitted a slight reduction (2 and 6 V) of switching voltages, it is not clear that this is worth the more complex electrode structure one must use.

The possibility of integrating a number of such switches on a single substrate to form a switching matrix has also been demonstrated.[12] Five switches were interconnected to form a 4 x 4 matrix with a cross-talk below -18 dB for each of the output ports. These accomplishments, coupled with the large fabrication tolerances permitted, indicate that the switched $\Delta\beta$ coupler could play an important role in optical communication systems, and, might even be of interest for electrical switching networks.

4.2.4 Integrated optical bistable devices

Two-port bistable device

Optical bistable devices have interested research workers for a number of years.[13,14] The early proposals for such devices, however, were based on the use of purely optically induced nonlinearities which require extremely high optical power densities. More recently, Smith and Turner[15] suggested the use of an electro-optically induced nonlinearity overcome this problem and have now demonstrated an integrated optical version of such a device, which is shown in Figure 4.10. The light at the output of the cavity formed by the silvered cleaved-crystal faces is detected and amplified by the avalanche detector. The electrical signal is then fed back to the electrodes which control the optical length of the cavity via the electro-optic effect. The hysteresis, and hence bistability, observed with this configuration is shown in Figure 4.10.

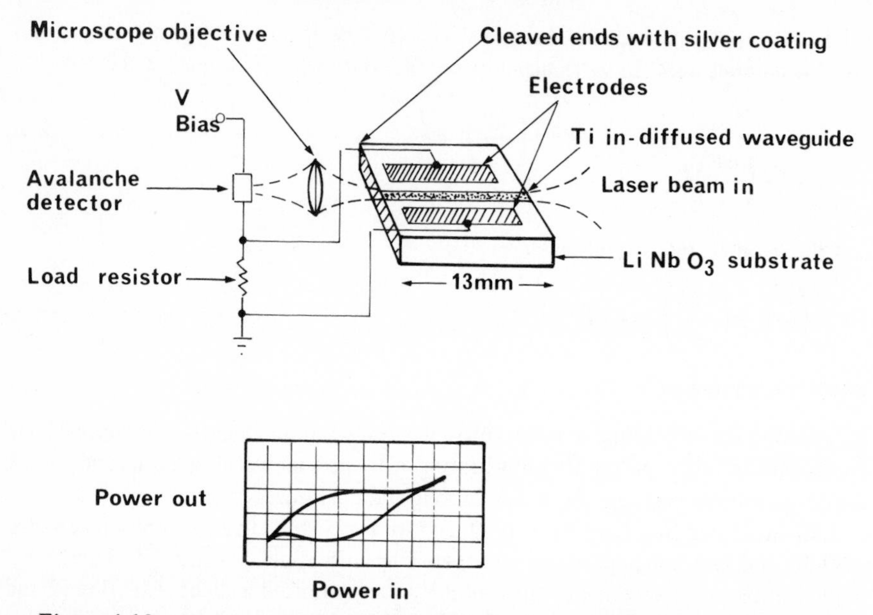

Figure 4.10. An integrated optical bistable device and the observed hysteresis

To analyse the operation of such a device we recall that with an interferometric structure such as this one, the ratio of the output optical power to the input power will be an oscillatory function of the effective cavity length. In our case the effective cavity length is a linear function of V, the voltage applied to the electrodes. Since in addition, $V = \gamma R P_{out}$, where γ is the current responsivity of the detector and R the load resistor, we can eliminate V to obtain a curve giving P_{out} versus P_{in} of the form shown in Figure 4.11. We can now see why optical hysteresis occurs. As one begins to increase P_{in}, P_{out} follows the lowest branch of Figure 4.11. Once P_{in} exceeds a certain critical value, P_{out} must jump to the second branch. If P_{in} is now reduced, P_{out} will follow this second branch until it is forced

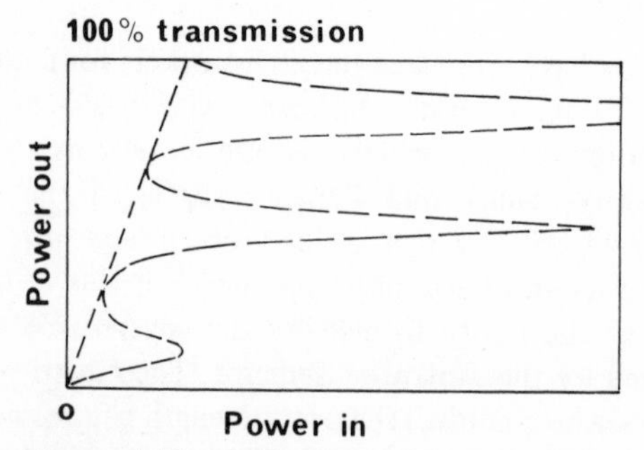

Figure 4.11. Optical power out versus optical power in for a bistable device

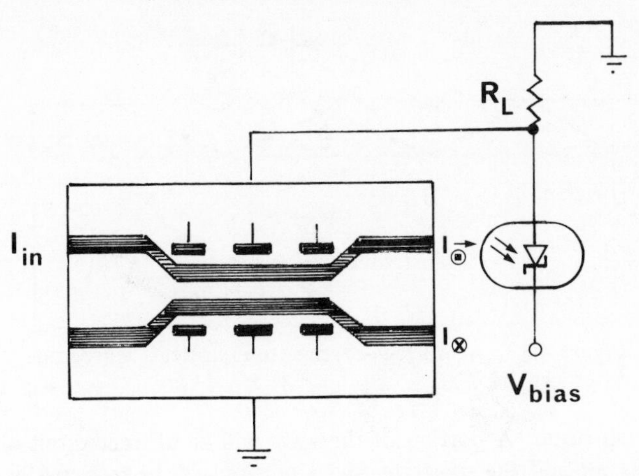

Figure 4.12. A four-port optical bistable configuration

to jump back to the lowest branch. This is the reason for the hysteresis shown in Figure 4.10. These devices can be designed to operate at nanosecond switching speeds, with the switching controlled optically or electrically, and with a power dissipation less than 1 pW.

Four-port bistable device

An interesting derivation of the previous device has also been reported.[16] In this device a switched directional coupler is used as shown in Figure 4.12, with one of its output ports connected to the detector which controls the drive voltage. The analysis of the device follows that used for the two-port bistable device but in this case there is no need for a resonator. Furthermore, in addition to performing the various functions associated with bistability (power limiter, etc.), this device permits remote optical switching of channels, which could prove to be an important function in optical communication systems.

4.3 PERIODIC STRUCTURES: APPLICATION TO FILTERS, DFB AND DBR LASERS

4.3.1 Introduction to corrugated waveguide filters

One of the interesting problems originally posed for integrated optics was a means of realizing filters and resonators in a planar form. These problems have now begun to be resolved by the incorporation of periodic structures in optical waveguides.[2,17,18]

Consider a planar optical waveguide with a corrugation etched into its surface (Figure 4.13). When a wave propagating in the guide reaches the perturbed area,

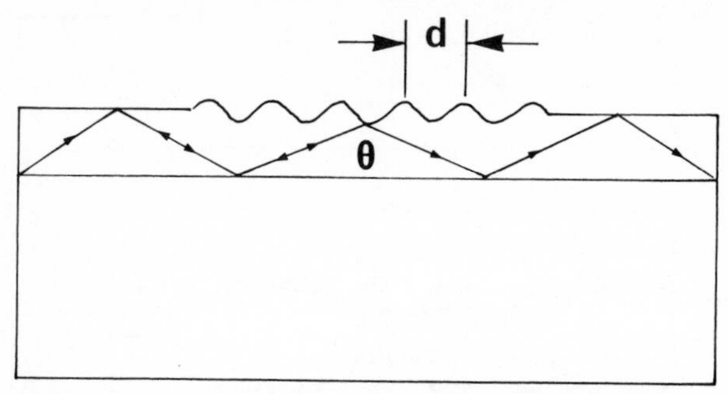

Figure 4.13. A surface-corrugation optical waveguide filter

two things will occur. A portion of the wave will be diffracted out of the guide, into both the air and the substrate, and a portion will be reflected back into the original mode but propagating at 180° to the original direction of propagation. If the Bragg condition is satisfied,

$$\lambda = 2N\Lambda \tag{4.8}$$

where λ is the vacuum wavelength, N the effective guide index of refraction given by $n_g \sin\theta$, and Λ the wavelength of the corrugation. After an interaction length L, determined by the coupling coefficient K, all of the incident power will be reflected. It is clear, therefore, that such structures can serve as frequency-selective rejection filters or mirrors. Such filters will have a bandwidth Δ given approximately by[18]

$$\frac{\Delta}{\lambda} = \frac{\Lambda}{L} \tag{4.9}$$

The device behaviour will be determined by the product KL where K, the coupling coefficient, is approximately given by:

$$K \approx \frac{\pi}{\lambda} \frac{\Delta h}{h_{\text{eff}}} \frac{n_g^2 - N^2}{N} \tag{4.10}$$

where Δh is the corrugation modulation depth and h_{eff} the effective guide thickness which takes into account the penetration depth of the evanescent waves. Filters having reasonable parameters (modulation depths of several hundred angstroms, periods of 100–3000 Å and lengths of the order of 5 mm) have yielded reflectivities on the order of 80% and bandwidths on the order of 1 Å, in good agreement with the previous formulas.[19]

More complex filter structures, including tapers and chirps, have also been studied. Tapers, a gradual change in the modulation depth, have permitted sidelobe levels of better than −70 dB to be achieved.[20] The gradual change in the corrugation period (chirps) permits a broadening of the filter response function.[19]

In the following sections we shall discuss the use of corrugated filters to form resonators for two types of injection lasers. This will lead us to a discussion of the technological problems posed when one attempts to fabricate filters have a specified performance.

4.3.2 Distributed feedback (DFB) lasers

In the distributed feedback (DFB) laser the period structure is distributed throughout the active lasing region.[17] The advantage of this type of resonator, compared with the usual cleaved-face resonator for injection lasers, is a reduced frequency sensitivity to variations in laser power and temperature. Nevertheless, the price to be paid for this frequency stability is a considerably more complex fabrication process.

Consider the cross section of a typical DFB laser shown in Figure 4.14. In such a structure it is necessary to stop the epitaxial growth, inscribe the appropriate grating, and then continue the epitaxial growth over the grating. Evidently this perturbs the crystalline structure in the region of the grating. By using a structure of the type shown in Figure 4.14 which separates the perturbed region from the active region where the recombination occurs, several groups have demonstrated DFB lasers which operate CW at room temperature.[21,22]

These lasers have demonstrated the desired reduction of frequency sensitivity to both temperature change and pump power as well as improved stability. A typical $d\lambda/dT$ reported for such lasers is 0–0.5 Å/°C, whereas in ordinary double-heterostructure lasers a typical value is 2–5 Å/°C.[23] Nevertheless, the *a priori*

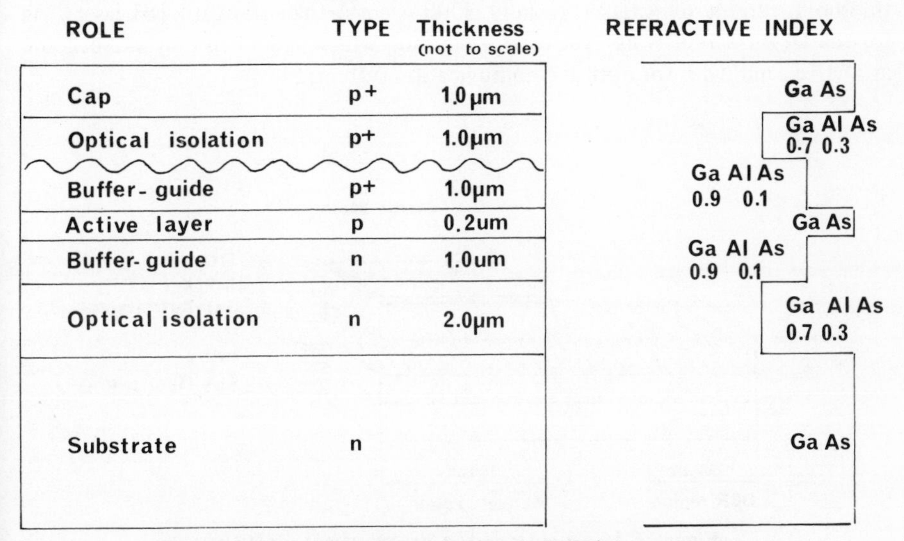

Figure 4.14. A distributed feedback (DFB) laser structure

determination of a precise lasing wavelength appears difficult with present-day fabrication techniques. The effective index is a sensitive function of guide thickness and for GaAs-like laser systems it would be necessary to control the absolute guide thickness to within 10–50 Å to control the output wavelength to 1 Å.[24]

Despite the technological difficulties which remain to be resolved, the DFB structure offers some interesting possibilities for wavelength multiplexing. Six DFB lasers, each having a different grating period and hence a different output wavelength, have been fabricated on a single substrate and interconnected to an output guide.[25] Another technique based on changing the angle between the beam and a single grating for multiple-frequency operation has also been proposed and demonstrated.[26] These lasers do, therefore, despite their structural complexity, offer interesting possibilities for *future* use.

4.3.3 Distributed Bragg reflector (DBR) lasers

In the DBR laser the periodic structures are placed outside the active region and replace the mirror function of the cleaved faces of an ordinary GaAs laser. The advantage of this structure, compared with the DFB laser, is that the perturbed region is completely separated from the active region. The disadvantage is that the beam is forced to propagate in the unpumped, and hence absorbing, distributed reflector. In order to retain the advantages of the distributed reflector and not introduce too much loss, several authors have used the hybrid structure shown in Figure 4.15.[27,28]

Again, these lasers have complex structures, and to date demand higher threshold current densities (typically 3000 A/cm^2) than standard DH lasers. As for the DFB lasers it is not yet clear, therefore, where and when they will become attractive candidates for optical communication systems.

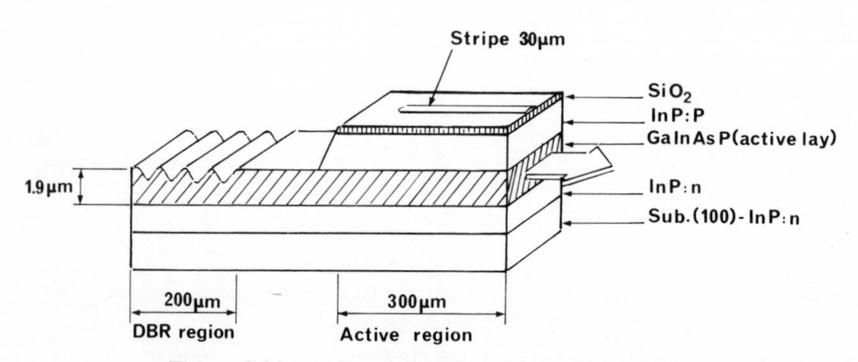

Figure 4.15. A distributed Bragg reflector (DBR)

4.4 OPTO-ELECTRONIC INTEGRATION

One of the problems that is often overlooked in integrated optics is the means of interconnecting optical and electronic circuitry. Two examples of integration of integrated optical devices and active electronics have already been demonstrated and we shall discuss them in these sections. The examples are opto-electronically integrated photodetectors and oscillator–laser integration.

4.4.1 Integrated optical photodetectors

In 1973 the possibility of fabricating optical waveguides on silicon substrates containing active electronic devices such as photodetectors was demonstrated.[29] In this work it was shown that it is possible to form low-loss (loss below 1 dB/cm) optical guides by RF sputtering high-index glass on an oxidized silicon substrate containing photodetector elements. A schematic of the device demonstrated is shown in Figure 4.16. In this device the guided light is coupled into the silicon substrate at the point where the SiO_2 layer stops. In this region a *p–n* junction converts the optical signal to an electrical signal. It was estimated that by proper design, the response time could be limited to the carrier drift time and would permit the detection of signals of up to 10 GHz.

A more complex version of such a device using a charge-coupled device (CCD) detector array has also been demonstrated.[30] The device contained 19 array elements and was successfully operated with a 100-kHz clock rate. While this device is intended for use in data handling, rather than transmission systems, it demonstrates the fact that it is rather easy to couple complex silicon circuits with optical waveguides.

Considerable work has also been performed on integrating photodetectors with guides in GaAs. By locally growing a region of InGaAs on to a GaAs waveguide, Schottky-barrier photodiodes were realized having avalanche gains of up to 250 and quantum efficiencies of around 50% for detecting light at 1.06 μm.[31]

Another interesting device that has been demonstrated is the electro-absorption integrated optics detector using reverse-biased *p–n* junctions.[32] In this device 1.06 μm light was guided within a GaAlAs double heterostructure and important photocurrents (~1 mA with quantum efficiency near 100%) were detected with applied voltages of the order of 10 V. An advantage of such a detector lies in the fact that it is based on a standard double heterostructure.

Proton-implanted integrated optical detectors have also been realized in GaAs.[33] These detectors, however, suffered from a degradation in their electrical characteristics due to defects produced by the implantation process. The quantum efficiencies reported were of the order of 16% for 1.06 μm light and the response times of the order of 200 ns. While this performance could be considerably improved by careful design, the basic damage problem remains and such detectors are probably of less interest for optical detection than those previously mentioned.

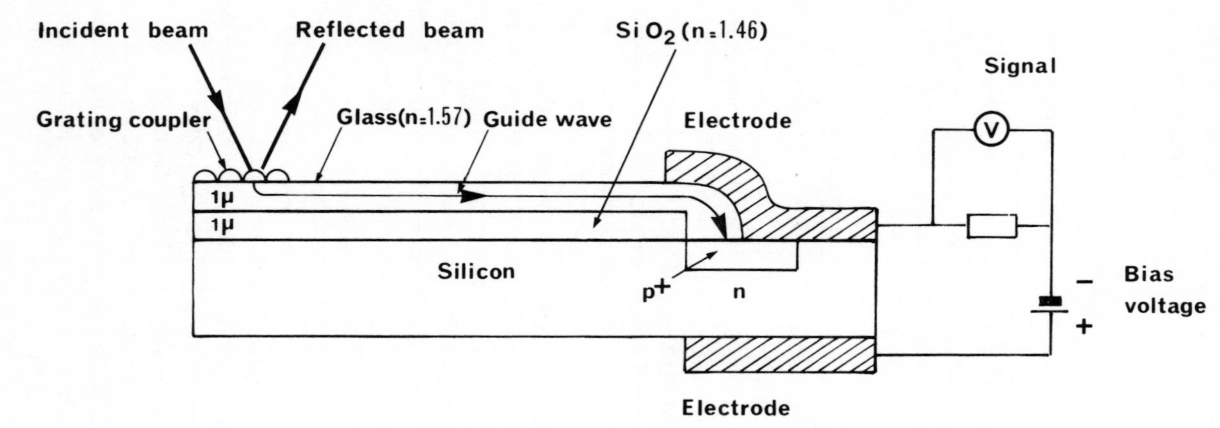

Figure 4.16. An integrated optical photodetector

In conclusion, one can say that two classes of materials have been used to demonstrate integrated optical photodetectors: silicon and the III–V semiconductors. Complex electrical circuitry can be realized in both, and the eventual choice between them, if one opts for opto-electronically integrated detection, will depend on the wavelength used in the transmission system. If long-wavelength (1.1–1.3 μm) emitters are used, the GaAs system will be necessary since Si sensitivity is very low in that region. If 0.9 μm emitters are used the Si system will be chosen since the III–V compounds do not permit the fabrication of low-loss guides in that region.

4.4.2 Oscillator–laser integration

The opto-electronic integration of a Gunn oscillator with a double heterostructure laser shown schematically in Figure 4.17 has recently been demonstrated.[34] This device is interesting since it provides an output pulse train having a repetition rate of up to 1 GHz. To fabricate this device the five epitaxial layers are grown on a semi-insulating substrate, which is necessary for the operation of the Gunn oscillator. After termination of the epitaxial growth, the device is etched to obtain the configuration shown. When the voltage applied to the Gunn device and the laser exceeds a threshold value, the Gunn device begins to oscillate, and the oscillating current pumps the laser. Due to the laser nonlinearity, the laser modulation can be considerably greater than the current modulation. In the experiment cited, values of 70% and 15% respectively were reported.

In conclusion, we can say that a number of opto-electronically integrated prototype devices have been realized and there is every reason to believe that a trend in this direction has been established and a number of other devices of this type should appear shortly. Despite their apparent technological complexity, some of these devices such as the oscillator–laser, or photodetector–laser structures could find use in multimode optical communication systems. This is because they are, actually, electronic devices with optical inputs or outputs. Their interior interconnections are purely electrical. It is probable that they are, however, the forerunners of a second generation of integrated optical devices in which data is handled, *within* the device in both an optical and electronic form.

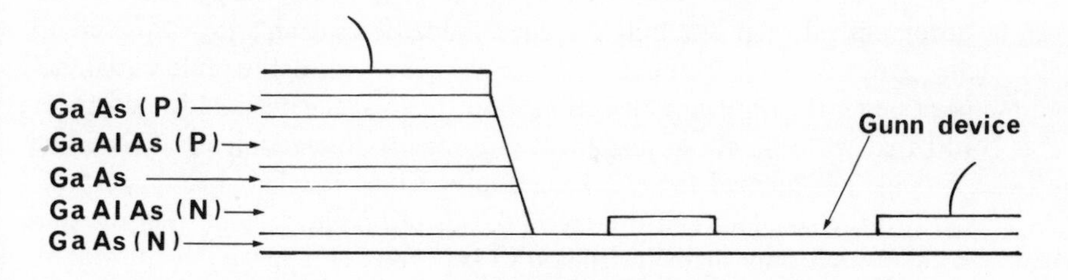

Figure 4.17. An integrated Gunn oscillator–injection laser structure

4.5 FIBRE–WAVEGUIDE COUPLING

One of the main problems which remain to be solved before integrated optical devices can find widespread use is the problem of connecting these elements to monomode fibres. This problem promises to be a difficult one in view of the sub-micron tolerances involved. Two types of approach have been attempted to date: end-fire, and evanescent field coupling. We shall discuss each of these techniques in turn.

4.5.1 End-fire coupling

The essential idea is simply to butt the core of the optical fibre against the end of the integrated optical channel waveguide, and, once alignment is attained, fix the elements in place. While conceptually simple, this technique remains rather difficult in practice. The practical difficulties revolve around the fact that both guide-structure extremities must be polished, and the structures must be very accurately aligned. In addition, even if 'perfect' alignment is attained the coupling cannot be total due to reflection effects and mode-structure mismatch between the essentially stripe-like modes of the channel waveguide and the circular modes of the fibre.

In a study of the coupling efficiency between a single-mode ($\Delta n = 0.07\%$) fibre with an 8 μm core and an approximately 4×2 μm Ti:$LiNbO_3$ waveguide, it was shown that a 3-dB insertion loss is attainable.[35] To maintain the coupling efficiency at 95% of this value, it is necessary to maintain the transverse dimensions to within approximately 2 μm, the guide separation to about 10 μm and the angular misalignment to below 2°. These results are rather encouraging. They indicate that by symmetrizing the channel-waveguide mode (by adding a cover film having the same index as the substrate, for example) and by reducing the reflections, insertion losses of the order of 1 dB should be attainable.

In the previous approach, all possible degrees of alignment freedom are maintained during the coupling process. If one wishes, however, to interconnect a number of channel waveguides to a fibre array (at the extremities of a switching matrix, for example) another approach is necessary. It has been suggested that this can be accomplished by positioning the single-mode fibres in an array of preferentially etched grooves in a silicon chip.[36] Since the precision of the preferential etch allows positioning the fibre core to within about 1.5% of the overall fibre diameter, it is hoped that with the aid of alignment grooves for observation a good coupling efficiency could be obtained and efficiencies approaching 30% have been reported.[37] As opposed to the standard end-fire method, this technique requires that the fibre core and the overall fibre diameter should be kept fixed. It appears, therefore, that whenever minimum insertion loss is a necessity the best existing method is the individual connection of guides and fibres.

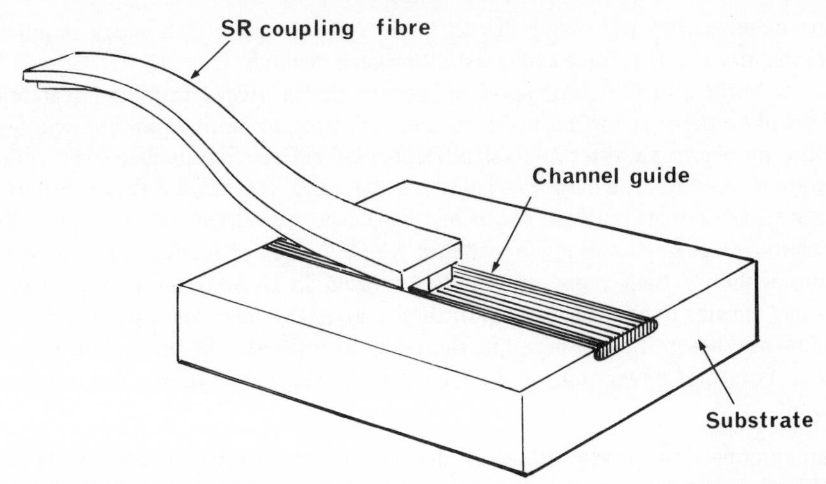

Figure 4.18. Coupling configuration for a sandwich ribbon (SR) fibre with a strip waveguide

4.5.2 Evanescent wave coupling

Another promising technique for coupling channel waveguides to single-mode fibres is evanescent wave coupling. In this method (Figure 4.18) a sandwich ribbon (SR) fibre, having its core on the surface of the fibre, is pressed against the channel waveguide to ensure an interpenetration of the optical waves.[38] By fabricating a glass channel waveguide and a glass SR fibre having synchronous modes, coupling efficiencies of over 90% have been demonstrated. Similar efficiencies have also been attained using waveguides tapered to zero thickness over the coupling region to allow for tapered velocity coupling,[39] which removes the stringent fabrication tolerances required for synchronous coupling.[40]

These techniques, which avoid the need for polishing the component extremities, are therefore a very promising means for connecting fibres and channel waveguides. Considerable work, however, remains to be done before such a technique will be sufficiently practical to permit field use in an optical communication system.

4.6 MATERIALS: HYBRID VERSUS TOTAL INTEGRATION VIEWPOINTS

It is clear that the materials which have played a major role in the development of the devices we have discussed can be divided into two main classes: the III–V semiconductors (essentially the GaAs family), and insulating crystals. The reason for the use of GaAs is evident. It is the only single type of material in which one can currently realize the vast majority of functions required for optical circuits: emission and detection of light as well as modulation and guiding. Furthermore, it offers the potential of integrating the required electronic circuitry. Therefore, if one believes that the major impact of integrated optics technology requires *total integration*, one is forced to work with such a material.

Many workers in the field, however, believe that a *hybrid* form of integrated optics will be the first to find practical use, and, perhaps more important, will for specific problems always retain significant performance advantages over fully integrated versions. The best examples, to date, are the metal-diffused lithium tantalate and niobate guides. These materials have permitted, for example, the fabrication of channel-waveguide switches whose figures of merit are an order of magnitude better than those of switches realized in GaAs. Important progress, however, remains to be made in both of these material systems, and their associated technologies, before they will permit the practical utilization of integrated optical devices. For total integration in the III–V semiconductor class, progress must be made in:

(a) integrating long-wavelength emitters (such as GaInAsP which emits at 1.1–1.3 μm) with GaAs structures, which have low absorption losses at these wavelengths);
(b) improving the individual device performances by an order of magnitude.

For hybrid integration with the metal-diffused guides, one can cite as major problems:

(a) incomplete knowledge of the guide formation and evolution processes in these materials;
(b) the increased complexity of the inter-element coupling posed by hybrid systems.

Which specific type of device and material systems will have the first applications will depend on both the progress made towards solving the above problems and on the evolution of the needs of the optical communications systems themselves.

4.7 CONCLUSIONS

Integrated optics technology has now made contributions to all the essential components for optical communication systems. In reviewing the current state of the art, however, it is clear that it is only in the domain of modulators and switches that integrated optics has led to devices whose performance and capabilities exceed those of standard, non-integrated, devices. If current trends continue, these elements, in a hybrid configuration based on metal-diffused guides, should be the first to find practical use in optical communication systems. Since such use will, however, be dependent upon the introduction of monomode-fibre systems, it should be at least 5–10 years before this could occur. With such a delay it is possible that totally integrated systems based on the III–V semiconductors will have become competitive, and currently unpredictable criteria such as overall cost-effectiveness will come into play to determine the system to be employed. It is evident therefore that it will be a number of years before the way in which integrated optics will influence optical communications systems will begin to become clear.

REFERENCES

1. D. B. Anderson, *Optical and Electrooptical Information Processing*, MII Press, 1965, pp. 221–234.
2. S. E. Miller, 'Integrated optics: an introduction', *Bell Syst. Tech. J.*, **48**, 2059–2069, 1969.
3. H. Kogelnik, *Integrated Optics* (ed. T. Tamir), Springer-Verlag, 1975, pp. 15–83.
4. J. Hammer, *Integrated Optics* (ed. T. Tamir), Springer-Verlag, 1975, pp. 140–201.
5. I. P. Kaminow, L. W. Stulz and E. H. Turner, 'Efficient strip-waveguide-modulator', *Appl. Phys. Lett.*, **27**, 555–557, 1975.
6. M. Papuchon, 'Recent progress in electrooptic modulation and switching using $LiNbO_3$ waveguides', *Frequenz*, **32**, 75–78, 1978.
7. M. K. Shams, H. Namiziki and S. Wang, 'Monolithic integration of GaAs–(GaAl)As light modulators and distributed-Bragg-reflector lasers', *Appl. Phys. Lett.*, **32**, 314–316, 1978.
8. J. McKenna and F. K. Reinhart, 'Double-heterostructure $GaAsAl_xAs_{1-x}$ pn junction-diode modulator', *J. Appl. Phys.*, **47**, 2069, 1976.
9. H. Kogelnik and R. V. Schmidt, 'Switched directional couplers with alternating deltabeta', *IEEE J. Quantum Electron.*, **QE-12**, 396–401, 1976. R. V. Schmidt and H. Kogelnik, 'Electro-optically switched coupler with stepped deltabeta reversal using Ti-diffused $LiNbO_3$ waveguides', *Appl. Phys. Lett.*, **28**, 503–506, 1976.
10. M. Papuchon, Y. Combemale, X. Mathieu, D. B. Ostrowsky, L. Reiber, A. M. Roy, B. Sejourne and M. Werner, 'Electrically switched optical directional coupler: COBRA', *Appl. Phys. Lett.*, **27**, 289–291, 1975.
11. R. V. Schmidt and P. S. Cross, 'Efficient optical waveguide switch/amplitude podulator', *Opt. Lett.*, **2**, 45–47, 1978.
12. R. V. Schmidt and L. L. Buhl, 'Experimental 4 × 4 optical switching network', *Electron. Lett.*, **12**, 575–577, 1976.
13. A. Szoke, U. Daneu, J. Goldhar and N. A. Kurnit, 'Bistable optical element and its applications', *Appl. Phys. Lett.*, **15**, 376–379, 1969.
14. H. M. Gibbs, S. L. McCall and T. N. C. Venkatesan, 'Optical bistability', *Phys. Rev. Lett.*, **36**, 1135, 1976.
15. D. J. Smith and J. Turner, 'A bistable Fabry–Perot Resonator', *Appl. Phys. Lett.*, **30**, 280–281, 1976.
16. P. S. Cross, R. L. Schmidt, R. L. Thornton and P. W. Smith, 'Optically controlled two channel integrated-optical switch', *IEEE J. Quantum Elect.*, **QE-14**, 577–580, 1978.
17. H. Kogelnik and C. V. Shank, 'Stimulated emission in a periodic structure', *Appl. Phys. Lett.*, **18**, 152–154, 1971.
18. D. C. Flanders, R. U. Schmidt, H. Kogeinik and C. V. Shank, 'Grating filters for thin-film optical waveguides', *Appl. Phys. Lett.*, **24**, 194–196, 1974.
19. H. Kogelnik, 'Filter response of nonuniform almost-periodic structures', *Bell Syst. Tech. J.*, **55**, 109–126, 1976.

20. P. S. Cross and H. Kogelnik, 'Sidelobe suppression in corrugated-waveguide filters', *Opt. Lett.*, **1**, 43–45, 1977.
21. H. C. Casey, Jr, S. Somekh and M. Ilegems, 'Room-temperature operation of low-threshold separate-confinement heterostructure injection laser with distributed feedback', *Appl. Phys. Lett.*, **27**, 142–144, 1975.
22. M. Nakamura, K. Aidi, J. Umeda and A. Yariv, 'CW operation of distributed-feedback GaAs–GaAlAs diode lasers at temperatures up to 300 K', *Appl. Phys. Lett.*, **27**, 403–405, 1975.
23. A. Yariv and M. Nakamura, 'Periodic structures for integrated optics', *IEEE J. Quantum Electron.*, **13**, 233, 1977.
24. D. B. Ostrowsky and J. Sevin, 'Contribution a l'etude des lasers a structure distribuée', *Photonics*, Gauthier-Villars, 285–308, 1973.
25. K. Aiki, M. Nakamura and J. Umeda, 'Frequency multiplexing light source with monolithically integrated distributed-feedback diode lasers', *Appl. Phys. Lett.*, **29**, 506–508, 1976.
26. M. A. Di Forte, M. Papuchon, C. Puech, P. Lallemand and D. B. Ostrowsky, 'Tunable optically pumped GaAs–GaAlAs distributed feed-back lasers', *IEEE J. Quantum Elect.*, **QE-14**, 560–562, 1978.
27. Th. I. Alferov, S. A. Gurevich, M. Mizerov and E. L. Portnoy, 'CW AlGaAs injection heterostructure lasers with second-order distributed Bragg reflectors', *Technical Digest, Topical Meeting on Integrated and Guided Wave Optics*, Salt Lake City, January 16–18, 1978.
28. H. Kawanishi, Y. Suematsu, Y. Itya and S. Arai, '$Ga_xIn_{1-x}As_yP_{1-y}$–InP injection laser partially loaded with distributed Bragg reflector', *Jap. J. Appl. Phys.*, **17**, 1439, 1978.
29. D. B. Ostrowsky, R. Poirier, L. M. Reiber and C. Deverdun, 'Integrated optical photodetector', *Appl. Phys. Lett.*, **22**, 463–465, 1973.
30. J. T. Boyd and C. L. Chen, 'An integrated-optical waveguide and a charge-coupled-device image array', *IEEE J. Quantum Electron.*, **QE-13**, 282–286, 1977.
31. G. E. Stillman, C. M. Wolfe and I. Melngailis, 'Monolithic integrated $In_xGa_{1-x}As$ Schottky-barrier waveguide photodetector', *Appl. Phys. Lett.*, **25**, 36–38, 1974.
32. S. Somekh, E. Garmive, A. Yariv, H. Garvin and R. Hunsberger, 'Integrated optical photodetector in GaAs', *Appl. Opt.*, **13**, 327, 1974.
33. H. S. Stoll, A. Yariv, R. G. Hunsberger and G. L. Tangonan, 'Proton implanted optical waveguide detectors in GaAs', *Appl. Phys. Lett.*, **23**, 664–665, 1973.
34. C. P. Lee, S. Margelit, I. Cery and A. Yariv, 'Integration of an injection laser with a Gunn oscillator on a semi-insulating GaAs substrate', *Appl. Phys. Lett.*, **32**, 866–867, 1978.
35. J. Noda, O. Mikami, M. Minakata and M. Fukuma, 'Single-mode optical-waveguide fiber coupler', *Appl. Opt.*, **17**, 2092, 1978.
36. H. Hsu and A. F. Milton, 'Flip chip approach to end-fire coupling between single-mode optical fibers and channel waveguides', *Electron. Lett.*, **12**, 404, 1976.
37. H. Hsu and A. F. Milton, 'Single-mode coupling between fibers and indiffused waveguides', *IEEE J. Quantum Electron.*, **13**, 224, 1977.
38. C. A. Millar and P. J. R. Laybourn, 'Coupling of integrated optical circuits using sandwich ribbon fibres', *Opt. Common.*, **18**, 80, 1976.
39. M. G. F. Wilson and G. A. Teh, 'Tapered optical directional coupler', *IEEE Trans. MTT*, **MTT-23**, 85–92, 1975.
40. C. D. Wilkinson, Private communication, 1978.

Optical Fibre Communications
Edited by M. J. Howes and D. V. Morgan

CHAPTER 5

Optical fibre and cables

C. K. KAO

5.1 PREFERRED FIBRE TYPES

The physical and optical characteristics of optical-fibre waveguides required in practical systems may differ widely, but a few designs represent the preferred solutions. Most of these are made with silica-based glasses and are generally suitable for operation in the near infra-red wavelength range.[1–12]

5.1.1 Single-mode waveguide

A single-mode waveguide has the ultimate wide-bandwidth capability and has a completely defined propagation characteristic. It is ideally suited for long-haul and high-capacity applications. A typical structure is as shown in Figure 5.1.

The core and cladding are made from glasses with high purity in order to avoid losses arising from impurity absorptions. The outer layer is also made from a glass. It can be the same material as the cladding. However, it can be a less pure material and have a different refractive index. Its function is to increase the outer diameter so that the fibre can be more easily handled and have improved rigidity to resist bending. The cladding thickness is required to be about 10 times the core radius, since significant field penetrates into the cladding region from the core.

Characteristics

– Attenuation
    Scattering limit ~1 dB/km at 0.85 μm for NA = 0.1.

– Bandwidth
    ~40 GHz km, limited by material and waveguide dispersions.

– Core
    One to several wavelengths in diameter. Core size can be increased by lower-

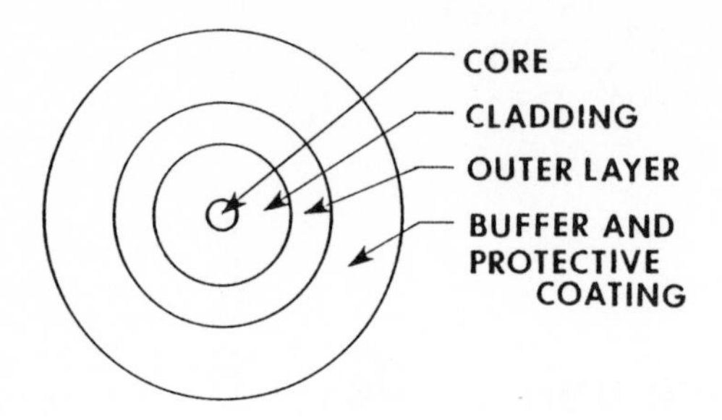

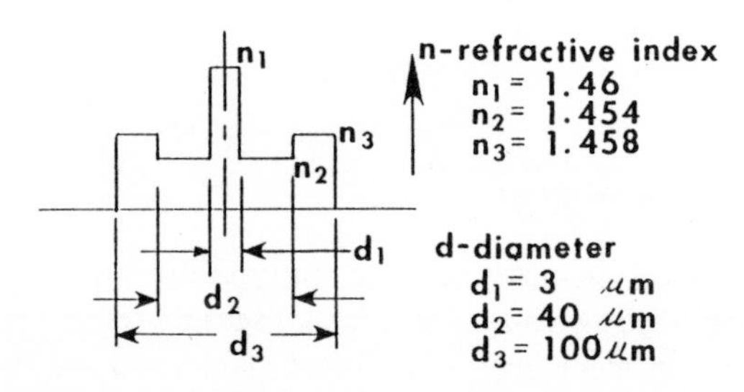

Figure 5.1. Single-mode waveguide structure

ing NA, but the radiative loss will increase at a bend and/or with dimensional variation.

– Cladding

Thickness must be at least 10 times core radius. Increased thickness eases handling and reduces susceptibility to microbending. The influence of lossy outer jacket on propagation will also be reduced.

– Coupling

Requires light-source emission spot size to be equal to or smaller than the fibre core area for maximum efficiency.

5.1.2 Multimode graded-index waveguide

A multimode waveguide with a relatively large core of profiled index, and a relatively high numerical aperture (multimode graded-index fibre) is a high-quality fibre suited for high-bandwidth, medium-haul applications. A typical structure is as shown in Figure 5.2.

As in the single-mode waveguide, the core and cladding are made from materials with high purity in order to avoid losses arising from impurity absorptions. The outer layer can be the same as the cladding. However, it can be a less pure material and have a different refractive index. Its function is to increase the outer diameter, so that the fibre can have improved rigidity to resist bending. The cladding layer can be thin. The function of the cladding layer is to act as a barrier to prevent

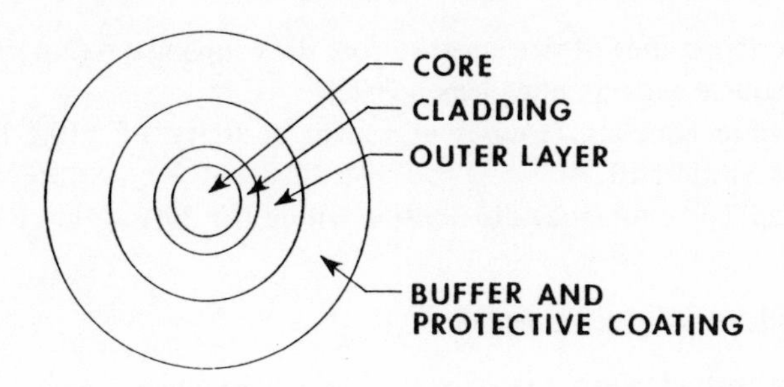

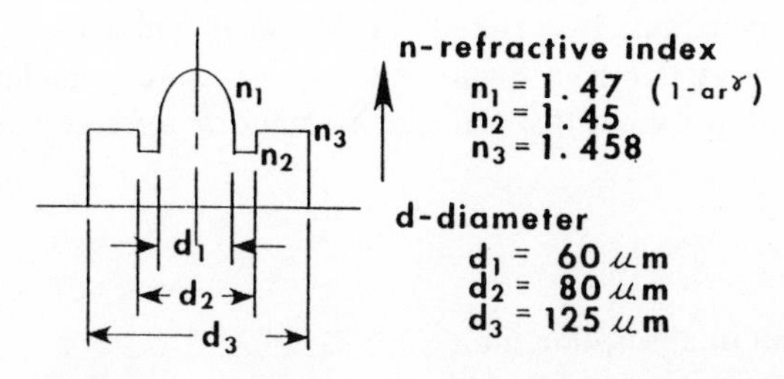

Figure 5.2. Multimode graded-index waveguide structure

impurity ions in the outer layer from migrating into the core region. If the outer layer has a different but higher refractive index, the cladding thickness is required to be about 10λ. This will ensure that the effective numerical aperture (NA) of the fibre is defined by the core/cladding refractive indices and that the higher-order propagating modes will not be unduly affected.

Characteristics

- Attenuation
  Scattering limit, increases with NA, ~2 dB/km
- Bandwidth
  >1 GHz km for a profile optimized for maximum bandwidth.
  ~300 MHz km for readily achieved profile.
  Limited by profile control and profile change with wavelength.
  Bandwidth decreases with an increase in NA.
- Core
  50–60 μm diameter with an NA of 0.2–0.3.
- Cladding
  10λ cladding thickness and 2:1 ratio of OD to core is adequate for low microbending loss.

– Coupling

To lambertian source of size equal to core the coupling $\eta = (NA)^2/2$.

To laser source with coupling lens $\eta > 90\%$.

Fibre-to-fibre coupling requires alignment accuracy of ±10% for coupling loss to be within 1 dB.

Loss is dependent on mode distribution within the fibre at the coupling point.

5.1.3 Multimode step-index waveguide

A multimode waveguide with a large core of step index and a large NA (multimode step-index fibre) can couple effectively to a light source such as an LED with a lambertian output radiation. It is suited for short-haul and low-cost applications. A typical structure is as shown in Figure 5.3. The structure is similar to the multimode graded-index fibre except that the core has uniform refractive index.

Characteristics

– Attenuation

Scattering limit or absorption limit.

– Bandwidth

~20 MHz km limited by modal dispersion.

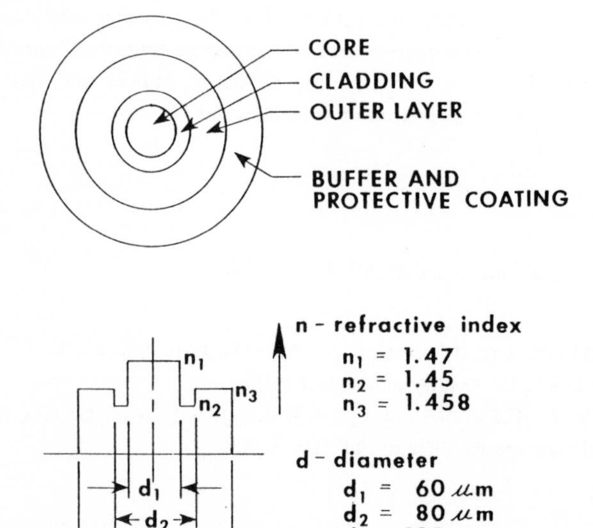

Figure 5.3. Multimode stepped-index waveguide structure

– Core

  $>80\ \mu m$ diameter with NA of 0.3 to 0.6.

– Cladding

  2:1 ratio of OD to core diameter is adequate for low microbending loss.

– Coupling

  To lambertian source of size equal to core the coupling $\eta = (NA)^2/2$.

  Fibre-to-fibre coupling is most sensitive to lateral misalignment (1 dB loss with 10% offset).

  Loss is dependent on mode distribution within the fibre at the coupling point.

5.1.4 Plastic-clad fibre

A plastic-clad fibre is a step-index multimode fibre with a glass core and a plastic cladding. It is a step-index fibre with a large core and a relatively high NA. It is a little less expensive than the multimode step-index fibre. A plastic-clad pure silica fibre suffers less radiation-induced loss and can be used for special applications where a resistance to radiation-induced loss is important. A typical structure is shown in Figure 5.4.

This type has somewhat higher loss and a limited operating temperature range (–40 to +125 °C). The plastic cladding is usually a silicone rubber. A protective plastic coating is required.

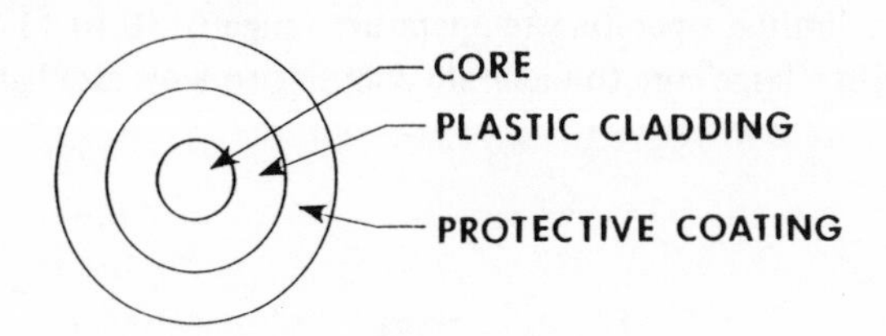

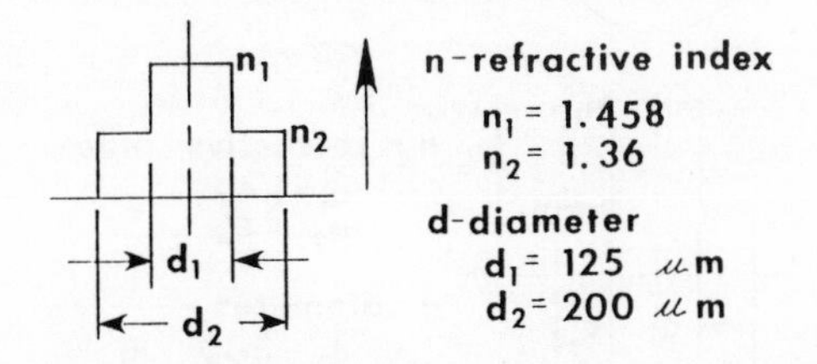

Figure 5.4. Plastic-clad multimode fibre waveguide

Characteristics

- Attenuation
  Absorption limit set by the plastic-clad material.
- Bandwidth
  ~20 MHz km limited by modal dispersion.
- Core
  100 μm up to 600 μm with NA of >0.3.
- Cladding
  Can be thin (several μm).
- Coupling
  As in multimode step-index case.
  Fibre-to-fibre coupling is complicated by the fact that the plastic is soft and mechanically difficult to hold to the tolerances required.
  The tolerance requirement is ±10% for 1 dB loss.

5.1.5 All-plastic fibre

An all-plastic fibre is a step-index multimode fibre with a much higher loss but can be handled without special care. It can be used conveniently for very short and low-cost links. A typical structure is as shown in Figure 5.5.

A good candidate plastic fibre is made with polymethyl methachrylate polymer and its copolymer. The mechanical properties of the polymer allow the fibre to be made without additional protective layers. These fibres have high loss (>400 dB/km) and have a limited operating temperature range (–30 to +125 °C). They are usually made with a large core to facilitate coupling to large size light sources.

Characteristics

- Attenuation

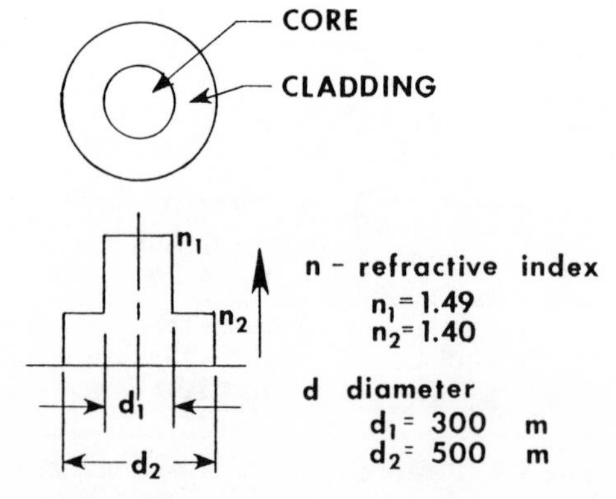

Figure 5.5. All-plastic multimode fibre waveguide

Molecular absorption bands and high scattering cause relatively low-loss transmission to be attained in the 0.6–0.8 m region. Typical attenuation 400 dB/km.

– Bandwidth
  Is of little concern over a short distance of a few metres.

– Core
  100 μm with NA of 0.5.

– Cladding
  Can be thin.

– Coupling
  To a source is similar to the multimode step-index case.
  Fibre-to-fibre coupling is relatively easy, since the plastic material involved is much harder than the plastic cladding material and end preparation can be achieved by simple cutting.

5.1.6 Fibre loss and bandwidth

Best fibre losses are shown in Figure 5.6 and the influence of loss and dispersion on repeater spacing for different information bit rate is shown in Figure 5.7.

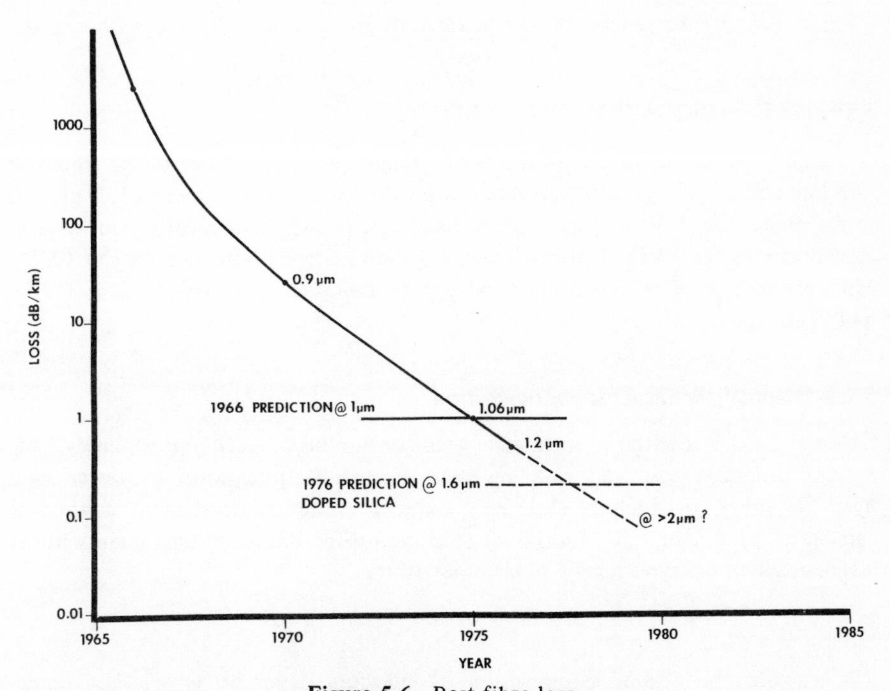

Figure 5.6. Best fibre loss

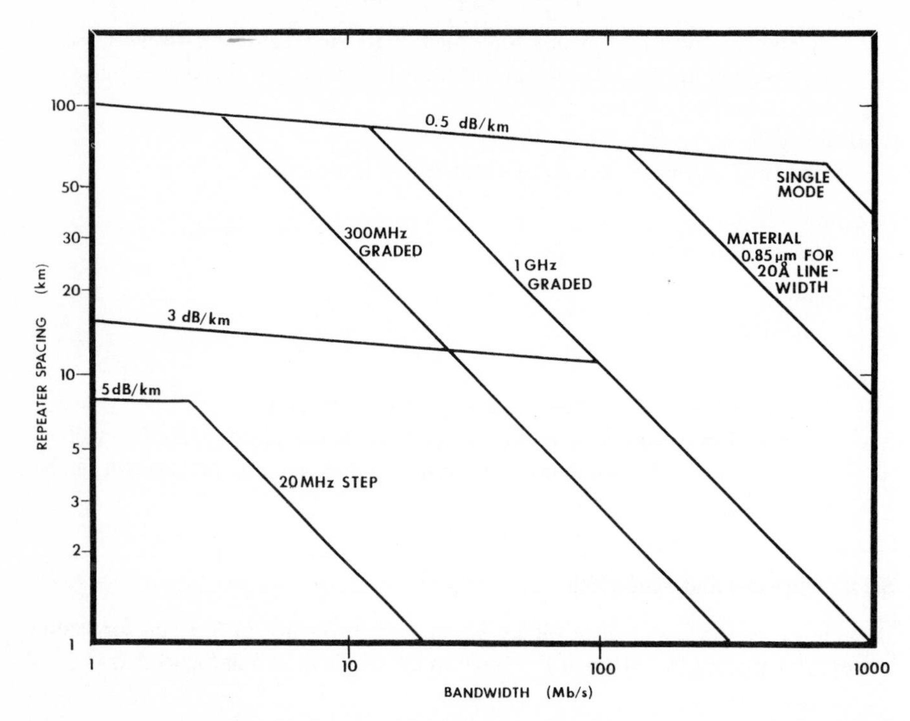

Figure 5.7. Repeater spacing and bandwidth application to different fibre types

5.2 FIBRE-FABRICATION PROCESSES

A number of different fibre-fabrication techniques have been under intensive development by many manufacturers worldwide. These will be described, in general terms, under the following descriptive headings: external chemical vapour deposition of soot (ext CVD); internal chemical vapour deposition of glass (int CVD); Multi-element glass; external chemical vapour deposition of glass (plasma CVD); and Phasil glass.

5.2.1 External chemical vapour deposition

External CVD is a batch process which has continuous-operation possibilities. This process produces core and cladding materials in ultra-pure form. However, care must be taken to exclude externally induced contaminations. This method is capable of producing both multimode step and profiled fibres of high quality but is less convenient for producing a single mode fibre.

Process 1

The glass of the desired composition is deposited, layer by layer, via a flame-hydrolysis process, in the form of a powder or soot, uniformly over the length of

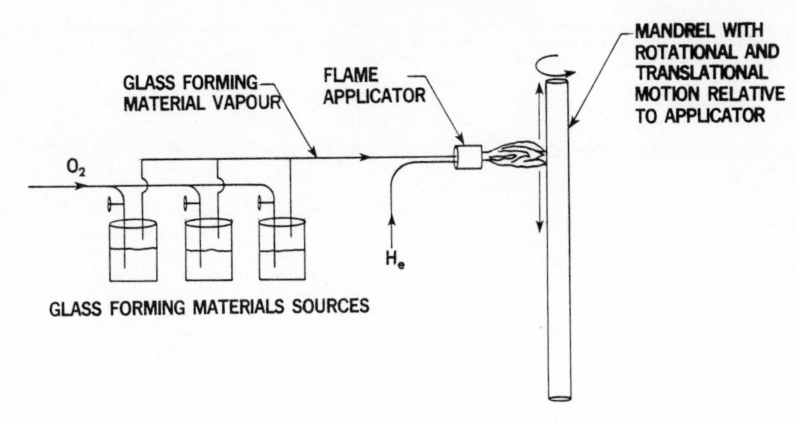

Figure 5.8. Principle of flame-hydrolysis deposition process

a mandrel. When deposition is completed the material is sintered and the mandrel removed. The consolidated tube is then fused and collapsed simultaneously into a preform rod. The principles are illustrated in Figure 5.8.

Process 2

The glass of the desired composition (radially varying) is deposited via a flame-hydrolysis process in the form of a powder or soot, over the end of a starting rod. The deposition is continued while the rod is slowly moved away from the flame in a vertical direction. A soot boule will be formed in that direction. The soot is sintered and fused simultaneously at a convenient position some distance away from the deposition end. This results in the continuous formation of a preform rod. The deposition schematic is as shown in Figure 5.9.

Using readily available glass-forming materials such as Ge, B, P and Si compounds, the processes have achieved the following performances.

– Maximum deposition rate
  ~2 g/min.

– Loss
  No absorption loss other than that due to OH^-, plus scattering loss.

– Maximum NA
  0.24.

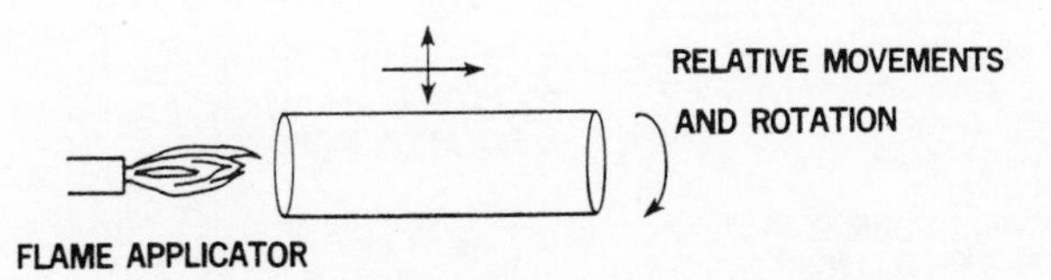

Figure 5.9. End-deposition schematic

– Dimensional restriction
  None.

– Dispersion
  300 MHz km readily achieved.
  1 GHz km possible.

– Strength
  May have internal flaws.

– Dopant material
  Ge, B, P.

– Maximum preform size
  10 km of 125 μm OD fibre.

5.2.2 Internal chemical vapour deposition

Internal CVD is a batch process. It forms the core and intermediate cladding materials in ultra-pure state within a substrate tube. It is capable of producing both single-mode and multimode step and profiled fibres of extremely high quality.

Process 1

The glass of the desired composition is deposited, layer by layer, from a vapour-phase reaction, in the form of glass uniformly over the length of the internal surface of the substrate tube (Figure 5.10). The heat to induce the chemical reaction is supplied by heating.

– Maximum deposition rate
  ~0.4 g/min.
  Since the bulk of the cladding layer is not deposited, this is equivalent of about 1.2 g/min compared with the external process.

– Loss
  Scatter loss limit.

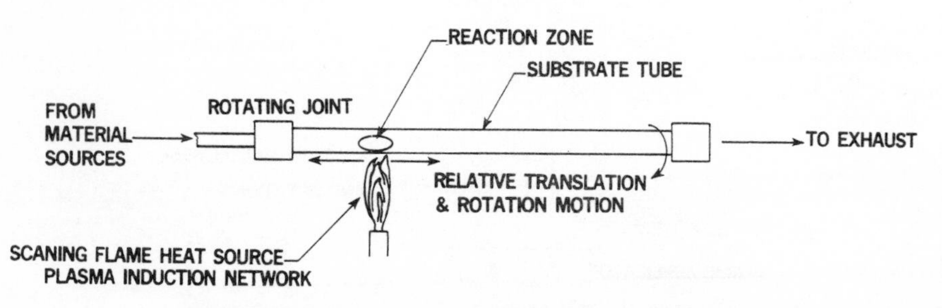

Figure 5.10. Substrate and heating arrangement

- Maximum NA
  0.38.
- Dimensional restriction
  OD to core diameter ratio unlikely to be less than 1.5 : 1.
- Dispersion
  1 GHz km achieved.
- Strength
  Few internal flaws in the deposited material.
- Dopant material
  Ge, B, P.
- Maximum preform size
  20 km of 125 μm OD fibre.

Process 2

The heat of reaction is generated by electric-current heating of the ionized gases in the form of an isothermal plasma (Figure 5.10). The gases are flowing at a similar rate as those in the indirect-heating case described in Process 1. The plasma temperature is extremely high but the tube temperature is below the glass-softening temperature. The deposition rate is expected to be higher but at the same time less controlled, while the tube is expected to be more stable during deposition, and hence scaling up to larger preform should be more readily achieved.

- Maximum deposition rate
  ~1.0 g/min.
- Loss
  Scattering loss.
- Maximum NA
  0.3.
- Dimensional restriction
  None.
- Dispersion
  100 GHz km possible.
- Strength
  Few internal flaws in the deposited material.
- Dopant material
  Ge, B, P.
- Maximum preform size
  20 km of 125 μm.

Process 3

The heat of reaction is generated by electric-current heating of the ionized gases at reduced pressure in the form of a non-isothermal homogeneous plasma. The reaction efficiency is 100% but the quantity of material flow is limited. The plasma temperature is well controlled. Deposition is made on the inner surface of a substrate tube heated to below softening point. The deposition rate is low but extremely precisely controlled. The tube is stable during deposition. Scaling-up is possible lengthwise but not easily in diameter. The arrangement is similar to the one shown in Figure 5.10.

– Maximum deposition rate
  ~0.3 g/min

– Loss
  Scatter loss limit

– Maximum NA
  0.3

– Dimensional restriction
  OD to core diameter ratio 1.5 : 1.

– Dispersion
  1 GHz km.

– Strength
  Few internal flaws in the deposited material.

– Dopant material
  Ge, B, P, F.

– Maximum preform size
  4 km of 125 μm.

5.2.3 External chemical vapour deposition of glass

Plasma CVD is the most highly developed process for making synthetic silica on an industrial scale. The addition of a dopant, however, gives rise to the need to control the differential vaporization characteristics. This process has been successfully employed to make a fluorine-doped silica. Very large boules can be fabricated. The arrangement is similar to the one shown in Figure 5.9 except that the flame torch is replaced by a plasma torch.

– Maximum deposition rate
  ~1.0 g/min.

– Loss
  Scatter loss limit.

– Maximum NA
  0.22.

– Dimensional restriction
  None.

– Dispersion
  300 GHz km marginally achievable.

– Dopant material
  F.

– Maximum preform size
  Unlimited.

5.2.4 Multi-element glasses

Multi-element glasses are made from ultra-pure basic oxides and carbonates. The purification of raw chemicals is carried out by wet chemical processes. The chemicals, to avoid recontamination, are mixed in a clean environment, and fired in pure-silica crucibles. The resultant glasses are drawn from crucible in rod form ready for remelting in double crucibles for fibre formation (see Figure 5.13). The purification, mixing and firing have also been achieved in a completely enclosed system.

The purity achievable is not as high as that produced by the CVD process, since the wet chemical methods are less effective in separating out the transition elements. (In the CVD process the transition-metal halides have very low vapour pressure and hence are not carried into the reaction zone.) The residue transition-metal ions give rise to an absorption-loss level of about 3 dB/km.

The use of multi-element glass is a natural candidate for large-volume continuous fibre fabrication and has cost advantages for large-scale production. Furthermore, the range of glass compositions are potentially larger.

– Production rate
  Very high.

– Loss
  Absorption 3 dB/km + scattering loss.

– NA range
  0.2 to 0.6.

– Dimensional restriction
  None.

– Dispersion
  300 MHz km.

– Strength
  Depends on glass systems but is lower than SiO_2 fibres.

– Composition
  Na–B–Si
  K–B–Si
  Ge–B–Si
  Na–Ca–Si
  K–B–Si
  Th–B–Si
  Pb–Si

5.2.5 Phasil system

The Phasil system is a fabrication process based on the vicor glass process. It is a batch process. The steps involved are lengthy but can be made efficient by processing a large number of preforms at once. The vicor rods are first leached so that only a honeycomb silica structure is left. Then, the desired pure dopant molecules are stuffed into the rod by a solution treatment. The clad layer is formed by a second partial leaching process. The structure is dried and sintered to form the fibre preform. By selective stuffing and leaching, a profiled core can also be achieved. Figure 5.11 is a flow chart of the Phasil process.

– Production rate
  High.

– Loss
  5 dB/km caused by absorption + scattering.

– Maximum NA
  0.3.

– Dimensional restriction
  None.

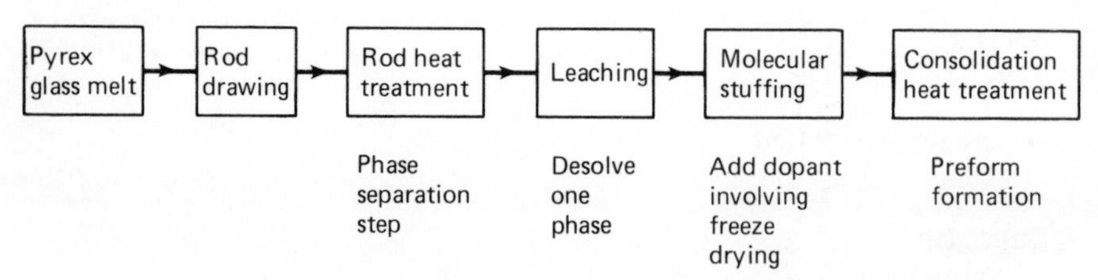

Figure 5.11. Flow chart of the Phasil process

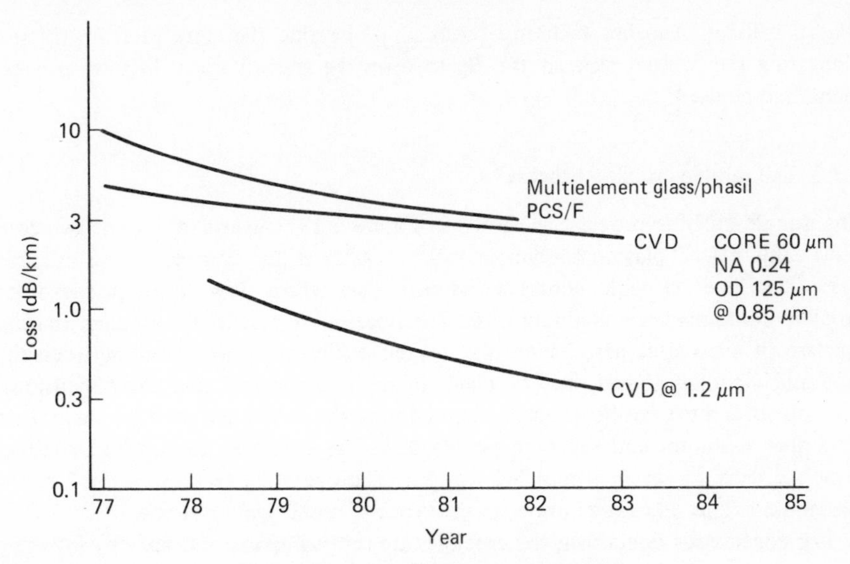

Figure 5.12. Fibre-loss specifications

– Dispersion
  300 MHz km possible.

– Strength
  Showing large number of internal flaws.

The five fibre fabrication techniques can be ranked as follows:

Internal CVD	Best performance but expensive.
External CVD	Good performance less expensive.
Plasma CVD	Moderate performance, moderate cost, at moderate volume.
Phasil	Moderate performance, low cost, at large volume.
Multi-element glass	Moderate performance, low cost, at very large volume.

Of these, the multi-element glass method produces glasses which are to be remelted in a double-crucible structure to produce fibres, possibly in continuous lengths. The other forms yield fibre preforms in a rod form. These rods are to be drawn into fibres in a batch mode. However, if the preform is sufficiently large the loss of efficiency and material wastages due to setting-up requirements may be minimal. The fibre-loss specifications expected from these processes are compared in Figure 5.12.

5.3 FIBRE-DRAWING PROCESSES

The double crucible process is a fibre-drawing technique where the fibre is drawn from the melts of the core and cladding glasses within a concentric crucible with central orifices. Another technique consists of heating the fibre preform tip and elongating the molten glass at the tip to form the fibre. Various heating arrangements can be used.

5.3.1 The double crucible process

The double crucible process, as shown in Figure 5.13, consists of a set of two concentrically placed platinum crucibles with nozzles at the centres of the crucible base. The inner crucible contains the core glass which flows through the outer crucible containing the cladding glass. The nozzles are arranged with such spacing and are of such sizes as to allow the desired diffusion to take place between the core and cladding glasses, and to yield the desired core and cladding dimensions. The crucibles must not be a source of contamination and hence must be made from ultra-pure platinum and must be pre-leached. The crucibles are heated by induction, d.c. current, or in a muffled furnace. The operating temperature is kept to a minimum. This sets a restriction on the types of usable glass systems.

For continuous operation, the crucibles are topped up continuously by lowering glass canes into the melt from the top of the crucibles.

– Maximum operating temperature
 ~800 °C.

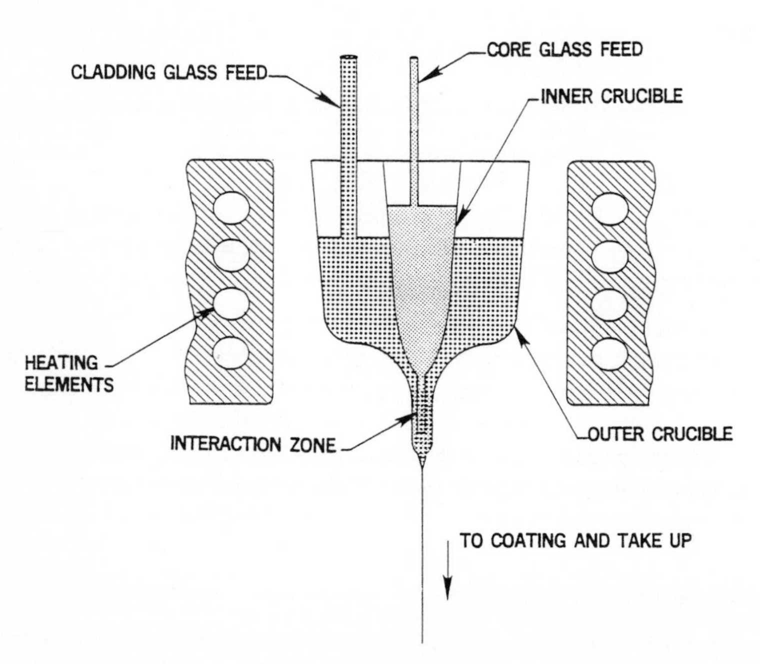

Figure 5.13. Schematic of double-crucible fibre-drawing apparatus

– Maximum fibre-pulling rate
  At least several hundred metres per minute.

5.3.2 Fibre drawing from preforms

Fibre drawing from fibre preforms involves heating the preform tip. Several heating sources are suitable but each has its limitations.

Oxyhydrogen flame

Flame stability depends on torch design and inter-torch interference. Flame fluctuations result in dimensional changes in the fibre. Achievable tolerance is about ±2%. The quantity of heat available is adequate for heating large preforms without causing preform vaporization due to intense local heating. Flame, however, generates H_2O as a by-product and this could affect fibre surface composition. Drawing temperature affects optical and physical properties and is also a function of drawing speed. Drawing speed of 60 m/min is possible. The strength of fibre produced is excellent.

Furnaces

Graphite furnaces can be energized by d.c. current or by a.c. induction. The operating temperature required for silica fibre drawing is sufficiently high to cause SiO_2 and C to interact. If a shield tube is used, it is likely to become another source of contamination. Most furnaces are not well insulated and would not produce sufficient heat for fibre drawing from large preforms. Drawing speed of 60 m/min for 11 mm diameter preforms is possible. With feedback control, ±1% tolerance is easily achievable. The strength of the fibre produced is good, if careful purging is executed.

Inductively heated zirconium furnaces have been successfully operated for SiO_2 fibre drawing. The heat capacity appears to be greater than the graphite furnace and result in less contamination.

The CO_2 laser is another heating source successfully used for SiO_2 fibre drawing. The absorption coefficient of SiO_2 of the 10.6-μm radiated power from the laser unfortunately is extremely large, and the heat conductivity of SiO_2 is relatively low. As a result, the fibre preform surface reaches extremely high temperatures and evaporates rapidly. This limits the size of preform which can be drawn into fibre by using laser heating and also causes a minor problem of SiO_2 dust disposal.

The critical parameters of fibre drawing are summarized in Table 5.1.

Note: The as-drawn fibre has extremely high strength. However, the pristine fibre surface must immediately be protected from abrasion and interaction with the environment. The present technique is to coat the fibre with a plastic coating by a solution-coating method. Several plastics have been used. These are Kynar, epoxy,

Table 5.1 Critical fibre-drawing parameters

● Drawing temperature	
× Silica/doped silica	
○ Multi-element glass	
● Heating method	
× Graphite furnace	d.c. or inductive
× Zirconia furnace	d.c. or inductive
× CO_2 laser	
× O_2–H_2 flame	
○ Resistance furnace	
○ Crucible heating	
● Coating material	
● Kynar	
● Epoxy	
● Silicone RTV	
● UV-cured epoxy	
● Preform size	
● Drawing rate	

uv-cured epoxy, and silicone RTV. Application techniques include the use of slightly tapered solid nozzle and flexible nozzle made from silicone RTV. Provided the fibre is well centred, the fibre surface is not damaged and relatively uniform and concentric coating can result.

5.4 FIBRE STRENGTH AND DURABILITY

5.4.1 Introduction

With the rapid progress being made in fibre-optic technology, attention is shifting from the more theoretical aspects of operational and commercialization problems.

When optical-fibre waveguides are employed in a practical environment, the durability is of major concern. In a fibre cable design, the mechanical requirements called for are more easily met if the fibre is strong and will not fail under stress during its operational life. A strong and durable fibre is required. Towards this end it is necessary to understand the strength and fatigue failure modes of glassy material and to be able to assess long-length and long-term strength of fibres. It is also necessary to understand the effects of environmental conditions, such as relative humidity and temperature, on fibre strength.

The determination of fibre strength, however, is by no means simple since fibre strength is governed by the existence of occasional flaws along the entire length of the fibre under consideration. When subjected to tensile stress, a fibre will fail if the stress concentration at a dominant flaw reaches the critical fracture stress of the material. When an applied stress lower than the critical fracture stress is applied, the surface flaws will tend to enlarge in such a way as to cause stress concentrations at the flaw tips to increase. This causes the flaws to propagate at progressively higher speed until the stress concentration again reaches the critical fracture stress, and failure occurs. This phenomenon is caused by a stress-induced corrosion involving water in the environment, and is known as a static fatigue phenomenon. It governs the expected life of the fibre under load.

5.4.2 Strength of glass

Structural strength of oxide glasses

Glass contains atoms located at the centre of oxygen tetrahedra which are connected at their corners and arranged in space in a random continuous network. Cations such as Si^{4+}, Ai^{3+} and B^{3+} that tend to form tetrahedra or other coordinated units with oxygen are known as network formers. Other cations such as alkali ions are called network modifiers and lie in open space in the network

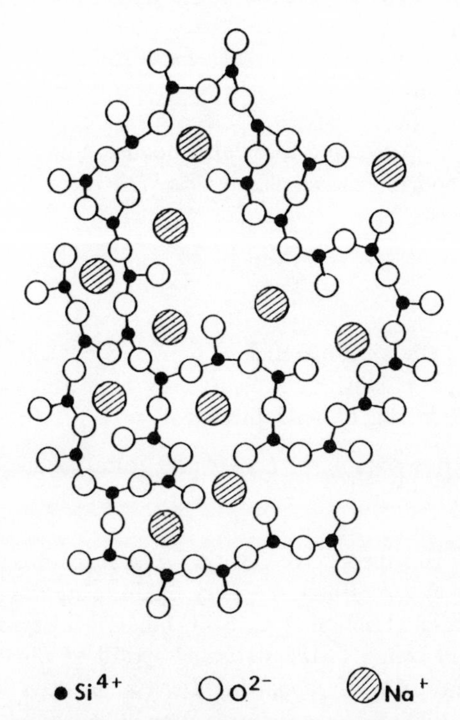

Figure 5.14. Schematic representation of a sodium silicate glass[15]

structure (Figure 5.14). Glass normally forms a loosely packed structure as a result of the random arrangement of the oxygen tetrahedra.

Glass is a strong, elastic material (apparently obeying Hooke's Law over a wide range of stresses), but it is also brittle. Its ultimate strength is determined by the bonding forces within its material structure. The variety and complexity of glass micro-structures preclude accurate determination of total strength by calculating the aggregate of bonding forces even for pure fused SiO_2 or simple binary glasses. However, with the knowledge of the dissociation energies of the bonds and of the modecular packing densities, it is possible to predict with reasonable accuracy the Young's modulus for most simple and a few complex glasses.

The ultimate strength of glass is only of theoretical interest. The practical strength is reached when a glass sample is subjected to stress, and failure occurs when the ultimate strength limit is exceeded at some portion of the structure. This depends on the stress distribution. Glass strength is thus dependent on the micro-heterogeneity and/or phase separation which are controlled by the glass composition, the thermal history and the melt atmosphere. The existence of micro-heterogeneity in glass reduces its strength since stress concentration at this region may occur. Even more dominant are stress concentrations on external surfaces where there are micro-flaws or potential micro-flaws caused by the natural process of surface formation.

Theories of glass strength

Flawless glass The stress required to break a bond can be estimated for individual materials. Various models yield similar results. The theoretical cohesive strength, σ_t, can be expressed as:

$$\sigma_t^2 = \frac{2\gamma E}{8a}$$

where γ = the surface energy of material
E = the Young's modulus
a = the atomic spacing or bond distance

For the Si–O bond this gives $\sigma_t = 2.6 \times 10^6$ psi* corresponding to a bond distance of 1.6 Å.

Stress-induced flaw This theory assumes that flaw appears as a direct result of the application of stress. A representative theory of this class was proposed by Cox.[13] He assumes that a certain fraction of the Si–O bonds in a glass are in a broken state at any given moment owing to the statistical spread of vibrational energy. Since a broken bond is incapable of supporting a stress, an extra load is momentarily placed on the bonds that are in the vicinity of the broken bond, which increases the

* 1 psi ≡ 6894.76 N m^{-2}.

probability of their breaking. If a critical number of neighbouring bonds are simultaneously in a broken condition, a fracture will propagate. The interstitially migrating ions are assumed to play an important role. The strain in the vicinity of the interstitial ion is assumed to reduce to an important extent the activation energy for abortive breakage of nearby Si–O bonds. Time-dependence, temperature-dependence, surface and bulk strength, pristine and practical strength, composition, environmental and moisture effects are all predicated. This theory, however, is somewhat controversial.

Glass with flaws This theory regards the existence of flaws as given and concerns itself with the condition under which the flaw will propagate. It is assumed that the flaws are narrow cracks with small radii of curvature at their tips, where applied stresses are concentrated. If the flaws are assumed to be fine cracks, the stress at their tips can be calculated from elasticity theory. The calculation has been carried out for a straight crack with an elliptical cross section with the applied stress perpendicular to the crack, as shown in Figure 5.15.

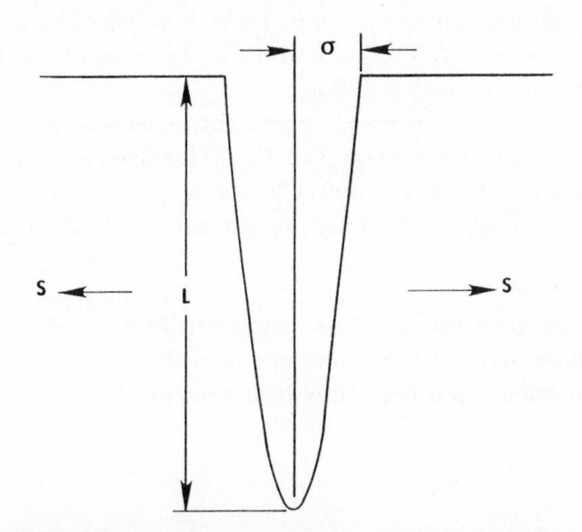

Figure 5.15. Elliptical cross section of a crack with applied tensile strength S

The stress σ at the crack tip in the direction of applied stress S is:

$$\sigma = S\left(1 + \frac{2L}{a}\right)$$

where L = the crack length (semi-major axis of the ellipse) and a = half the crack width (semi-minor axis of the ellipse). With the radius of curvature $\rho = a^2/L$ at the crack tip, and assuming $L \gg a$, this gives:

$$\sigma = 2S\sqrt{\frac{L}{\rho}} \tag{5.1}$$

It is difficult to apply Equation 5.1 to actual fracture problems because the radius ρ of the crack tip is probably of atomic dimensions, and has therefore not been measured directly. Many experimenters have assumed that ρ is constant for cracks of different lengths (though Hillig and Charles[16] found that ρ changed much more than L as corrosion proceeded). From Equation 5.1 the fracture stress S_f should then be inversely proportional to the square root of the crack length. This dependence was found by Griffith[14] in his well-known theory. The Griffith equation,

$$\sigma = \sqrt{\frac{2E\gamma}{\pi L}}$$

gives the fracture stress for a crack of elliptical shape with semi-major axis length L. E is the Young's modulus and γ the surface energy. The stress at the crack tip with load exerted along the minor axis of the ellipse is then equal to the cohesive or bond strength. Since σ is proportional to $1/\sqrt{L}$, σ decreases by a factor of 10 for a hundredfold increase in L. For fibres with σ in the region of 5×10^5 psi, the crack size responsible is of the order of 100 Å.

When a glass surface is exposed to the environment and has been handled, a large number of flaws will develop. The sizes of the flaws will vary: for a carefully protected surface the flaw size is generally smaller. The fracture strength is always dependent on the flaw at which the highest stress concentration develops (the dominant flaw).

When mechanical flaws are eliminated, the remaining flaws will be due to chemical and structural causes. These may generally be smaller than mechanical flaws. At the lower limit, the flaw sizes approach the constituent atomic spacings. The Griffith equation is not necessarily valid when the flaw size approaches atomic dimensions.

Glass surface structures

Surface structure of oxide glasses depends mostly on the reaction of nonbridging oxygen at the surface, which could be described as unpaired electrons. In a freshly prepared silicate glass there are unsatisfied Si–O and Si– bonds. These bonds react rapidly with atmospheric water to form silanol (SiOH) groups. The surface of a glass is, therefore, normally composed of metal hydroxyl groups. The structural arrangement and the thickness of the hydrated layer depend on glass composition, thermal history, humidity and surface treatment after melting and cooling.

The infra-red spectra of silica surfaces with different heat treatments have been studied. The presence of physically absorbed molecular water was indicated by the presence of broad absorption bands at about 3450 cm^{-1} and another at about

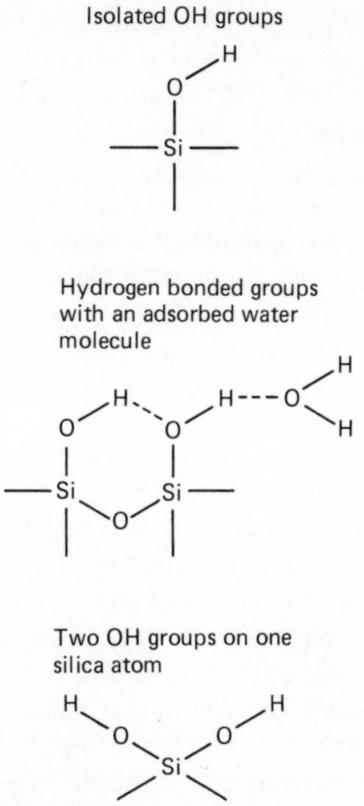

Isolated OH groups

Hydrogen bonded groups with an adsorbed water molecule

Two OH groups on one silica atom

Figure 5.16. Hydroxyl groups on silica surfaces

1250 cm^{-1}. Other bands indicate that some hydroxyl groups are close enough together to be hydrogen-bonded. The density of isolated SiOH groups on a silica surface has been calculated as 1.4 groups/100 Å and that of hydrogen-bonded groups as 3.2 groups/100 Å. Figure 5.16 indicates the types of hydroxyl groups existing on silica surfaces.

Altogether, four types of hydroxyl groups are known to exist on silica surfaces: isolated silanol groups, two OH^- groups on one silica atom, adjacent OH^- groups bonded with a hydrogen bond, and molecularly absorbed water. The relative concentration of each of the groups depends on the melt atmosphere and thermal history of the glass, and the temperature and humidity during the spectroscopic measurement.

Internal silanol groups near the surface have been observed. Exchange reactions studies of D_2O on glass surfaces indicate the presence of silanol groups, which results from the diffusion of water molecules into silica and their subsequent reaction with the silica lattice to form two SiOH groups. These internal groups do not form at room temperature, because of the low diffusion coefficient of water in bulk silica. The formation of these internal groups is evident above 100 °C.

Multi-component glasses will have the following modified cations, in addition to the four types present in silica. Other glass forms will provide sites for hydroxyl groups, and a monovalent cation R^+ in silicate glass can undergo an exchange reaction with water as follows:

$$H_2O + SiO{-}R^+ \longrightarrow SiOH + ROH$$

As hydrogen is smaller than the monovalent cations, this reaction results in a tensile stress at the glass surface which can increase the reactivity and further hydration or crack formation in some cases.

Fatigue strength

When subjected to tensile stress, glass will fail if the stress concentration at a dominant surface flaw reaches the critical fracture stress. When an applied stress lower than the critical stress is applied, and in the presence of moisture, there is a time delay to failure. The flaws will tend to enlarge in such a way as to cause stress concentration at the flaw tip to increase. This results in the propagation of the flaw at progressively higher speed until the stress concentration reaches the critical value and fracture occurs. This is known as a fatigue phenomenon and it governs the long-term strength of glass. Static fatigue is caused by water in the environment. Glass that is baked out and tested in vacuum shows little fatigue.

Hillig and Charles[16] have developed a theory to explain delayed fracture in glass in terms of stress-induced corrosion by water at flaw tips where there should be a preferential attack due to severe stretching of bonds. It is assumed that the rate of reaction of water with glass controls the rate of change in the shape of the crack tip. This reaction is

$$-\overset{|}{\underset{|}{Si}}-O-\overset{|}{\underset{|}{Si}}- + H_2O \dashrightarrow 2(SiOH) \qquad (5.2)$$

This reaction of water with the silicate network leads to a corrosion of the load-bearing elements of the glass. Under stress, reaction 5.2 is accelerated. The time to fracture is thus related to the flaw growth rate.

Hillig and Charles assumed that crack propagation was an activated process in which the activation energy was stress-dependent. The basic equation for crack velocity can be developed from the absolute-rate theory of chemical reactions. If the rate-limiting step for crack propagation is assumed to be an attack of the silicon–oxygen bonds by hydroxyl ions, then the crack-propagation equation is

$$v = v_0(OH^-)(A_g)\exp[(-\Delta E^{\ddagger} + \sigma\Delta V^{\ddagger}/3 - V_M\gamma/\rho)/RT] \qquad (5.3)$$

where (OH^-) is the hydroxyl-ion activity at the crack surface. (A_g) represents the chemical activity of a flat glass surface in contact with the corrosive environment. The first term in the exponential, $\Delta E^{\ddagger}$, represents the activation energy for the chemical reaction. σ gives the stress at the glass liquid interface and $\Delta V^{\ddagger}$ the

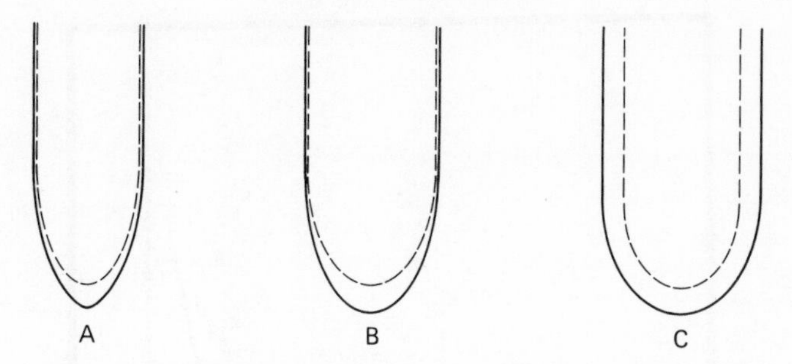

Figure 5.17. Hypothetical changes in flaw geometry due to corrosion and dislocation (S. M. Weiderhorn, *J. Am. Ceramic Soc.*, **50**, 407, 1967)

activation volume for the chemical reaction. The final term in the exponential accounts for the changing chemical activity of the glass surface with surface curvature. V_M is the molar volume of the glass, γ is the interfacial surface tension at the glass – medium interface and ρ is the radius of curvature of the glass surface.

Hillig and Charles used Equation 5.3 as a starting point to develop the relationship between the rate of crack propagation and the geometry of the crack surface. For a given glass composition and chemical environment, the radius of curvature and stress at the crack surface are the only variables in Equation 5.3. They control crack geometry and the rate of crack motion. Changes in flaw geometry due to chemical attack could be demonstrated by considering the relationship between ρ and σ in Equation 5.3 and the stress-concentration equation at the flaw tip, $\sigma = S(1 + 2(L/\rho)^{1/2})$, where S is the applied stress and L is the half-length of an elliptical crack (Figure 5.15). At stresses greater than a characteristic threshold value, flaw sharpening occurs and ρ decreases as the crack advances (Figure 5.17).

Conversely, at stresses below the threshold stress, crack blunting occurs. The threshold stress was interpreted to be an eventual static fatigue limit. This qualitative theory of static fatigue is in agreement with most experimental observations on the subject.

Equation 5.3 is sometimes written more concisely as follows:

$$v = v_0 \exp(\beta\sigma) \tag{5.4}$$

where v is the velocity at which the glass surface corrodes under a tensile stress σ, v_0 is the rate of corrosion with no stress, and β measures the effect of stress on the corrosion reaction. β is written as equal to V/RT where V is the 'activation volume'. The experiments of Weiderhorn[17] on the velocity of crack propagation show the exponential form of Equation 5.4. He found that for a given system (environment, temperature, glass composition) there is a unique relationship between the crack velocity v and the crack-tip stress-intensity factor K_I. A trimodal log V–K_I curve is obtained (Figures 5.18 and 5.19).

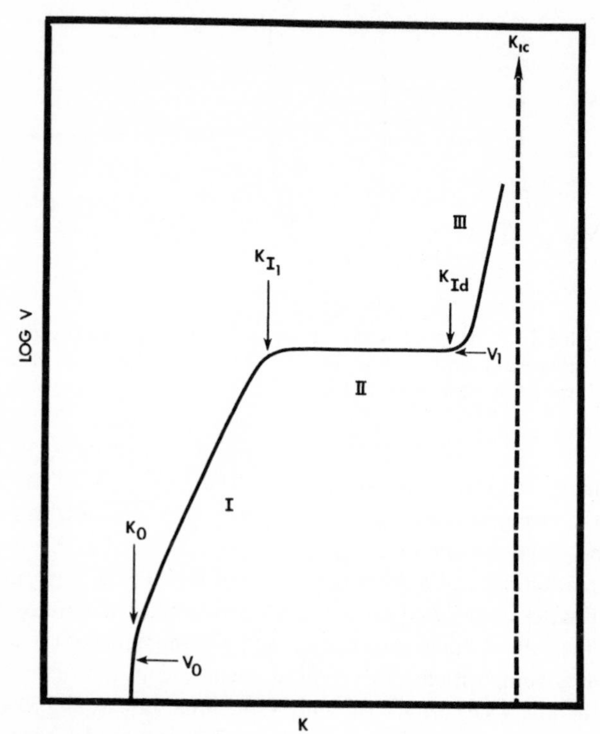

Figure 5.18. Schematic representation of the effect of crack-tip strss intensity, K on crack velocity, v, during slow crack growth[28]

At lower stresses (region I) the velocity is exponentially dependent on stress, as required by Equation 5.4. There is also a dependence on water in the environment. More generally, in this region, the crack velocity can be expressed as a power function of the stress-intensity factor:

$$v = AK_I^n \tag{5.5}$$

where n and A are constants (typically $15 < n < 50$ for glass).

At higher stresses (region II) there appears to be a levelling-off of the velocity-dependence on stress, until the velocity no longer depends on stress. In this transition region the velocity still depends on humidity. It is suggested by Weiderhorn that in this region the crack-propagation velocity is limited by the rate of transport of water vapour to the crack tip. At still higher stresses (region III) the velocity again increases exponentially with stress and no longer depends on environment. For most glass systems, region II occurs at a sufficiently high crack velocity that the crack propagation time in region I represents the time delay to failure. When K_I approaches the critical value K_{IC}, catastrophic failure follows. It should be

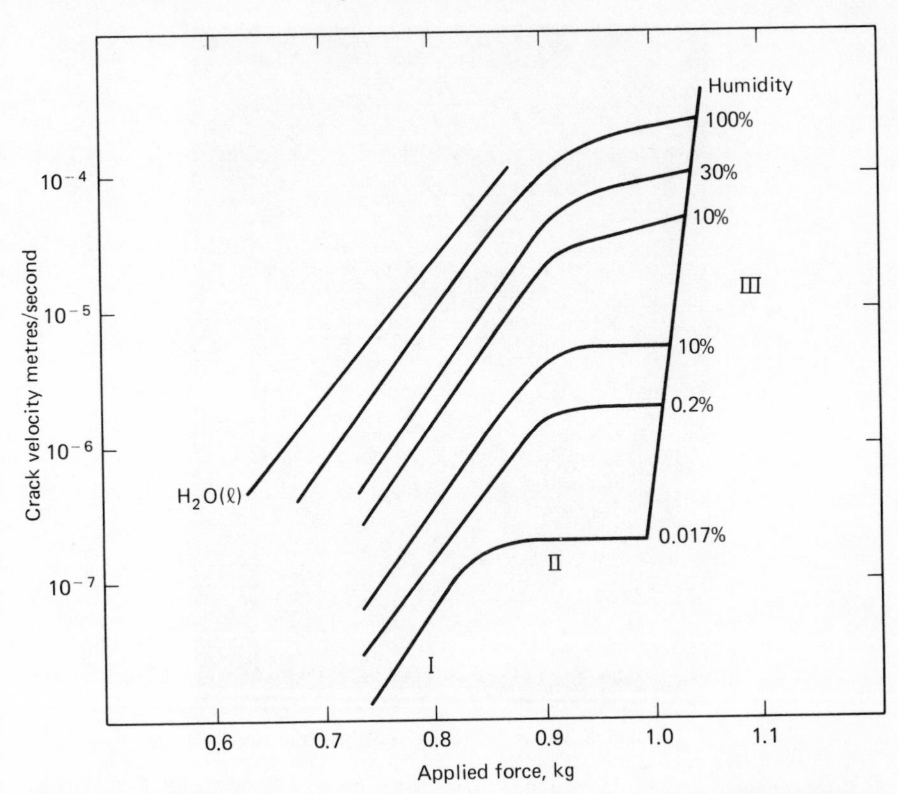

Figure 5.19. Crack velocity as a function of applied force in a soda-lime glass at room temperature (Dr R. J. Charles, General Electric Research High Strength Materials)

noted that the slow crack-growth limit K_0, if it exists, occurs at such low velocities (10^{-10} m/s) that its existence has not been generally proven. (It has been suggested that static fatigue could also be due to water absorbed on glass surfaces. In this theory, water is assumed to diffuse on the glass surface to a site at which breakdown of the silica network is initiated. As the molecularly absorbed water is bonded with weak hydrogen bond, a weakened glass network results and this enables the flaws to propagate.)

The direction of propagating cracks has also been studied. In any practical situation the orientation of cracks with respect to the stress is a random variable. Thus the actual stress at the crack is modified by factor $\cos \alpha$, where α is the angle between the crack length and the direction of the tensile stress. Furthermore, the crack may penetrate into the glass at some angle with its surface, again reducing the effective stress at its tip. On the other hand, for randomly oriented cracks of the same depth, the highest stress occurs at the one oriented perpendicular to the stress, and this crack should be the first to propagate.

Figure 5.20. A typical glass-fibre fracture

The surface of glass after fracture shows certain regular characteristics. The radial region closest to the initiating flaw is smooth and is called the mirror region. Beyond it there is a misty region where the surface roughens, and further away is a region of gross roughening called 'hackle'. The mist results from the beginnings of branching of the crack, and the hackle results from more extensive branching (Figure 5.20). Experimental evidence illustrates that crack propagation ceases when stress is removed. Comparison of cyclic and static fatigue tests demonstrates that the time to failure depends on the magnitude and total duration of the applied stress, but not on its cyclic nature. This implies that crack healing does not appear to take place.

The data obtained by various experimenters show the following for glass samples (some data are summarized in Table 5.2):

(a) Crack growth at low stress is exponentially dependent on the stress. There is also a dependence of water in the environment.
(b) At a higher stress, the stress dependence is almost nonexistent. The rate is dependent on the amount of water present. The rate of growth is suggested to be governed by the rate of transport of water to the crack tip.

Table 5.2 Fracture mechanics data for glasses

	n		Strength (kpsi)*		
Glass	Dynamic fatigue	Velocity measurement	Abraided	Etched	Etched liquid N_2
Soda lime silicate	16.0	16.6	13	250	450
Fused silica	37.8	36.1	15		2000 (pristine)
Borosilicate	27.4	34.1	12	200	400
Polymeric coated soda lime silicate	15		13	300	

* 1 psi = 6894.76 N m^{-2}.

(c) At sufficiently high stress the propagation rate is much higher and is dependent only on the material composition.
(d) An acidic or basic environment is seen to influence the crack growth. Hillig observed that immersing in 10% NaOH solutions greatly reduced the fatigue resistance of fused silica rods.
(e) The effect of alcohol is related to water content in the alcohol.
(f) Crack-propagation studies in vacuum indicate that for SiO_2 and some glasses, slow crack growth does not take place, but for others, crack growth is possible. Activation energies range from 70 to 200 kcal/mole, indicating that chemical reaction is unlikely to be the cause of the crack growth. The cause is the subject of various hypotheses yet to be confirmed, associated with the material's elastic constants and cohesive forces.
(g) The correlation of data from the direct and indirect methods has been good when identical test conditions have been maintained.
(h) The factor n appears to be independent of the surface condition of the samples. This implies that the fatigue mechanism is probably a single process.

5.4.3 Fibre strength

It is reasonable to start by assuming that the glass properties in fibre form are not significantly different from those found in bulk or sheet forms, despite the greatly changed size and surface-to-volume ratio. Hence, the fibre strength is determined by the same mechanisms which govern glass strength. In particular:

1. Fracture originates from flaws (microcracks) (Figure 5.21), formed as a result of: (a) inhomogeneities; (b) phase separation arising from thermal conditions employed in fibre fabrication; (c) interaction with the environment of the newly formed surface; (d) absorbed ions; (e) crystallization, (f) mechanical damage. The larger flaws are caused largely by mechanical means. However, as *in situ* pro-

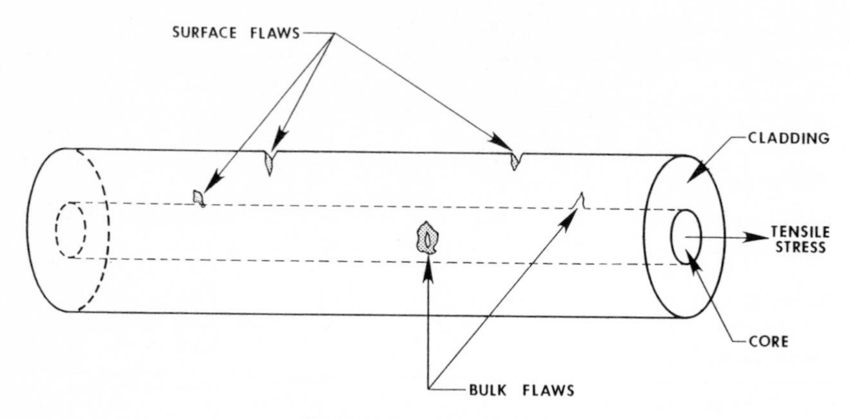

Figure 5.21. Fibre flaws

tection methods are perfected, the largest flaw may well be the random surface structural differences.

2. Under load, subcritical surface microcracks will propagate. When a critical stress, which equals the surface energy, is reached at a crack tip, the crack will propagate at a high velocity, resulting in fracture. The critical stress is most likely to be reached first at the tip of the deepest flaw since stress applied transversely to such a crack will produce a very high stress concentration at the tip of the crack through leverage action. Fatigue and ageing effects are associated with the propagation of subcritical surface cracks under load.

3. Internal stress within the fibre tends to put the core region in tension and the outer surface in compression, thus improving the strength.

5.4.4 Fibre durability

The existing theory of fatigue constructed from studies on bulk glass should be applicable to fatigue phenomena in fibres. Experimental investigations of fibre durability have been carried out. Interpretations of these results indicate that the existing theory of fatigue applies well to fatigue phenomena in fibre.

Water-vapour effect on fatigue

Hillig and Charles have proposed a fatigue failure theory for bulk glass. The salient assumptions are as follows. The fatigue failure of glass is a stress-induced corrosion at crack tips. The rate of change in the shape of the crack tip is controlled by the rate of reaction of water with glass. This corrosion reaction is:

$$-\overset{|}{\underset{|}{Si}}-O-\overset{|}{\underset{|}{Si}}- + H_2O \longrightarrow 2(-\overset{|}{\underset{|}{Si}}OH)$$

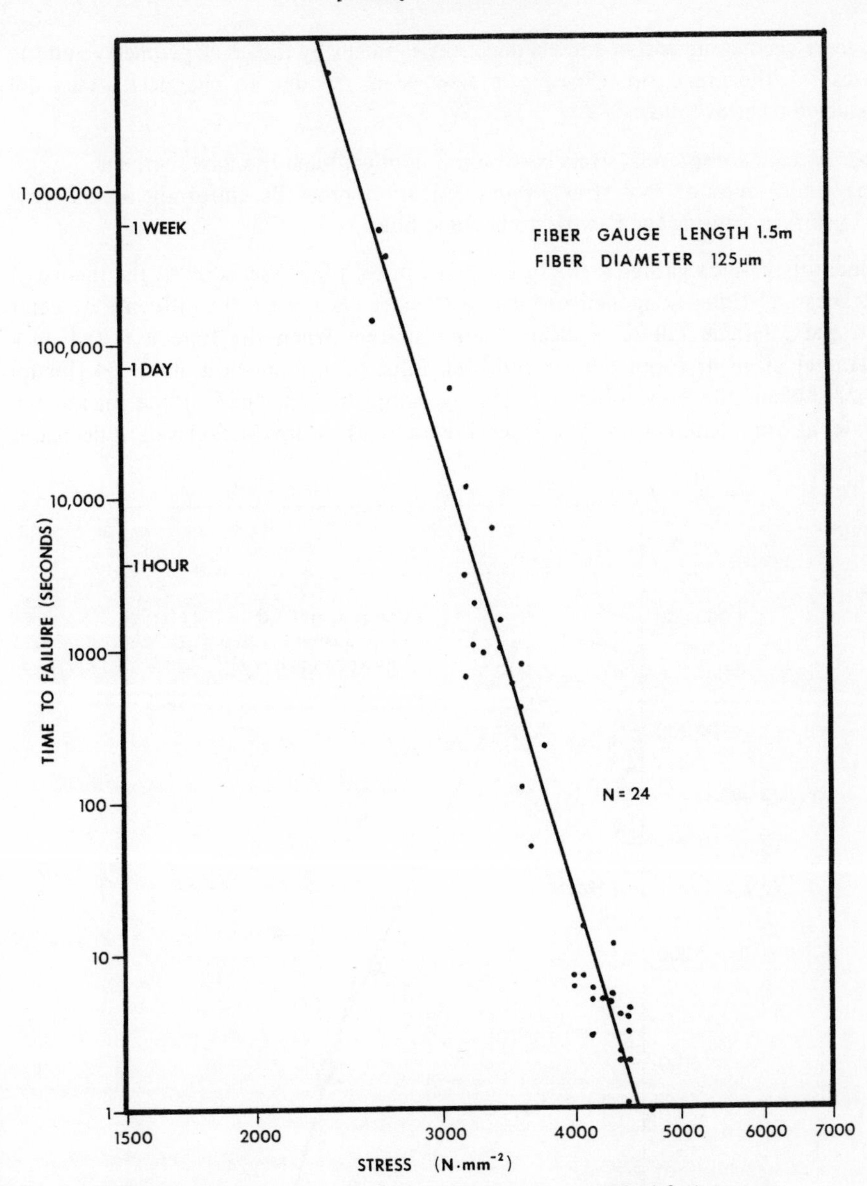

Figure 5.22. Time to failure vs. stress level – mandrel fatigue test

The crack propagation is assumed to be an activated process with an activation energy dependent on stress exerted at the crack tip. The crack propagation velocity is proportional to the activity of the water at the crack surface and to $\exp(-E/RT)$ which describes the activation energy E for the chemical reaction and the reaction-rate dependence on temperature T. Crack propagation velocity is thus controlled for a given glass composition and chemical environment by the crack geometry and the stress at the crack tip. Changes in flaw geometry due to chemical attack are assumed to be as follows:

(a) At high stress levels, stress corrosion is dominant and the flaw sharpens.
(b) Under zero or low stress, water corrosion proceeds uniformly on the flaw surface, causing the flaw tip to be more blunt.

Much of the data gathered for fused-silica optical fibres seems to fit the theory of Charles and Hillig as applied to bulk fused silica glass. First, the influence of water on static fatigue failure is clearly demonstrated. When the fibre is tested on a mandrel in air at room temperature, the value of n obtained is about 24 (Figure 5.22). When the same fibre is tested by complete immersion of the mandrel in water at room temperature, n changes (Figure 5.23). At low stress levels, n decreases

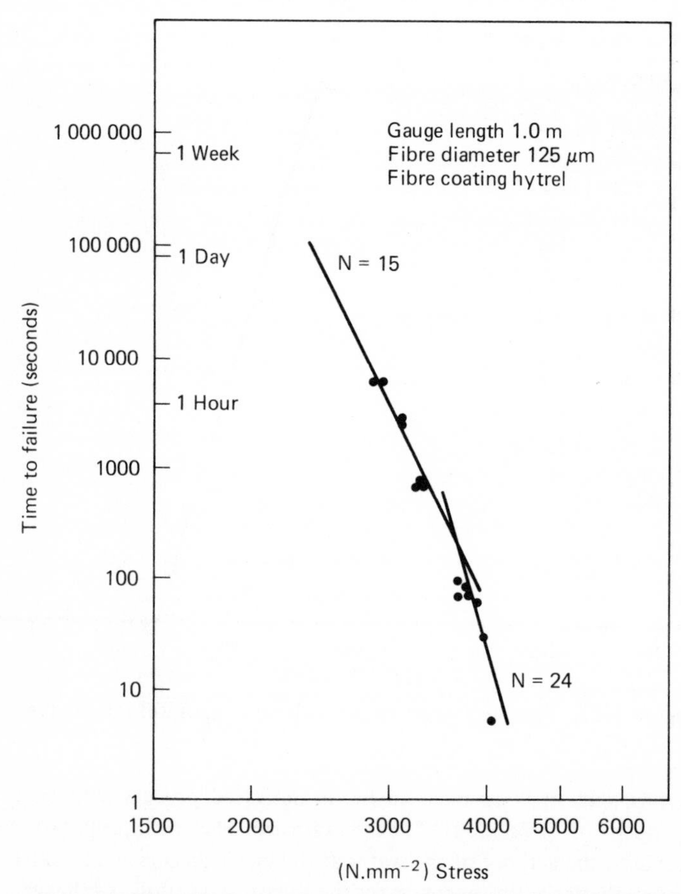

Figure 5.23. Effect of water on time to failure vs. stress

to about 15 and beyond the threshold stress (3700 N/mm^2) n changes back to 24, the value found for tests in air (Figure 5.22). This indicates that:

(a) Below 3700 N/mm^2 the increase in water concentration at the crack tip (from atmospheric water) involves a decrease of n from 24 to 15. This results in a decrease in time to failure t for the same stress σ, according to the equation $\log t = -n \log \sigma + A$, where A is a constant.
(b) Above 3700 N/mm^2 there no longer seems to be any dependence on water concentration. However, other experiments have indicated that it takes about an hour for water to penetrate through the plastic jackets. Since the time to failure at these stress levels is less than one hour, it is likely that a high concentration of water did not reach the glass prior to failure. For this reason, an n-value similar to that of fibres tested in air was found. If, however, the fibre is pre-soaked, the value $n = 15$ then prevails even at this high stress region.

General environmental effects

Silica is a weakly acidic compound. Since the time to failure is shorter in water than in air, it is possible that silica reacts with water to form silicic acid. The actual corrosion mechanism may involve transport of the silicic acid away from the crack site through dissolution in the water, thus allowing formation of more silicic acid unless an equilibrium is reached. Above the fatigue limit, application of stress and an increase in water concentration both accelerate the reaction rate at the crack tip.

In considering Figure 5.24, it can be seen that time to failure is roughly the same over the range pH = 0 to pH = 15.4 at both stress levels. However, there is a sharp increase in time to failure at pH = −1.1. This result indicates that there would seem to be a threshold of pH above which time to failure or crack propagation velocity is independent of pH, and below which it is dependent on pH. The increase in time to failure below the pH threshold may be explained as due to the lack of available water in a strongly acidic environment.

In other investigations on bulk glass, it has been observed that time to failure increased at low pH and decreased at high pH, for uncoated silica. This decrease at high pH is not seen in the data of Figure 5.24. This is attributed to the presence of the plastic coating over the fibre surface. The plastic coating effectively serves as a filter. Only H_2O penetrates the plastic in appreciable quantity within the time duration of the test.

Activation energy and temperature dependence

From the data gathered on fused-silica optical fibres, the activation energy for the chemical reaction can be derived. This has been found to be dependent on tip stress as assumed by Hillig and Charles. Times to failure at different temperatures were measured – a typical set of results can be seen in Figure 5.25. The activation

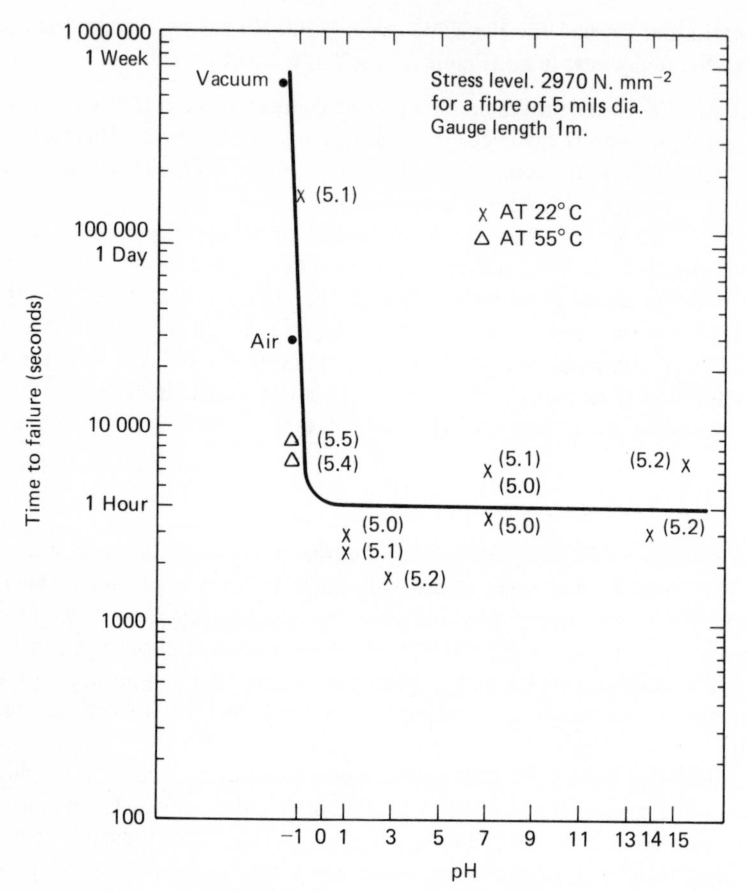

Figure 5.24. Time to failure vs. pH

energy calculation is derived from the slope of the straight line obtained by plotting $\ln(t)$ versus $(1/T)$. If λ is the slope of the line, then

$$\ln(t) = \lambda/T$$

or

$$t = t_0 \exp(\lambda/T)$$

This, compared with the usual Arrhenius equation of an activated process, gives:

$$\lambda = E/R \longrightarrow E = R\lambda$$

where E = activation energy
R = the gas constant

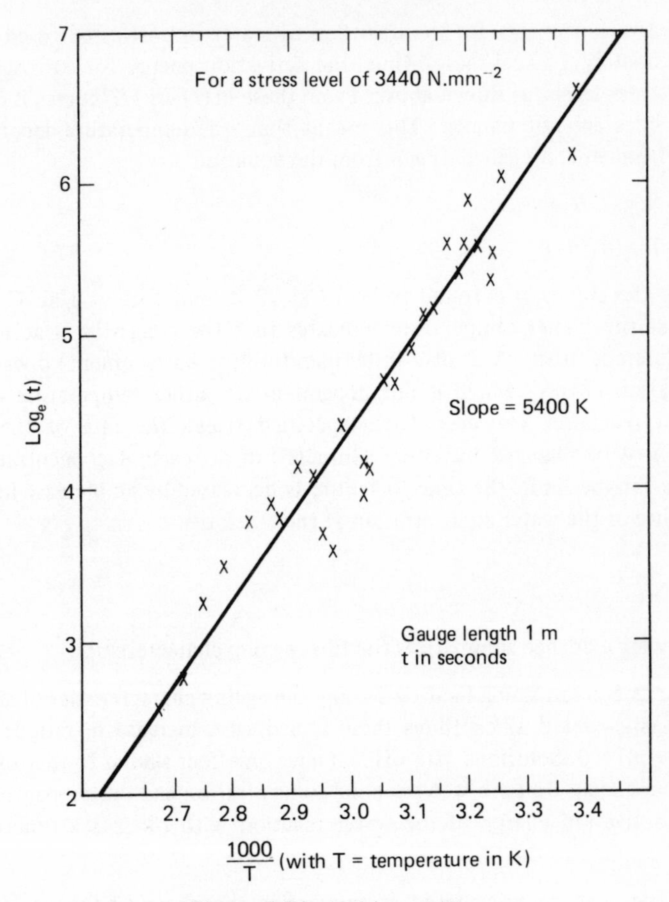

Figure 5.25. ln (t) vs. $1/T$

This gives, for water:

$E = 10.7$ kcal/mole at $\sigma = 3440$ N/mm^2
$E = 11.1$ kcal/mole at $\sigma = 3200$ N/mm^2
$E = 13.2$ kcal/mole at $\sigma = 2860$ N/mm^2
$E = 18$ kcal/mole at $\sigma = 0$

and for air:

$E = 10.5$ kcal/mole at $\sigma = 3440$ N/mm^2

The activation energy in air is approximately the same as that in water for the same applied stress. This must be so, if the activation energy is dependent on tip stress as stated above.

The activation energy for the diffusion of water in unstressed fused silica glass is approximately 18 kcal/mole. Thus, the activation energy for corrosion is lower for high stress levels, as shown above. From these ln (t) vs. $1/T$ curves it can be seen that the lines are not parallel. This means that n is temperature-dependent for a given environment. By calculating n from the equation

$$n = \frac{\log (t_2/t_1)}{\log (\sigma_1/\sigma_2)}$$

and using the curves, n is found to be 17 at 22 °C and 13.5 at 100 °C. The value of 13.5 at the higher temperature indicates that the reactivity is accelerated by increased temperature. A change in temperature (or environment) does not affect the activation energy, which is not dependent on either temperature or concentration of reactants. However, for a specified stress, the rate of the corrosion reaction may be changed by either temperature or reactant concentration. Thus, above the fatigue limit, the time to failure is decreased by an increase in either the temperature or the water concentration at the crack tip.

Conclusions

The following evidence summarizes the fibre-ageing characteristics:

(a) Water is a determining factor affecting the ageing characteristics of silica fibres.
(b) In plastic-coated silica fibres there is a drastic increase in fatigue resistance below pH = 0. Solutions with pH > 0 have an effect similar to that of water.
(c) n is dependent on both water-vapour concentration and on temperature.
(d) The activation energy of the water reaction with silica is a function of the stress applied to the fibre.

This evidence can be interpreted as supporting the chemical theory of fatigue in glass proposed by Hillig and Charles. The near-perfect specimens enable very consistent results to be obtained. These results contribute towards a better understanding of fatigue.

5.4.5 Minimum lifetime

The minimum lifetime of a fibre under load can be estimated from the dynamic fatigue test results and by using a proof test (see Section 5.6.2). After the dynamic fatigue test, the median strength $\bar{S}$ and standard deviation at a number of constant stressing rate $\dot{\sigma}$ are determined. A least-square analysis will yield the equation:

$$\ln \bar{S} = a - b \ln \dot{\sigma} \qquad (5.6)$$

An example is given in Figure 5.26 with the dynamic test results given in Figure 5.27. In order to predict minimum lifetime, the inert strength S_i (when fatigue is absent) must also be determined. This can be obtained by carrying out the dynamic

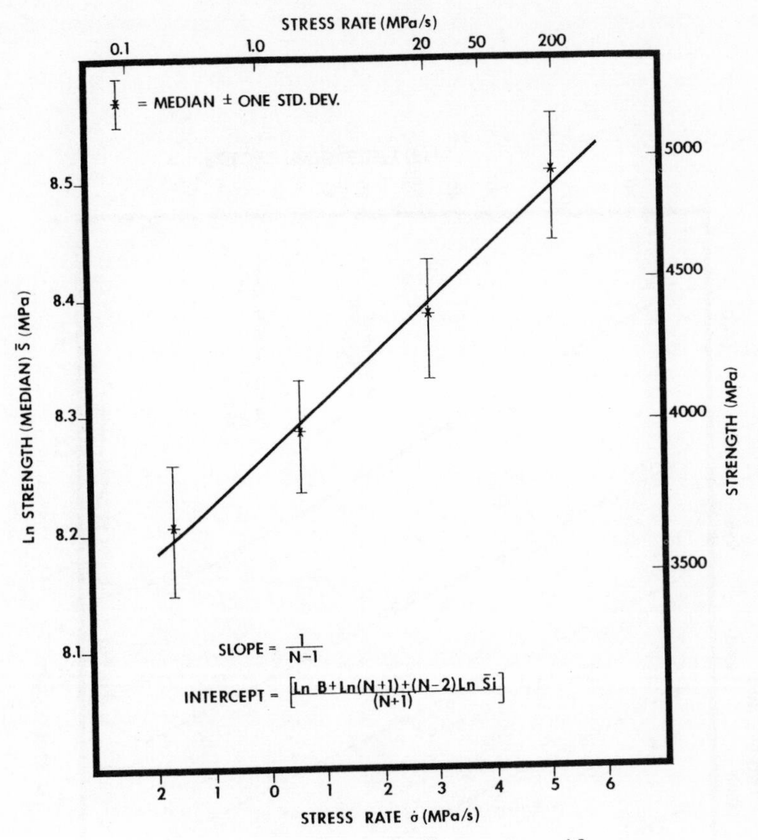

Figure 5.26. Typical ln $\bar{S}$ vs. ln $\dot{\sigma}$ plot[18]

fatigue test at a high stress and in a dry environment (a typical result is shown in Figure 5.27). This gives the Weibull parameter m from the equation with S_0, a scaling parameter.

$$\ln\{\ln[1/(1-F)]\} = n \ln(S_i/S_0) \tag{5.7}$$

From fracture mechanics theory, fracture strength is related to stress rate by

$$\ln S = \frac{1}{n+1}[\ln B + \ln(n+1) + (n-2)\ln S_i + \ln\dot{\sigma}] \tag{5.8}$$

where S_i is the strength in an inert environment and B and N are constants for a given environment and material composition. In Equation 5.7 the inert strength is a measure of the initial flaw size and B and n are parameters that characterize subcritical crack growth. By comparison of Equation (5.8) with Equation (5.7)

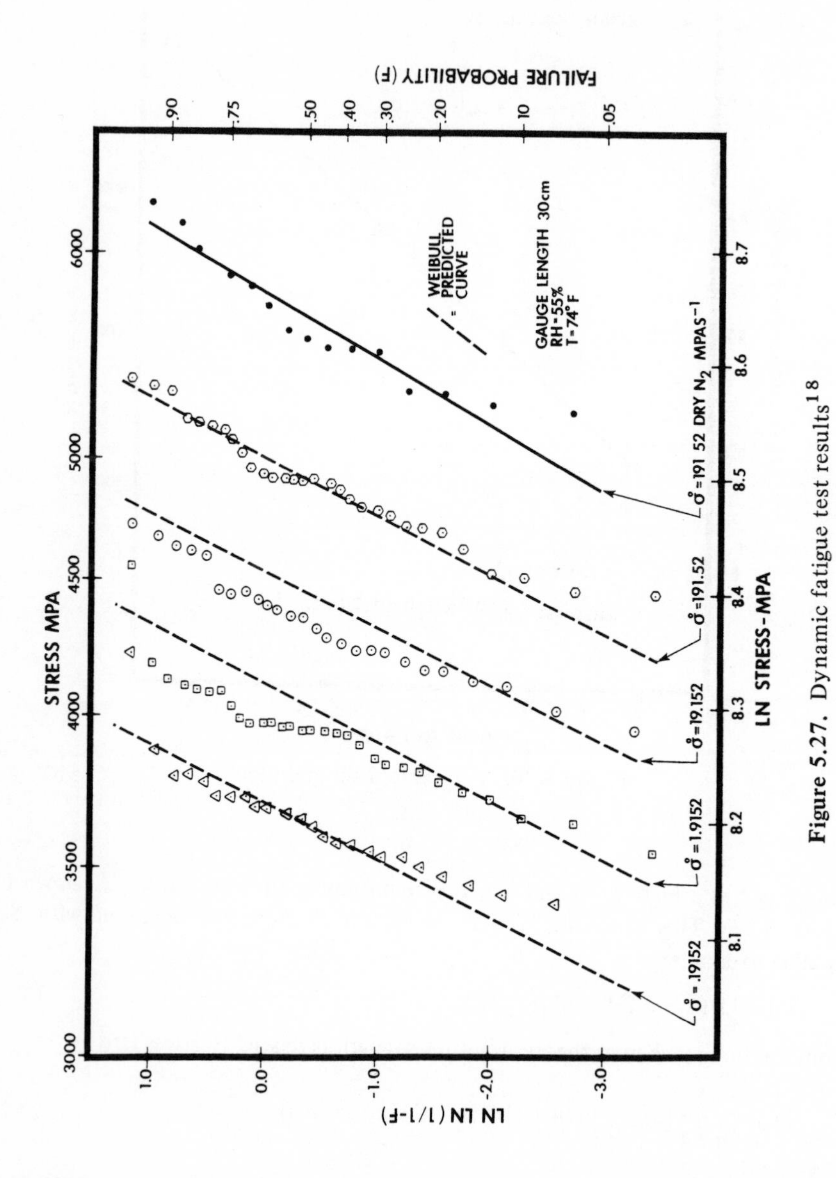

Figure 5.27. Dynamic fatigue test results[18]

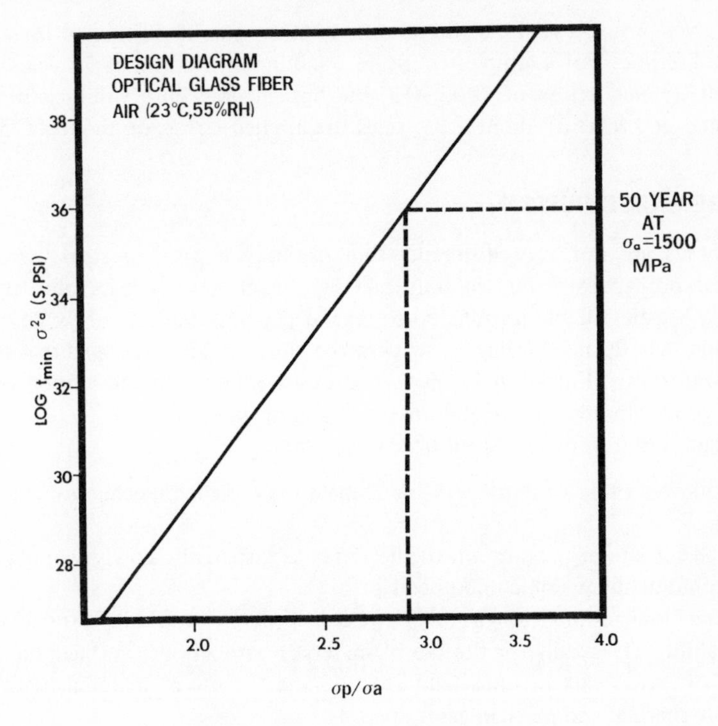

Figure 5.28. Minimum lifetime[18]

the crack-growth parameters n and B can be obtained from

$$\frac{1}{n+1}[\ln B + \ln(n+1) + (n-2)\ln \bar{S}_i] = a$$

$$\frac{1}{n+1} = b$$

where $\bar{S}_i$ is the median inert strength.

When n and B are known the minimum lifetime after a proof test can be estimated.

To illustrate lifetime predictions after proof testing, a design diagram has been constructed in Figure 5.28 based on the minimum lifetime after proof testing (t_{min}) which is given by:

$$t_{min}\sigma_a^2 = B(\sigma_p/\sigma_a)^{N-2} \tag{5.9}$$

From Equation 5.9 it is seen that a plot of ln ($t_{min}\sigma_a^2$) vs ln (σ_p/σ_a) is a straight line with a slope of $N-2$ and an intercept of B. Such a design diagram can be used to determine the required proof stress to guarantee a minimum lifetime in service, or for a given proof-test stress, the allowable stress can be calculated for a given minimum lifetime. For example, to assure a minimum lifetime of 50 years under a constant applied stress of 1500 MPa, the optical-glass fibre would have to be proof tested at a level of about 2.85 times the applied stress, or about 4275 MPa.

5.5 FIBRE CABLE DESIGN

Optical fibres are not very different from ordinary copper or steel wires. The mechanical properties differs in only one significant way. A glass fibre remains completely elastic for elongations up to several percent and has hardly any yield. This means that during cabling, a twist in the fibre could cause perturbations on the lay uniformity. Furthermore, as the fibres are rather thin the tension control has to be good, otherwise loose and uneven stranding will occur.

The major criteria for successful fibre cabling are:

(a) to achieve a cable structure which will meet the designed mechanical characteristics;
(b) to transfer lateral pressure on to the fibres as a spatially slowly varying force, thus minimizing possible microbends;
(c) to have low residual longitudinal stress in order to minimize fatigue failure probability. This calls for the use of materials with smooth surface, the avoidance of cross-overs in fibres and spacers, and the use of materials with compatible thermal and mechanical properties.

5.5.1 Application requirements

Fibre cable types range from lightweight, high-strength cable, to fully armoured cable for a variety of applications. Some examples are: weapon-guidance cables and shallow-sea armoured cables. For cables with very stringent requirements in overall size and load strength, special fibres with ultra-high strength and resistance to microbending may be called for.

The ideal strength member for optical-fibre cable should have similar elongation characteristics as the optical fibres. This means that pristine fibres could be used as strength members. Alternatively, materials with high Young's modulus, such as high-tensile steel and Kevlar, are good possibilities. They will withstand 2–3% elongation. Carbon fibre can be used advantageously, although the limited elongation could cause the cable to fail on high-impact loading when significant elongation is expected. The reason for using a material with high modulus and a high elongation as a strength member is to achieve a cable of minimum cross section. If the overall dimension of the cable is not critical then the choices of strength member become less critical.

Fibre cables may be designed with the fibres closely supported or deliberately laying loose within the cable. The former provides a stable structure which can be made gas-tight, while the latter ensures little or no microbends but has a more

Table 5.3 Fibre cables – technology issues

• Without metal parts	• Strength member	– Kevlar
		– Glass fibre
		– Elastomer
	• Moisture shield	– None
	• Rodent protection	– None
• With metal parts	• Strength member	– Steel
	• Moisture shield	– Al tape
	• Rodent protection	– Steel tape
• Construction	• Tight	– Stable configuration
		– Homogeneous material
		– High-NA fibre
	• Loose	– Unstable configuration
		– Not gas-tight
		– Larger cross section
	• Armoured	

unpredictable dynamic characteristics. The use of a metal outer shield and/or pressurization can reduce the rate of moisture penetration into the fibre, thus reducing fatigue effects. The metal sheath also acts as a protection against attacks by rodents. Cable design issues are listed in Table 5.3.

5.5.2 Strength members

Tensile strength is of special importance in long cables (for example, cables installed in ducts) where the tension required for pulling-in increases throughout the process as a function of the coefficient of friction between the cable sheath and adjacent surfaces. For straight ducts the tension increases linearly with the length, but for curves in ducts an additional exponential increase occurs which can substantially raise the pulling-in tension. A figure of merit for tensile properties is provided by the ratio of the tension at a standard strain to the weight per unit length and the tensile strength achievable per unit cross-sectional area: this applies to individual components as well as the completed cable. Preferred features of a strength member are therefore:

(a) high Young's modulus;
(b) strain at yield greater than the maximum designed cable strain;
(c) low weight per unit length;
(d) flexibility, to minimize restriction of the bending capability of the cable.

Other features that may be relevant include friction against adjacent components, transverse hardness, and stability of properties over a range of temperature including those encountered during cable manufacture.

High-modulus materials are inherently stiff in the solid form, but flexibility can be improved by employing a stranded or bunched assembly of units of smaller cross section, preferably with an outer coating of extruded plastic, helically applied tape, or a braid. Such a coating is particularly necessary if the strength member comes into contact with coated fibres, since a resilient or smooth contact surface is required to avoid optical losses due to microbending, a phenomenon commonly observed in fibres subject to localized mechanical stress.

Some main types of material for the construction of strength members on account of their high Young's moduli are steel wires, plastic monofilaments, multiple textile fibres, glass fibres and carbon fibres. Some significant features of these materials are summarized below:

Steel wires

These have been widely used in conventional cables for armouring and longitudinal reinforcement. Various grades are available with tensile strengths to break ranging from 540 to nearly 3100 MNm^{-2}. All have the same Young's modulus (19.3×10^4 MNm^{-2}) and the choice is guided by the preference for a high strain at yield compatible with that of optical fibres. The main disadvantage of steel is its high specific gravity which substantially adds to the cable weight.

Plastic monofilaments

These are available commercially in several basic materials. Specially processed polyester filament exists which combines a high elastic modulus (up to 1.6×10^4 MNm^{-2}) with good dimensional stability at elevated temperatures, and a smooth cylindrical surface. This type of strength member is of particular interest where low weight or absence of metals are prime requirements for the cable, but is not technically competitive with steel for cables to be installed in long ducts (500 to 1000 m).

Textile fibres

Commercial forms normally consist of assemblies of many small-diameter fibres laid up in twisted or parallel configurations. Typical examples in conventional cables are polyamides (Nylon) and polyethylene terephthalate (Terylene, Dacron, etc.), with elastic moduli which may be as high as $1.5 \times 10^4 MNm^{-2}$ for the individual fibres. Due to the large number of individual fibres, they are resilient in a transverse direction and are useful as cable fillers and binders as well as providing improved tensile properties in optical-fibre cables, but are more bulky than monofilaments of equivalent strength. An exceptional member in this class which has been widely employed in optical cables is Kevlar, an aromatic polyester. The individual fibres have the exceptionally high modulus (for an organic material) of up to $13 \times 10^4 MNm^{-2}$ which, coupled with its specific gravity of 1.45, gives it an effective strength-to-weight ratio nearly four times that of steel. Commercial forms of Kevlar suitable for cable reinforcement consist of composites of large numbers of single filaments assembled by twisting, stranding, plaiting, etc., and/or resin bonding, and retain a high proportion of the single-fibre modulus.

Glass fibres

For some applications the optical fibres may supply sufficient tensile strength, but additional nonactive fibres can be used, generally in a manner similar to textile fibres, if higher strength is required. Elastic modulus is high, typically 9×10^4 MNm^{-2}.

Carbon fibres

These have been successfully employed in rigid and semi-rigid plastic or metal composites, and have a modulus of up to 20×10^4 MNm^{-2} in single filaments. Relevant properties of the above materials, with the exception of carbon fibre, are summarized in Table 5.4.

Table 5.4 Properties of strength-member materials (after Foord)

Material	Specific gravity	Young's modulus (MN m^{-2})	Tensile strength (MN m^{-2})	Strain at break (%)	Normalized modulus-to-weight ratio
Steel wire	7.86	19.3×10^4	5 to 30×10^2	25 to 2	1.0
Polyester monofilament	1.38	1.4 to 1.6×10^4	7 to 9×10^2	15 to 6	0.3
Nylon yarn	1.14	0.4 to 0.8×10^4	5 to 7×10^2	50 to 20	0.3
Terylene yarn	1.38	1.2 to 1.5×10^4	5 to 7×10^2	30 to 15	0.3
Kevlar *49* fibre	1.45	13×10^4	30×10^2	2	3.5
Kevlar *29* fibre	1.44	6×10^4	30×10^2	4	1.6
S-glass fibre	2.48	9×10^4	30×10^2	3	1.4

In practical cables, the other components may add considerably to the weight but relatively little to the strength, and this most strongly affects the modulus-to-weight ratio when the strength member has a low specific gravity. The net effect is that the optimum modulus-to-weight ratio is generally achieved with steel wires and Kevlar 49, followed some way behind by glass and Kevlar 29, the remainder coming a poor third. The ultimate selection of material for the strength member depends upon the relative importance of cost, mechanical performance and the acceptability of a metallic component.

5.5.3 Fibre packaging

Surprisingly small external forces can cause lateral deformations, mode coupling, and optical loss in clad fibres. The pressure exerted on the individual fibre in a cable will almost certainly have an effect on optical loss. The lowest loss values are measured almost invariably in connection with extremely small mode coupling and after carefully eliminating external forces on the fibre. Maintaining these loss values in a cable may require better fibre and, more importantly, effective jackets optimally designed to shield against external forces.

Elastic deformations

Consider the simple model of a fibre pressed against an elastic plane surface that is slightly rough (Figure 5.29). The pressure from above is uniform, but as a result of the roughness, the contact forces between the fibre and the surface are not uniform along the fibre. Thus, the fibre bends slightly, yielding to a force $f(z)$ per unit length. According to the theory of the thin elastic beam, the lateral displacement $x(z)$ of the fibre axis is related to $f(z)$ by

$$\frac{d^4x}{dz^4} = \frac{f(z)}{H}$$

where $H = EI$ is the flexural rigidity or stiffness; E is Young's modulus and I the moment of inertia. For the circular cross section of the fibre

$$I = \frac{\pi}{4} a_1^4$$

where a_1 is the radius of the fibre.

The force $f(z)$ not only causes a bending action, but also a deformation $u(z)$ of the surface. Provided that $f(z)$ does not change too drastically along z, $u(z)$ is a linear function of the applied force. This forms the basis of an analysis which leads to the comparison of four fibre coatings. The first is made entirely from a soft plastic, the second from a hard plastic, and the third and fourth are hybrid structures. We assume a modulus of 1 kg/mm^2 for a typical soft material and 100 kg/mm^2 for a typical hard material. In Figure 5.30, the outer-jacket radius a_2 is plotted versus the excess loss computed for each structure if the mean lateral pressure is 0.1 g/mm. The pressure obviously determines the choice between a soft or a hard material, if the jacket is to be made from one material alone. The decrease

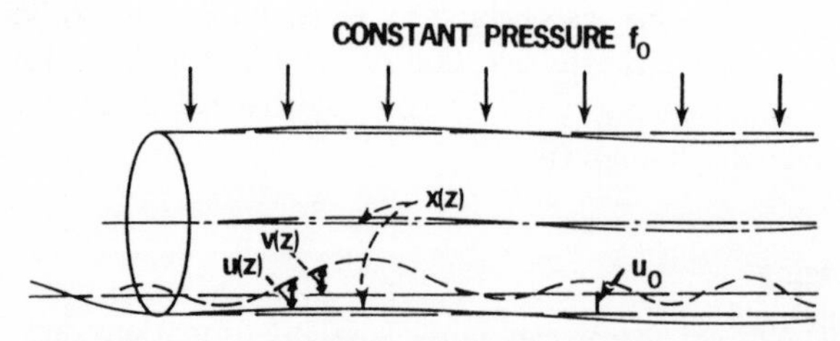

Figure 5.29. Fibre pressed against a rough surface[22]

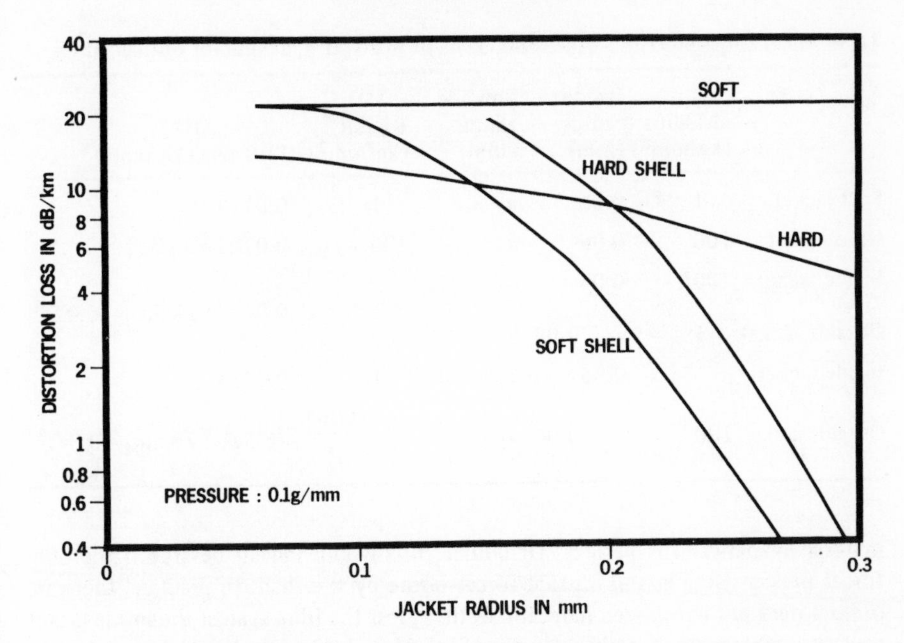

Figure 5.30. Distortion loss vs. outside-jacket radius

of the loss contribution with increasing jacket radius in case of the hard jacket comes about as a result of the increase in stiffness. The corresponding increase afforded by the soft jacket is negligible. The last two columns of Table 5.5 list the rigidity and stiffness parameters used in each case.

The third structure has a hard jacket padded with a soft outer layer. The layer thickness of 20 μm was chosen to avoid any deformation beyond its elastic limit. The fourth structure has a hard shell surrounding a soft material. The thickness of this shell should be approximately $0.02 a_2$. This optimum is a result of an increase both in stiffness and lateral rigidity as the shell thickness is increased. The two pairs of D–H values listed in the case of the fourth configuration refer to the two independently deforming structures (shell and fibre). The slight advantage of the soft over the hard shell, evident in Figure 5.30, is too small to be decisive. It may well be off-set by weight and cost considerations. The substantially improved fibre protection afforded by the hybrid structures as compared to simple jackets, however, is well worth considering. A jacket diameter of 0.5 to 0.6 mm permits a virtual elimination of the distortion loss in the example considered here. A similar reduction by a single hard jacket requires at least twice this jacket diameter.

The excess loss computed for the structure with a hard shell vanishes when the modulus of the inner jacket is reduced to zero. This implies that the protection provided by a stiff shell that surrounds the fibre in a loose way without any

Table 5.5 Characteristics of several types of protective jackets for optical fibres

	Modulus (kg/mm²)	Inside radius (mm)	Outside radius (mm)	(D) Rigidity (kg/mm²)	(H) Stiffness (kg/mm²)
Soft jacket	1	0.06	a_2	1	0.0713
Hard jacket	100	0.06	a_2	100	$0.0702 + 78.5a_3^4$
Inside jacket	100	0.06	$a_2 - 0.02$	1	$0.0702+78.5(a_2 - 0.02)^4$
Outside jacket	1	$a_2 - 0.02$	a_2		
Inside jacket	1	0.06	$a_2 - 0.04$	1	0.0713
Outside jacket	100	$a_2 - 0.04$	a_2	$1 + \frac{0.0064}{a_2^3}$	$78.5a_2^4 - 78.5(a_2 - 0.04)^4$

material in between is perfect. Of course, this would indeed be true if the only forces present were lateral outside forces borne by the shell. In practice, there are other forces not considered here; forces that press the fibre against the inside jacket wall in a cable bend, for example. Such forces could determine the distortion loss of the loosely jacketed fibre.

5.5.4 Cable designs[19–28]

Sufficient detail is now available to define the principal features of a practical cable design as follows:

(a) Strength member that enables the cable to be held under a high tension at low strain (0.5% to 1.0%) and with sufficient flexibility to allow bending round small radii.
(b) Plastic-coated optical fibres of high intrinsic strength, arranged within the cable so that when bent or stretched the fibre strain remains within the limits set for the cable.
(c) Other components to protect the coated fibres from microbends and environmental influences which may cause deterioration of their optical performance.

The ideal position for either of the first two components to provide maximum flexibility with minimum longitudinal strain is at the neutral axis of the cable, and consequently two types of construction are possible with one component arranged around the other. Both systems have specific advantages. A centrally located strength member provides maximum flexibility, but if placed around the fibres may protect them from radial crushing forces. Both types of construction are shown in Figure 5.31, and a number of variations have been developed.

Some structures of fabricated optical-fibre cables are shown in Figure 5.32.

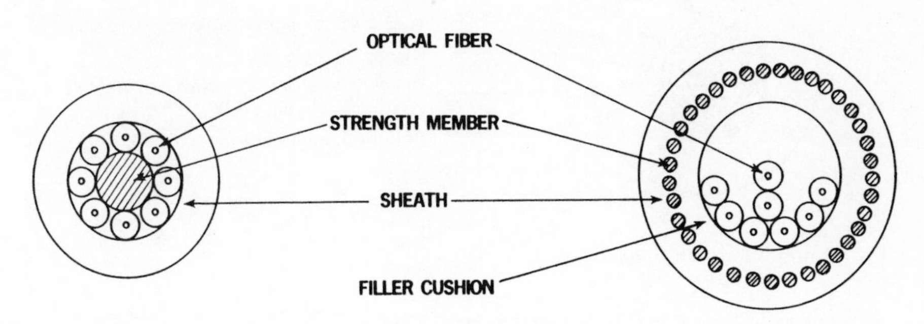

Figure 5.31. Typical constructions used for optical-fibre cables (R. J. Baskett, S. G. Foord)

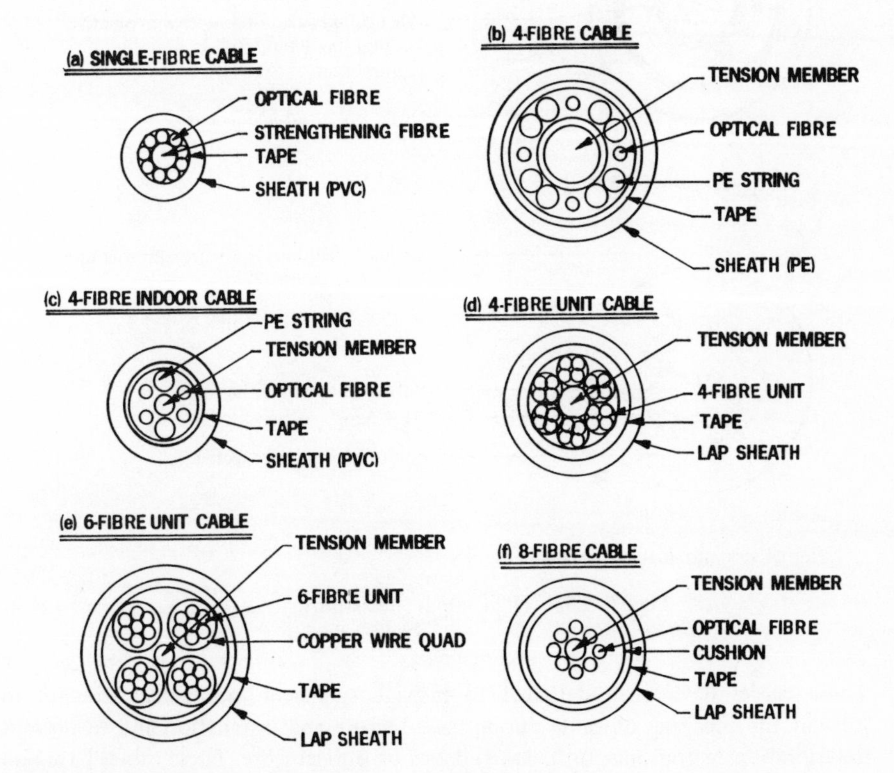

Figure 5.32. Cable structure (Dr M. Hoshikawa, Sumimoto Industries, Ltd, 'A Study on Optical Fiber and Fiber Cable')

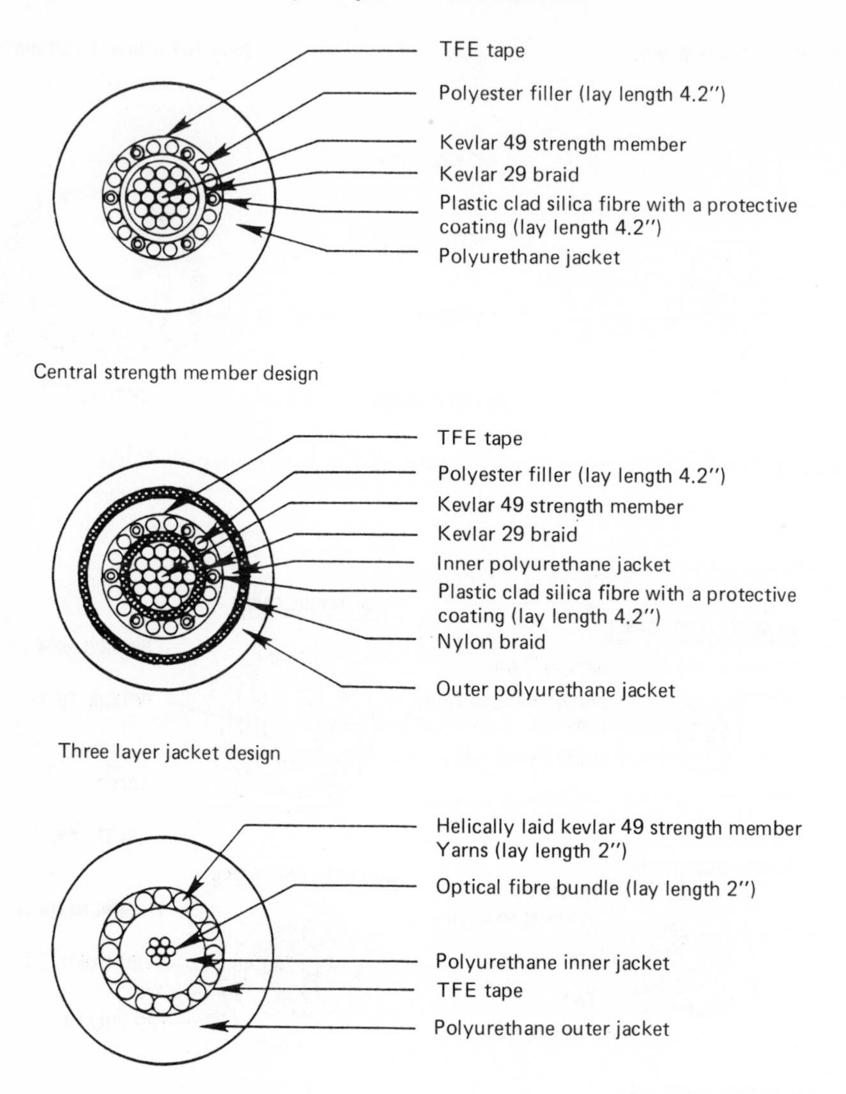

Figure 5.33. High impact-resistant cables

These cables have been designed to provide sufficient mechanical strength to prevent the breaking of cable during manufacture and installation and to prevent deterioration of transmission characteristics of optical fibre. These cables have also been designed to reduce the stress or strain on fibres which could be caused by bending, tension and side pressure.

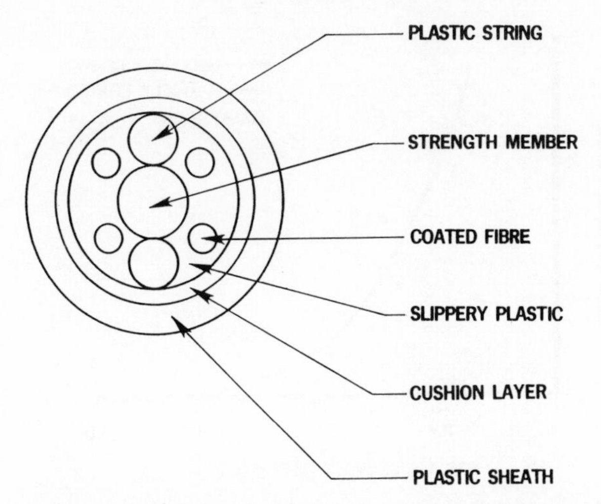

Figure 5.34. Cable structure (Dr Masao Hoshikawa, Sumimoto Electric Industries, Ltd)

For good impact resistance, two basic approaches can be followed: fibres laid against a hard surface; and heavy plastic-jacketed cable core. Typical designs are as shown in Figure 5.33.

To illustrate in more detail the microbending effects in cable the following example is given. Step-index fibres with the following parameters were fabricated into a cable to study the influence of microbending and refractive index difference:

Core diameter	60 μm
Cladding thickness	15 μm
Fibre diameter	150 μm
Diameter of coated fibres	1.0 mm

The cross-sectional view of the cable is shown in Figure 5.34. A double-coated and triple-coated fibre shown in Figure 5.35 was used. The additional losses due to

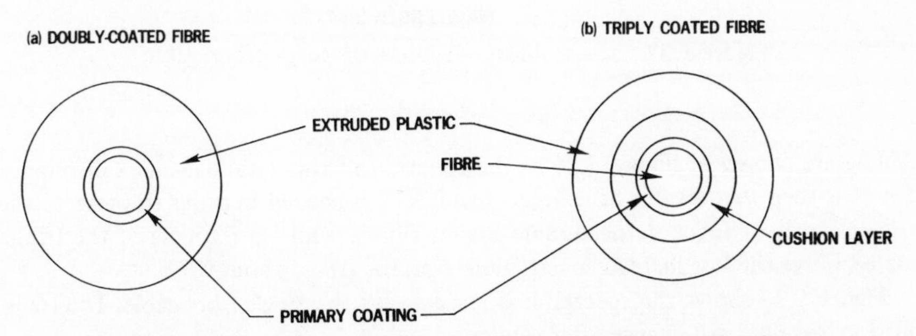

Figure 5.35. Cross-sectional views of two kinds of coated fibres

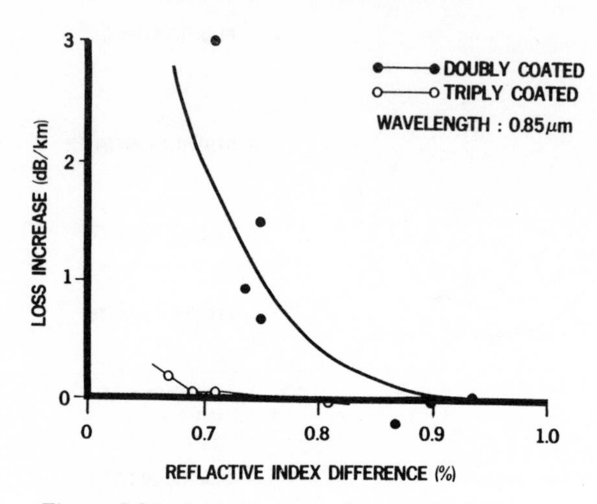

Figure 5.36. Loss increase due to cabling vs. refractive index difference

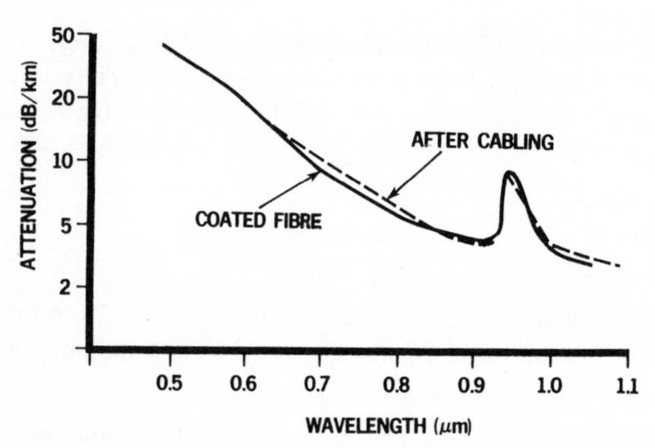

Figure 5.37. Spectral-loss response of single-fibre cable

cabling are shown in Figure 5.36 as the function of the refractive-index difference Δn. It is seen that a value of n larger than 0.85% is needed in order to preserve the original level of loss for the double-coated fibres, while in the case of the triple-coated fibres the loss increase is negligible even for Δn of about 0.7%.

Figure 5.37 shows the spectral-loss response of the single-fibre cable. The transmission loss does not change after cabling.

5.6 FIBRE AND CABLE TESTING

Optical transmission characteristics and mechanical properties of fibres and fibre cables are measured by a variety of techniques. The methods for evaluating optical characteristics are applicable for both fibre and fibre cables while the methods for evaluating mechanical properties are distinct.

5.6.1 Measurements of optical characteristics of fibre waveguides

Attenuation measurements

Techniques are available for measuring total attenuation in fibres. One is to measure the optical power transmitted through two lengths of a fibre with input-launching and output-detection conditions held invariant. The second technique employs an optical time-domain reflectometer.

Insertion loss

The optical power at a propagation distance X in the fibre waveguide can be expressed as

$$P(X) = P(X_1)\exp\left[-\int_{X_1}^{X} \alpha(X)\,\mathrm{d}X\right] \tag{5.10}$$

where $\alpha(X)$ is the loss coefficient and $P(X_1)$ is the power in the fibre at position X_1. If the loss coefficient is constant, the attenuation coefficient is given by

$$\bar{\alpha}(X_1, X_2) = \frac{1}{X_2 - X_1}\int_{X_1}^{X_2} \alpha(X)\,\mathrm{d}X \tag{5.11}$$

where X_1 and X_2 are the two fibre lengths used in the measurement. This loss equals the constant steady-state attenuation of the fibre provided that each of the modes has the same attenuation. Individual modes, however, have different losses. High-order modes, with higher penetration in the generally lossier cladding and more coupling to unguided modes via guide imperfections, have more loss than lower-order modes. This differential loss tends to be statistically compensated along the fibre by modal coupling. In fact, when complete compensation is achieved, a steady-state mode distribution is established which propagates with a characteristic attenuation constant.

Optical time-domain reflectometry

The loss-measurement techniques discussed above provide the averaged insertion loss for a given fibre length, but give no information concerning the length dependence of the loss. An optical time-domain reflectometer can be used to measure attenuation and allows the length dependence of the fibre attenuation to be displayed. The technique, based on the analysis of back-scattered light in the fibre, requires neither cutting the fibre nor access to both ends of the fibre. It is, therefore, convenient for use with cabled fibres. In the back-scattering experiments, a pulse of light is launched into the fibre in the forward direction using either a directional coupler or a system of external lenses and a beam splitter.

The waveform of the return light pulse is detected by a photodetector and processed with a boxcar integrator. The return waveform consists of three distinct segments:

(a) an initial pulse, which results from any non-directionality in the input coupling mechanism;
(b) a long tail caused by the distributed Rayleigh scattering that occurs as the input pulse propagates down the fibre; and
(c) pulses caused by the discrete reflections that may occur along the fibre length as a result of fibre imperfections, in-line connectors, a break, or the Fresnel reflection incurred from the end of the fibre.

The Rayleigh-scattered return can be used to extract the attenuation coefficient. The time dependence of the detected back-scattered power can be converted to a length dependence by multiplying by the velocity of light in the fibre core. The detected back-scatter power may be expressed as

$$P(X) - kP(0) \exp[-2\bar{\alpha}(X)X] \tag{5.12}$$

where X is a position along the fibre length, k is a constant, $P(0)$ is the power launched into the fibre at the input, and $\bar{\alpha}(X)$ is the average total attenuation coefficient of the forward and back-scattered signals.

Delay distortion

Delay distortion can be measured by one of several techniques, either in the time domain (impulse response measurements) or in the frequency domain (transfer function measurements).

Single-pass impulse response measurements A narrow pulse of light is injected into one end of a fibre, and the broadened output pulse is detected at the other end by a fast detector that is sensitive and linear, and computational facilities are required for deconvolving the received pulse obtained with the unknown fibre from the pulse that is received with a short fibre substituting for the unknown fibre. Launch conditions are as important in dispersion measurements as in loss measurements. Since the launch affects the distribution of energy in the various modes, it affects the impulse response measurement. The impulse response should ideally be measured under various launch conditions (different angles, spot sizes, etc.) to obtain a complete fibre characterization. In practice, however, this is too difficult and time-consuming to implement. A viable alternative that gives realistic and repeatable results is to use mode mixers at the transmit end in an attempt to launch the steady-state distribution.

Direct measurement of the transfer function An alternative to producing narrow pulses of high light is to modulate a CW light signal sinusoidally around a fixed level. The reduction in modulation index associated with propagation through the unknown fibre represents the roll-off of the transfer function at the modulation frequency. There are advantages to this technique: the transfer function is obtained directly without Fourier-transforming time-domain data. Sources that operate CW, but with limited output into a fibre (like narrowband incoherent sources), are more efficiently utilized (higher signal-to-noise ratio at the receiver). This technique has been used to measure the fibre-transfer function at various wavelengths using an incoherent light source, narrowband optical filters and an optical modulator to generate the sinusoidally varying light power.

In principle, both the amplitude and the phase characteristics of the transfer function can be obtained with this technique. But, in practice, only the amplitude variation is recorded.

5.6.2 Measurement of mechanical characteristics of fibre waveguides

Fibre-strength measurement

The measurement of fibre strength is complicated by the statistical variation of fibre strength along the length of a fibre (due to the random nature of flaw depth and spatial distribution). If sufficient data can be gathered that the statistics of the large but rare flaws are characterized, then extrapolation of statistical values of fibre tensile strength of one test gauge length to that of another gauge may be defined with known confidence limits. Otherwise, extrapolation is at best fortuitous. On the other hand, if the nature of the rare flaws are known, then more suitable testing techniques may be evolved. What is desired is to have test procedures which enable reliable long-length strength to be predicted, while the procedures should be simple and should not destroy more fibre than necessary.

Strength-test plan

The mechanical strength-test plan for high-strength fibre defines the procedures of performing a series of strength tests and reduction of data obtained from these tests so that fibre strength may be determined and appropriately specified. The test plan calls for dynamic fatigue tests, static fatigue tests and proof tests. Data are gathered and processed (using Weibull statistics) to provide a coherent and correlated set of results from which fibre strength specifications may be drawn.

Dynamic fatigue test The dynamic load test measures the mechanical tensile strength of a fibre at a defined constant rate of loading. The loading rate is expected to affect the average strength and strength distribution of a fibre, since fibres in general suffer strength degradation due to stress-induced corrosion (fatigue). For a high-strength fibre, the fibre strength spread is expected to be essentially unimodal and narrowly confined. These tests could be carried out with different specimen gauge lengths. In order to obtain statistically relevant data, at least 20 and preferably more than 40 samples should be tested. Different rates of loading spanning several orders of magnitudes, e.g. 200 MPa/s to 0.2 MPa/s, would provide the required data.

The test equipment is illustrated in Figure 5.38. The fracture load is applied through a constant-speed drive motor at a constant rate until fracture occurs and the load is recorded, and the diameter of the broken fibre end is measured and recorded.

Static fatigue testing Static fatigue testing is a method of studying the fatigue phenomenon in which the fibre is subjected to a constant stress and time to failure is measure. Two techniques may be used: axial loading, and bending stress.

The test equipment for axial loading is illustrated in Figure 5.39b. An ensemble

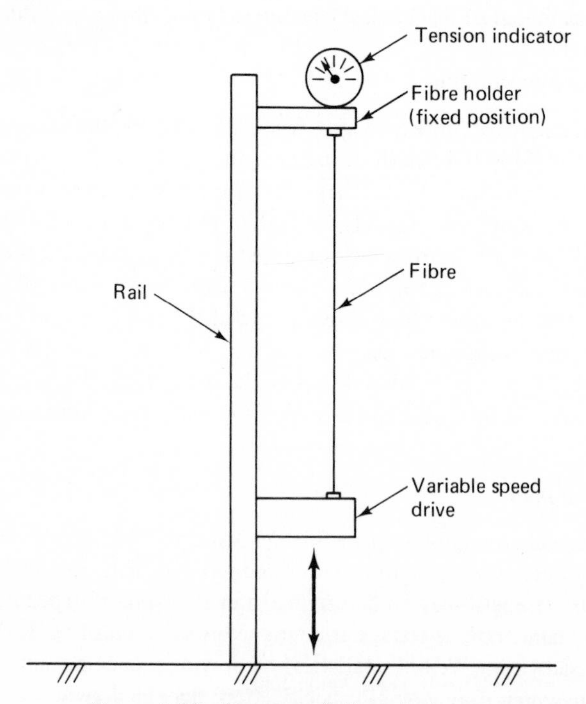

Figure 5.38. Dynamic fatigue test

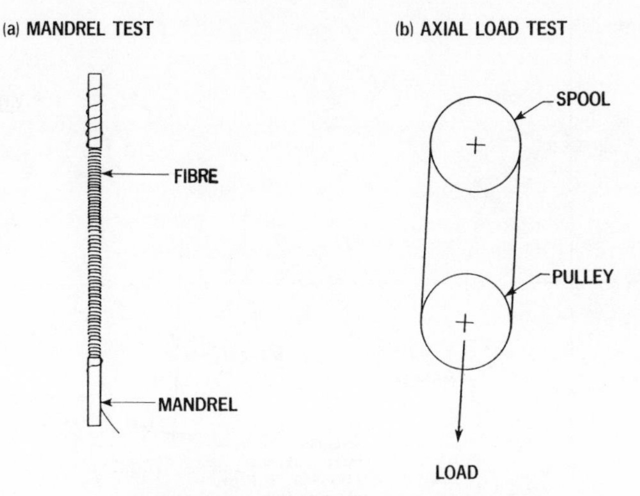

Figure 5.39. Fatigue-testing techniques

of specimens is used for each fibre sample. Each fibre is attached by wrapping each end several times around the top cylindrical support, and secured. A load is applied to the lower cylinder. The top cylinder and the cylinder to which the weight is attached have a sufficiently high diameter to avoid introducing excessive additional bending stresses. All fibre loops are made first and then weights are placed in position gradually while avoiding impulsive loads. The times to failure can be conveniently recorded by monitoring the time to fracture of the fibres.

The test equipment for bending stress consists of precision mandrels of different diameters. Plastic-coated optical fibres are subjected to flexural stresses by winding around a mandrel (Figure 5.39a). The stress level can be varied by the proper choice of the mandrel size, and is calculated as follows. Take R, the radius of the mandrel, r the radius of the fibre core and t, the thickness of the plastic cladding (Figure 5.40), then the elongation at the outermost portion of the core surface is calculated as follows: if L_0 is the neutral axis length, and L the maximum length of the fibre under bending stress, then the elongation is $(L - L_0)$. Hence, the strain is

$$\epsilon = \frac{L - L_0}{L_0} = \frac{(R + t + 2r)\theta - (R + t + r)\theta}{R + t + r)\theta} = \frac{r}{R + t + r} \tag{5.13}$$

and the stress is

$$\sigma = E\epsilon = \frac{Er}{R + t + r} \tag{5.14}$$

where E is the Young's modulus (typically, $7.2 \times 10^{10}\,\mathrm{N/m^2}$ for fused silica).

Convenient mandrel sizes for testing and the corresponding flexural stresses are

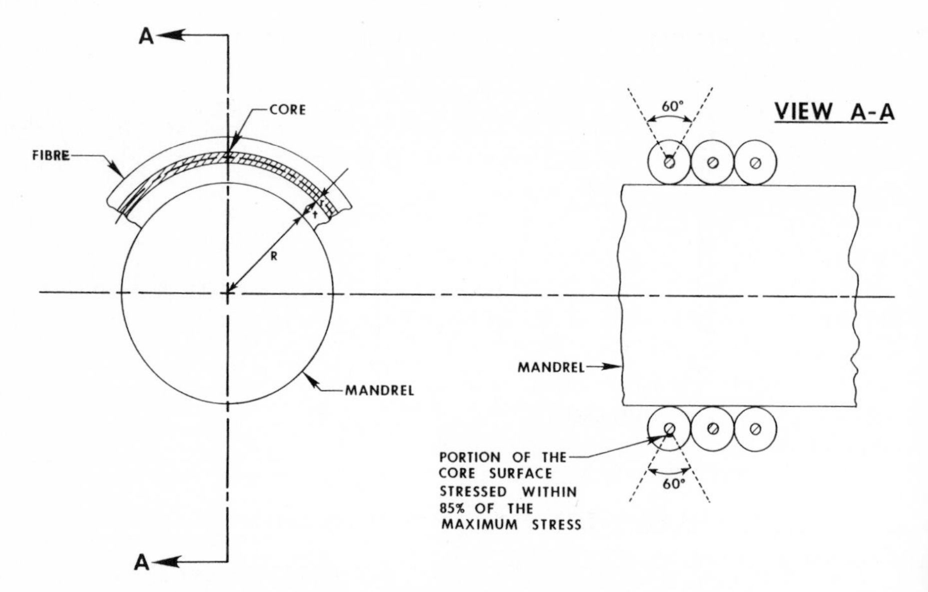

Figure 5.40. Fibre on mandrel test

shown in Table 5.6, for a fibre diameter of 125 μm (5 mils). In this test, only a fraction of the fibre surface (one-sixth, equivalent to an angle of 60°, see Figure 5.40) is subjected to tensile stress within 85% of the maximum stress. Hence, the surface of the specimen fibre of length L under test is equivalent to a specimen of $L/6$ in length. However, despite this reduced length, the shorter specimen length is of little consequence to enable meaningful fatigue results to be established, since it appears that the dominant flaws of a strong fibre are distributed all along the surface.

An ensemble of several specimens of a fibre is used for each mandrel size. These are wound over the mandrel to form a single-layer winding. Care is taken to avoid

Table 5.6 Mandrel sizes and equivalent stresses for the bending-stress fatigue test

Mandrel diameter		Equivalent stress		
millimetres	inches	N/mm^2 for 125 m fibre diameter	psi for 0.005 in. fibre diameter	Equivalent % elongation
1.55	0.062	4152	6.02×10^5	6
2.08	0.083	3311	4.8×10^5	4.8
2.38	0.095	2972	4.31×10^5	4.3
2.73	0.109	2655	3.85×10^5	3.8
3.13	0.125	2365	3.43×10^5	3.4

introducing unwanted tensile stress during winding through axial tension. Time to failure of the first fracture observed visually or monitored through transmitted light of each fibre is recorded and the diameter of the fibre at the fracture measured. The occurrence of fracture is contained by the plastic clad and stress is maintained at all parts of the sample except perhaps the immediate vicinity of the fracture (~1 mm). This differs from the axial load test, where all stresses are removed after fibre fracture. The mandrel size and the fibre diameter enable the maximum stress at the surface to be calculated. The environmental conditions surrounding the mandrel can be readily established to a range of temperature, humidity and concentration of chemical reactants and these can be established incomparably more easily than using the axial load method. However, especially with the smallest mandrel diameter, there is a small error margin in determining the time to failure, since at a high stress, time to failure is comparable with the necessary time to wind the fibre around the mandrel.

The principle of this technique is illustrated by the following theory:

Time to failure is governed by the stress concentration at a flaw site. This can be expressed:

$$t_2 = (\sigma_1/\sigma_2)^n t_1$$

where t_1 is the time to failure at a reference stress concentration σ_1, and t_2 is the time to failure at the applied stress concentration σ_2. It is assumed that surface flaws only play a part in the mandrel test, because bulk flaws are subjected to lower stresses (see Equation 5.14) and may not suffer from stress-induced corrosion. n has been found to be anywhere from 15 to 25, depending on the environmental conditions.

From the theory of Griffith, the stress required to produce a fracture is inversely proportional to the square root of the depth of the flaw. If the fibre has a certain class of flaw the stress intensity for fracture to occur is defined as follows. Let σ_1 be the stress intensity for the first failure to occur in time t_1. If the flaw size changes by 20%, σ_1 will change by 10%.

$$\sigma \propto 1/\sqrt{L}$$

$$\frac{\Delta\sigma}{\sigma} = \frac{1}{2}\frac{\Delta L}{L}$$

Let $\sigma_2 = \sigma_1/1.1$ be the stress intensity for failure to occur in time t.

Then

$$t_2/t_1 = (\sigma_1/\sigma_2)^n = (1.1)^n$$

For

$$n = 24, t_2/t_1 \approx 10$$

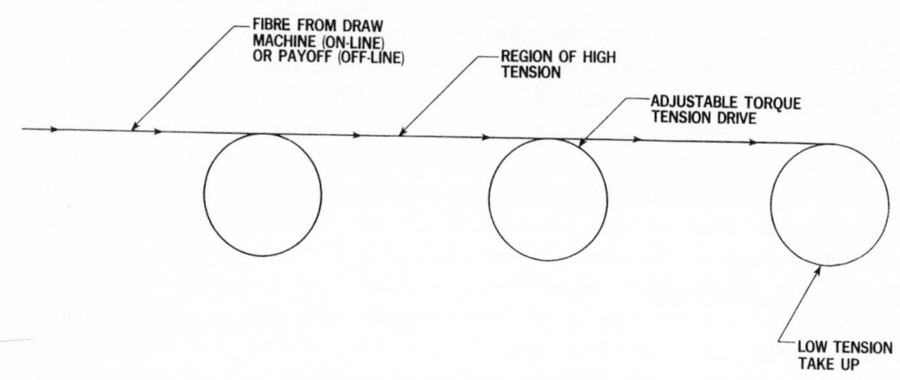

Figure 5.41. Proof-test device

Thus, if $t_1 = 1$ hour, then the failures occurring within a time period of 10 hours will be flaws with stress concentrations within 10% of each other. Thus, a particular class of flaws will fail on an average within the time period $t_2 = 10t_1$.

For a strong fibre, the prevalent flaw sizes have a very small spread from a mean size, which can be seen from dynamic load test results. The stress to failure spread is under 10%. This justifies the use of the time to failure of the first fracture as a measure since this time is at least accurate to an order of magnitude and should, in general, be more accurate. Results of static fatigue measurements have also substantiated this approach.

Proof test The long-length strength of fibres is governed by the largest flaws of the weakest point along the entire length of the fibre. The statistical results from the short-length strength tests allow long-length strength to be predicted if the strength distribution is unimodal and the Weibull parameter is accurately determined. The proof test subjects the entire fibre to a given stress. It can be used effectively to truncate the flaw distribution.

A proof-test device is illustrated in Figure 5.41. The region of tension indicated can be of any desired length and the level tension is set by the adjustable-torque tension drive wheel. This is to be set at a chosen proof-test load. A friction wheel allows load to be released quickly.

5.6.3 Cable testing

The optical characteristics of the cable are evaluated by using the same measurement methods for fibres. These measurements are required to be carried out when the cable is subjected to the conditions called for by the physical environmental specifications. The mechanical tests are concerned with survivability of fibre when the cable is subjected to stresses at various temperatures and humidities. These tests are generally based on standard test procedures used in the copper cable industry.

A list of such tests is given below:

- Mechanical evaluation
    - Tensile strength test
    - Bend test
    - Twist test
    - Impact resistance
- Environmental evaluation
    - Temperature-cycling test
    - Moisture resistance
    - Fungus testing
- Nuclear survivability test
    - Measured radiation-induced optical losses after short exposure
    - Radiation-induced optical losses after long exposure

Typical examples are listed below:

Mechanical evaluation

As a part of the mechanical evaluation, the cables are subjected to tests for tensile strength, bend, twist and impact resistance.

(a) Tensile-strength test. Three samples of each type of cable are tested for fibre breakage utilizing a 400-pound* load over a 24 inches† gauge length. The total time to reach the required tensile strength is 26 to 30 seconds. The load is maintained for one minute, then released in 19 to 28 seconds.
(b) Bend test. Three samples of each type are tested to 2000 cycles around a diameter of 5 x cable diameter.
(c) Twist test. Three samples of each cable type are subjected to the twist test of 2000 cycles of 360° twist over 10 cm gauge length.
(d) Impact resistance. The impact-resistance test is performed by dropping a 1-cm radius spherical impact tool of 5 kg from a height of 10 cm for 200 times.

Environmental evaluation

The environmental evaluation consists of a heat-cycling test, a moisture-resistance test and fungus testing.

(a) Temperature-cycling test. One sample of each type of cable is subjected to temperature cycling from –55 °C to +85 °C.

* 1 lb = 0.453 kg.
† 1 inch = 24.4 mm.

(b) Moisture resistance. Moisture-resistance testing includes attenuation measurement after moisture cycling, immersing in 98% humidity at 50 °C.
(c) Fungus testing. Fungus test in accordance with Procedure 1, Method 508 of Mil-Std-810B is an example.

Nuclear survivability test

(a) Cable irradiation. A short section of the cable is exposed to a dose of 1.27×10^5 rads of gamma and 4.4×10^{10} neutrons/cm^2 and fibre-attenuation increase monitored.
(b) Radiation-induced optical losses are measured after 10 seconds (neutrons and gamma mixed).
(c) Radiation-induced optical losses are measured after 24 hours (gamma irradiation).

Cable testing has demonstrated the following:

(a) Attenuation and dispersion. The cabling process has negligible adverse effect on fibre attenuation and dispersion over the temperature range.
(b) Tensile test. The cable has safely withstood tensile loading of the designed value. Continuous loading, under static conductions, has caused no attenuation increase.
(c) Impact test. The cable has safely withstood typically 200 impacts at designed impact loading at a single point on the cable.
(d) Environmental tests. The cable has performed in a highly satisfactory manner at high and low temperature extremes and under conditions of vibration, moisture, and sometimes nuclear radiation.

REFERENCES

1. K. C. Kao and G. A. Hockham, 'Dielectric-fibre surface waveguides for optical frequencies', *Proc. IEE*, **113** (7), 1151–1158, 1966.
2. R. D. Maurer, 'Glass fibers for optical communications', *Proc. IEEE*, **61**, 452–462, 1973.
3. F. W. Dabby *et al.*, 'Borosilicate clad fused silica core fiber optical waveguide with low transmission loss prepared by a high efficiency process', *Appl. Phys. Lett.*, **25** (12), 1974.
4. T. Nakahara *et al.*, 'Optical fiber made by new rod-in-tube method', Fall Meeting The Electrochemical Society – Pittsburg, Penn., Oct. 1978 – Abstract No. 138, 373–375.
5. P. B. O'Connor *et al.*, 'Preparation and structural characteristics of high silica, graded index optical fibers', *Am. Ceramic Society Bulletin*, **55** (5), 513, 1976.
6. K. J. Beales *et al.*, 'Preparation of sodium borosilicate glass fibre for optical communication', *Proc. IEE Special Issue on Optical Fiber Technology*, **123** (6), Paper 7684 E, 1976.

7. P. B. Macedo *et al., Proceedings Second European Conference on Optical Fiber Communication*, Paris, 1976, pp. 37–39.
8. K. Rau *et al.*, 'Progress in silica fibre with fluorine dopant', Tropical Meeting Williamsburg, Feb. 1977, P TuC4-1 to TuC4-4.
9. P. W. Black, 'Fabrication of optical fiber waveguides', *Electrical Communications*, **51** (1), 4, 1976.
10. M. C. Paek, 'Laser drawing of optical fibers', *App. Op.*, **13**, 1383, 1974.
11. M. Nakahara, 'Drawing techniques for optical fibre', *Rev. Elec. Comm. Lab. Japan*, **26** (3-3), 476, 1978.
12. B. A. Proctor *et al.*, 'The strength of fused silica', *Proc. Roy. Soc. (London)*, **A297**, 534, 1967.
13. S. M. Cox, 'Glass strength and ion mobility', *Physical & Chemical of Glasses*, **10** (6), 286–289, 1969.
14. A. A. Griffith, *Phil. Trans. Roy. Soc.*, **A221**, 163, 1921.
15. B. E. Warren, 'Surface structure of glass', *J. Am. Ceramic Soc.*, **21**, 259–265, 1938.
16. W. B. Hillig and R. J. Charles, *High Strength Materials* (ed. V. F. Zackey), Wiley, 1965, pp. 682–705.
17. S. M. Weiderhorn, 'Environmental stress corrosion cracking of glass', *Corrosion Fatigue NACE*, 2, 731–742, 1972.
18. J. E. Ritter, Jr, *et al.*, 'Application of fracture mechanics theory to fatigue failure of optical glass fibers', *J. Appl. Phys.*, 1978 (to be published).
19. S. G. Foord and J. Lees, 'Principles of fibre-optical cable design', *Proc. IEE Special Issue on Optical Fiber Technology*, **123**, (6), Paper No 7697E, 1976.
20. M. Hoshikawa *et al.*, 'A study on optical fiber and fiber cable', *Sumitomo Electric Tech. Review*, **17**, 77–88, 1977.
21. R. E. J. Baskett and S. G. Foord, 'Fiber optic cables', *Electrical Communications*, **52** (1), 49, 1977.
22. D. Gloge, 'Optical-fiber packaging and its influence on fiber strengthness and loss', *Bell System Technical Journal*, **54** (2), 245, 1975.
23. C. M. Miller, 'Laminated fiber ribbon for optical communication cables', *Bell System Tech. J.*, **55** (7), 929, 1976.
24. R. L. Lebduska, 'Fiber optic cable test evaluation', *Optical Engineering*, **13** (1), 49, 1974.
25. J. C. Smith and M. Pomerantz, 'Fiber optic cables for local distribution systems', *IWCS Proceedings of 25th International Wire & Cable Symposium*, 226–234, 1976.
26. M. Barnoski and S. Personick, 'Measurements in fiber optics', *Proc. IEE*, **66** (4), 429, 1978.
27. R. J. Freiburger *et al.*, 'Mechanical environmental performance of rugged lightweight fiber optic cables from technical field applications', *Proc. Cable & Wire Synopsis 1978* (to be published).
28. A. C. Evans and S. M. Weiderhorn, 'Proof testing of ceramic materials', *Int. J. of Fracture*, **10** (3), 379, 1974.

Optical Fibre Communications
Edited by M. J. Howes and D. V. Morgan

CHAPTER 6

Optical communication systems

I. GARRETT and J. E. MIDWINTER

6.1 INTRODUCTION

An optical communication system is essentially very simple. If we consider initially a binary digital system, then we can imagine an incoming signal which will consist of a series of zeros and ones, randomly distributed. These signals are used to switch a light source off and on. Light from the source is coupled into the optical fibre, transmitted through it by total internal reflection at the core cladding interface, to emerge attenuated at the far end. Here, it impinges upon a photodetector which converts the light (photons) to electron–hole pairs in the semiconductor material. These in turn, under the influence of the bias field applied to the detector, give rise to an electric current which then activates a decision circuit whose task is to decide whether the received signal corresponds to a zero or a one. A simple system of this type is illustrated in Figure 6.1.

Also shown in Figure 6.1 is a power penalty. We will see from the following analysis that our simple system can be generalized to include the effects of such things as bandwidth limitation in the fibre, transmit pulse width and extinction ratio, all of which degrade the quality of transmission, simply by adding a notional attenuator to the ideal system. The setting for the attenuator can be calculated in decibels for any of the impairments likely to be met so that any real system can be compared to the ideal system by this single notional element. This is a very useful way of modelling a fibre transmission system since it reduces the model to a form that can be simply calculated or in which simple programmes can be used to optimize the system for a given task.

Before we launch into an analysis of the optical system, we will examine some of the underlying mechanisms which control its performance. Light behaves when detected by a photodetector as discrete energy packets, quanta or photons. The energy of a photon is given by hf where h is Planck's constant and f is the optical frequency. At radio or microwave frequencies, since $hf \ll kT$, quantum effects are normally drowned by thermal noise fluctuations. However, in the optical system,

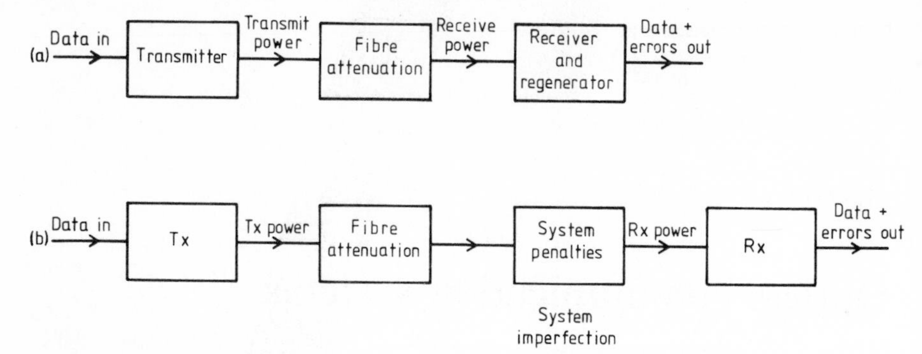

Figure 6.1. A simple block schematic of a fibre communication system: (a) a perfect system, and (b) with allowance for system imperfections expressed in terms of a notional attenuator whose setting is controlled by the degree of imperfection. In this way an imperfect system can be modelled by a perfect system by the addition of a simple attenuator. Optimization of the design involves minimizing the notional theoretical attenuation setting

the signal arriving at the receiver should be viewed as composed of a discrete number of quanta. For a received ONE, the design target is typically 200 to 2000 quanta. The probability of arrival of a quantum is described by Poisson statistics, so that we may say broadly that if a mean number μ are expected, there will be a random fluctuation of order $\sqrt{\mu}$ about that mean from pulse to pulse on arrival. This signal-dependent noise must be added to the other receiver noise sources, such as occur in the following amplifier or in the photodetector itself, to calculate the total noise level. But notice that during the reception of a ZERO, there will be less receiver noise than during the reception of a ONE, so that the decision threshold to obtain equal error probability for both ZERO and ONE reception will not be at $\mu/2$ but somewhere below that. Notice also that because of the random and quantum nature of the arriving light signal, some minimum received power is required if an acceptable error rate is to be maintained.

With this picture in mind, it is now easy to see where some of the power penalties arise. If there is no spill-over of signal during transmission from the allotted bit time, and the receiver operates at a very low electrical noise level, then during a ZERO pulse reception, the noise will be very low. The decision threshold can be set close to the zero level, and the value of μ for reception of a ONE also set low for an acceptable error probability. However, as soon as optical power from an adjoining ONE spills over into the bit interval occupied by a ZERO, not only will the received signal waveform become distorted, an effect that can be corrected by suitable equalizers, but the noise level will inevitably rise during the zero interval. The decision threshold will have to be raised and in turn μ must also be raised to hold the error rate constant. Thus, despite perfect equalization (correction of the distorted mean waveform), more mean signal power will be required to hold the error rate constant. This effect is fundamental to an optical system and leads to heavy power penalties from any form of signal degradation, as we will see in the following analysis.

It follows from the above that an ideal optical system would be one in which all the optical power associated with a given transmitted bit arrived in a discrete package at the receiver, well free of any adjoining time slot. In reality this is unlikely to happen, so that most systems show a power penalty relative to the 'ideal' system, a penalty which may vary from 1 to 10 dB or more. In our analysis of these effects, we will examine the receiver first to establish the basic results governing its performance, consider how typical detectors might be used in optical receivers and then move on to examine how the transmitter and the transmission medium might affect the overall system performance.

6.2 THE DIGITAL OPTICAL RECEIVER

The functions of a digital optical receiver are identical with those of the receiver on any other type of digital system: detection of the incoming signal, amplification with agc, pulse-shaping (equalization) if necessary, timing extraction, decision and error detection or correction. Optical signal-processing is still in its infancy, and at present all these functions are carried out on the electrical signal produced by the detector in response to the received optical signal. This section is concerned with the theory of the performance of optical receivers, and how that performance is affected by the characteristics of the received signal. Several authors have analysed this problem in the past.[1–6] Personick's analysis[4] is the most comprehensive but complicated. In this section we use the analysis by Smith and Garrett,[5] which is simpler and nearly as accurate.

Figure 6.2 is a schematic diagram of a digital optical receiver, up to the input to the decision stage. The detector is assumed to be a photodiode of either the avalanche type (APD) with quantum efficiency of η electrons per photon, mean avalanche gain $\bar{g}$, and dark current I_d, or else of the PN or PIN type, in which case $\bar{g} = 1$. The photodiode has capacitance C_d, and R_b is the bias resistor generating thermal noise current $i_b(t)$. The amplifier has input resistance R_a and capacitance C_a, and noise current and voltage sources $i_a(t)$ and $e_a(t)$ of spectral density S_I and S_E respectively.

We consider an incident digital optical signal:

$$p(t) = \sum_{-\infty}^{\infty} b_n h_p(t - nT)$$

where $p(t)$ is the received optical power, T is the bit time and $h_p(t)$ the pulse shape. Let:

$$\int_{-\infty}^{\infty} h_p(t) = 1$$

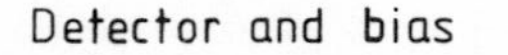

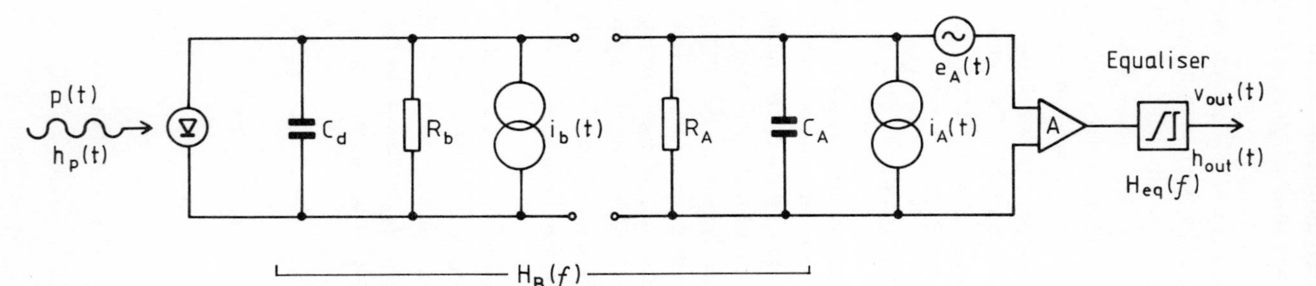

Figure 6.2. The equivalent circuit of an optical receiver, showing the noise sources to be taken into account. The analysis of the receiver must include both the detector and the immediately following amplifier and equalizer. The equalizer may well be used only to correct for the electrical response of the detector/amplifier circuit and may not need to correct for the transmission-medium response

then b_n is the energy in the nth pulse, and in a binary system can take two values b_0 and b_1 corresponding to ZERO and ONE bits. The lower limit to b_1 is set by the quantum nature of light. The arrival of a photon at the detector is an event with a Poisson probability distribution: if μ is the expectation number of photons arriving in a given time interval, say one bit time, then the probability of m photons arriving is:

$$p(m) = \frac{e^{-\mu}\mu^m}{m!}$$

If we had a noiseless receiver, we could say that if one or more photons were detected in a bit period the pulse was a ONE, and otherwise it was a ZERO. The error probability is then the probability of detecting no photons when μ are expected, and we can calculate μ from the desired maximum error probability; for example, an error probability of 10^{-9} demands $\mu = 20.7$.

Any practical receiver is noisy. In addition to the quantum noise on the incoming optical signal, which manifests itself as shot noise on the primary photo-current from the detector, there are various sources of thermal noise to be considered, as well as shot noise on the dark current of the input stage and excess noise from avalanche multiplication in an APD.

The mean output current from the photodiode at time t is given by:

$$\overline{i_p(t)} = \frac{\eta q}{\hbar\Omega}\, \bar{g} p(t)$$

where $\hbar\Omega$ is the photon energy. This current results in a mean voltage at the output of the equalization network given by:

$$\overline{v_{\text{out}}(t)} = \frac{A\eta q}{\hbar\Omega}\, \bar{g} p(t) * h_{\text{b}}(t) * h_{\text{eq}}(t)$$

where $h_{\text{b}}(t)$ and $h_{\text{eq}}(t)$ are the impulse responses of the bias and equalization networks and '*' denotes convolution. Superimposed on this output voltage are the various noise contributions. The shot noise contribution has been discussed in detail by Personick.[4] We make the approximation that the mean-square shot-noise voltage at the decision time $\langle v_{\text{s}}^2 \rangle$ is related to the mean unity-gain photo-current over the bit time $\langle i_0 \rangle_T$ by the usual expression:

$$\langle v_{\text{s}}^2 \rangle = 2q \langle i_0 \rangle_T B_N R^2 A \overline{g^2}$$

where R is the total resistance of R_{b} and R_{a} in parallel and $\langle i_0 \rangle_T$ contains possible contributions from neighbouring pulses if they overlap into the bit time under decision, and B_N is the noise equivalent bandwidth, defined as usual for positive frequencies only:[6]

$$2B_N = \frac{1}{R^2} \int_{-\infty}^{\infty} |H_{\text{eq}}(f) H_{\text{b}}(f)|^2 \, \mathrm{d}f = \frac{1}{R^2} \frac{\hbar\Omega}{A\eta q} \int_{-\infty}^{\infty} \left| \frac{H_{\text{out}}(f)}{H_{\text{p}}(f)} \right|^2 \mathrm{d}f$$

Here $H(f)$ is the Fourier transform of $h(t)$.

The mean photocurrent over the bit time depends on whether the pulse is a ONE or a ZERO. We consider the worst-case shot noise, when all neighbouring pulses are zero. Then[5] assuming $b_0 = 0$,

$$\langle i_0 \rangle_{T,1} = \frac{\eta q}{\hbar \Omega} \frac{b_1}{T} + I_d \text{ for a ONE pulse} \tag{6.1}$$

$$\langle i_0 \rangle_{T,0} = \frac{\eta q}{\hbar \Omega} \frac{b_1}{T}(1 - \gamma) + I_d \text{ for a ZERO pulse} \tag{6.2}$$

where

$$\gamma = \int_{-T/2}^{T/2} h_p(t)\, dt$$

is the fraction of the pulse energy contained within the bit time. The factor $(1 - \gamma)$ thus takes account of intersymbol interference due to pulse dispersion in the fibre.

The thermal noise contributions come from the bias resistor:

$$\langle v_R^2 \rangle = \frac{4k\theta}{R_b} B_N R^2 A$$

and from the amplifier noise current and voltage sources:

$$\langle v_I^2 \rangle = 2 S_I B_N R^2 A$$

$$\langle v_E^2 \rangle = 2 S_E B_N' A^2$$

where

$$2B_N' = \int_{-\infty}^{\infty} |H_{eq}(f)|^2\, df = \frac{\hbar \Omega}{A \eta q} \int_{-\infty}^{\infty} \left| \frac{H_{out}(f)}{H_p(f)} \frac{1}{\frac{1}{R} + 2\pi j f C} \right|^2 df$$

Here C is the total capacitance of C_d and C_a in parallel. Following Personick,[4] we normalize $h_{out}(t = 0)$ to unity, putting $A\eta q/\hbar\Omega = 1$, and normalizing the bandwidth integrals with respect to the bit time. Then we find, for the total mean-square noise voltage:

$$\langle v_N^2 \rangle = \left(\frac{\hbar \Omega}{\eta} \right)^2 \left\{ \frac{\bar{g}^X}{q} \langle i_0 \rangle_T T I_2 + \frac{Z}{\bar{g}^2} \right\} \tag{6.3a}$$

where

$$Z = \frac{T}{q^2} \left(S_I + \frac{2k\theta}{R_b} + \frac{S_E}{R_2} \right) I_2 + \frac{(2\pi C)^2}{T q^2} S_E I_3 \tag{6.3b}$$

which is a dimensionless parameter characterizing the thermal noise of the receiver, and

$$I_2 = \int_{-\infty}^{\infty} \left| \frac{H'_{\text{out}}(\phi)}{H'_{\text{p}}(\phi)} \right|^2 d\phi$$

$$I_3 = \int_{-\infty}^{\infty} \left| \frac{H'_{\text{out}}(\phi)}{H'_{\text{p}}(\phi)} \right|^2 \phi^2 \, d\phi$$

which are the normalized dimensionless bandwidth integrals.

The worst case values of $\langle v_N^2 \rangle$ for ONE and ZERO pulses can be obtained by substituting Equations 6.1 and 6.2 for $\langle i_0 \rangle_T$ into Equation 6.3. Using the Gaussian approximation for the noise-voltage distribution, we obtain the expression for the receiver sensitivity:

$$b_1 - b_0 = Q(\sqrt{\langle v_N^2 \rangle_1} - \sqrt{\langle v_N^2 \rangle_0}) \tag{6.4}$$

where Q is a number depending on the maximum permitted error rate P_{E}:

$$P_{\text{E}} = \int_Q^{\infty} \frac{1}{\sqrt{2\pi}} e^{-x^2/2} \, dx = ½ \, \text{erfc}\,(Q/\sqrt{2})$$

For an error probability of 10^{-9}, $Q = 6.00$. Using the simple approximation for the excess noise due to avalanche multiplication, $\overline{g^2} = \bar{g}^{2+x}$, and assuming $b_0 = 0$, Equation 6.4 becomes:

$$b_1 = \frac{Q}{g}\left(\frac{\hbar\Omega}{\eta}\right)\left\{\sqrt{\bar{g}^{2+x}\frac{\eta}{\hbar\Omega} b_1 I_2 + Z} + \sqrt{\bar{g}^{2+x}\frac{\eta}{\hbar\Omega} b_1 (1-\gamma) I_2 + Z}\right\} \tag{6.5}$$

The next logical step in the analysis is to find the optimum value of avalanche gain g_{opt} by differentiating Equation 6.5 with respect to $\bar{g}$ and putting $db_1/d\bar{g} = 0$. It is useful first of all to interpret Equation 6.5 in terms of conceptually useful dimensionless parameters. Let N_{pe} be the number of primary photoelectrons within the bit period for a ONE pulse. Then

$$N_{\text{pe}} = \frac{b_1 \eta}{\hbar\Omega} = T \langle i_0 \rangle_{T,1}$$

and the shot noise can be represented as:

$$\langle i_s^2 \rangle = \frac{2 I_2 q^2}{T} N_{\text{pe}}$$

The thermal-noise terms can be represented as a notional number N_{t} of electrons in the bit time:

$$N_{\text{t}} = Z/I_2$$

For example, if the only source of thermal noise were a resistance R, we would have

$$N_t = \frac{2k\theta}{R}\frac{T}{q^2}$$

$$Z = \frac{T}{q^2}\frac{2k\theta}{R} I_2$$

from Equation 6.3b.

Let the variance of the voltage at the output of the equalization network be N_0 and N_1 for ZERO and ONE pulses, where N_0 and N_1 are proportional to the total noise which is made up of contributions from the thermal noise parameter N_t, the shot noise, etc. Taking the constant of proportionality to be unity, we have:

$$N_0 = N_t I_2$$

$$N_1 = N_t I_2 + N_{pe} I_2 \bar{g}^{2+x}$$

From the Gaussian approximation,

$$\bar{g} N_{pe} = Q(\sqrt{N_0} + \sqrt{N_1}) \tag{6.6}$$

and we get back to Equation 6.5 with $\gamma = 1$ (no intersymbol interference). We can use Equation 6.6 to investigate two limiting cases. The thermal noise limit arises when $N_1 \approx N_0 \approx N_t I_2$. Then

$$N_{pe} = 2Q\sqrt{N_t I_2}$$

and the decision threshold level $D = N_{pe}/2$.

In the opposite limit when shot (quantum) noise dominates, $N_0 \approx 0$, $N_1 = N_{pe} I_2$ and so

$$N_{pe} = Q^2 I_2$$

The minimum useful value of I_2 is 0.5, corresponding to the Nyquist bandwidth. Then for an error probability of 10^{-9}, $Q = 6$ and $N_{pe} = 18$ (using Poisson statistics we found $N_{pe} = 20.7$).

We can interpret Q loosely in terms of optical signal-to-noise ratio. The optical signal power is proportional to N_{pe}, while $\sqrt{N_1}$ and $\sqrt{N_0}$ are rms noise currents and may be thought of as proportional to some optical noise power. Then:

$$\left(\frac{S}{N}\right)_{\text{optical}} = \frac{N_{pe}}{\frac{1}{2}(\sqrt{N_1} + \sqrt{N_0})} = 2Q$$

Thus an error probability of 10^{-9} calls for an optical SNR of 12 (10.8 dB), hence an electrical SNR of 144 (21.6 dB). Substituting for N_0 and N_1, Equation 6.6 can be written

$$N_{pe} = Q^2 I_2 g^x + \frac{2Q\sqrt{N_t I_2}}{\bar{g}}$$

The first term is the quantum limit multiplied by the excess multiplication noise factor, while the second term is the thermal noise limit divided by the avalanche gain. From this we obtain the optimum value of gain:

$$g_{\text{opt}}^{1+x} = \frac{2}{Qx}\sqrt{\frac{N_t}{I_2}}$$

With this value of gain, the shot noise is $1/x$ times greater than the thermal noise. This equation is valid only for $\gamma = 1$ (no intersymbol interference). For other values of γ, Smith and Garrett derived the more general equation:

$$N_{\text{pe}} = \frac{b_1\eta}{\hbar\Omega} = Q^{\left(\frac{2+x}{1+x}\right)} Z^{\left(\frac{x}{2+2x}\right)} I_2^{\frac{1}{1+x}} L \tag{6.7}$$

where

$$L^{1+x} = \left(\frac{1-\gamma}{K(2-\gamma)}\right)\left\{\left[\tfrac{1}{2}\frac{(2-\gamma)}{(1-\gamma)}K+1\right]^{1/2} + [\tfrac{1}{2}(2-\gamma)K+1]^{1/2}\right\}$$

and

$$K = \sqrt{1 + \frac{16(1-\gamma)}{(2-\gamma)^2}\left(\frac{1+x}{x^2}\right)} - 1$$

L is a parameter which depends only on γ and x, and is typically between 2 and 3. It is plotted in Figure 6.3 as a function of γ for different values of x.

Equation 6.5 can be expanded to include a dark-current contribution represented by N_d, the number of leakage electrons in the bit time, and also intersymbol interference and non-zero extinction ratio ϵ (Hooper and White[7]):

$$(1-\epsilon)\bar{g}N_{\text{pe}} = Q[\sqrt{[N_t + N_d\bar{g}^{2+x} + \bar{g}^{2+x}(1-\gamma+\gamma\epsilon)N_{\text{pe}}]\,I_2} + \sqrt{(N_t + N_d\bar{g}^{2+x} + \bar{g}^{2+x}N_{\text{pe}})I_2}\,] \tag{6.8}$$

This equation is general, and permits one to find the optimum gain and the minimum value of N_{pe} in terms of the noise sources N_t and N_d and the parameters γ and ϵ. Nonzero extinction ratio ϵ is likely to be common in systems using laser sources, since as we shall see, it is frequently desirable to bias the laser just below its threshold for oscillation during a ZERO in the bit stream. This means that during such periods the laser is acting as an LED and consequently launches substantial power into the fibre.

At optimum gain, the multiplied shot-noise term $\bar{g}^{2+x}N_{\text{pe}}$ is a few times the thermal-noise term N_t, while $N_d + (1-\gamma+\gamma\epsilon)N_{\text{pe}}$ is usually much smaller than N_{pe}. So it is possible to approximate Equation 6.8, replacing both $N_d + (1-\gamma+\gamma\epsilon)$ N_{pe} and N_d by N, and we find:

$$N_{\text{pe}} = \left(\frac{Q}{1-\epsilon}\right)^2 I_2\bar{g}^x + \frac{2Q}{\bar{g}(1-\epsilon)}\sqrt{I_2N_t}\sqrt{1+\frac{N}{N_t}\bar{g}^{2+x}} \tag{6.9}$$

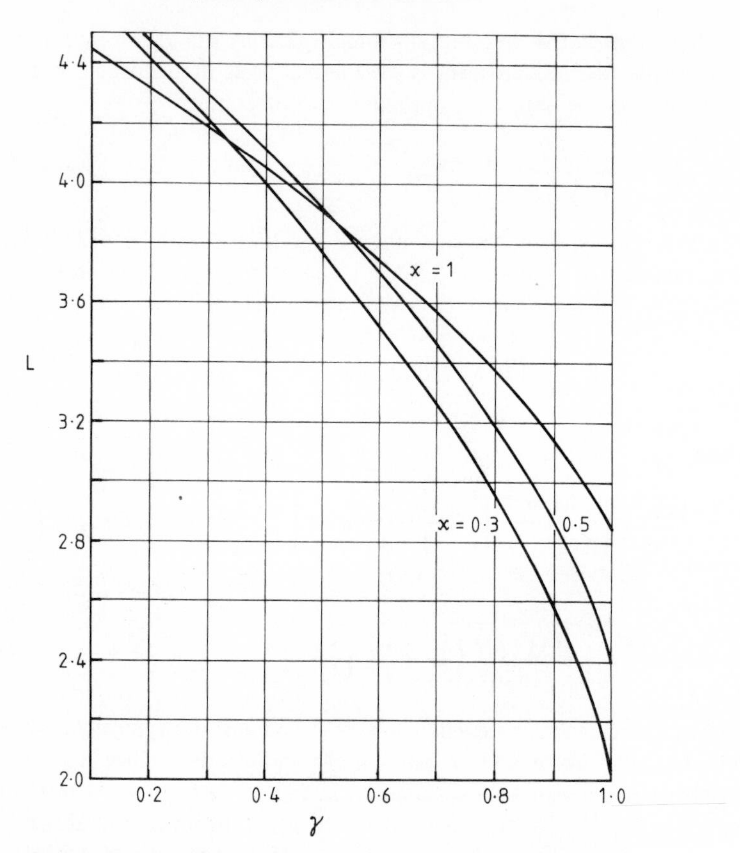

Figure 6.3. A plot of values of the parameter L against γ, the fraction of energy received in the relevant bit interval. L is used in the calculation of receiver power penalty and is a function of γ and x only

The thermal-noise term is increased by a factor which depends on the parameter N/N_t, the ratio of contributions from dark current, intersymbol interference and nonzero extinction ratio to the thermal-noise contribution. This parameter determines the optimum gain, and because it is multiplied by $\bar{g}^{2+x}$, it has a large effect when the gain is large. The dependence of g_{opt} and N_{pe} on N/N_t is shown in Figure 6.4 for various values of N_t, taking $x = 0.3$, typical of a good silicon APD, and $\epsilon = 0$. Conventional receivers using packaged components have N_t of around 10^6 for a bit rate of, say, 100 Mbit/s. In the absence of intersymbol interference, dark current, etc. (i.e. small N/N_t), the optimum gain is large and N_{pe} is around 240 electrons per bit time. However, the optimum gain decreases very rapidly for larger N/N_t, since N/N_t is multiplied by $\bar{g}^{2+x}$. We see, then, that if the receiver noise parameter is large, the receiver can be made sensitive by employing a large

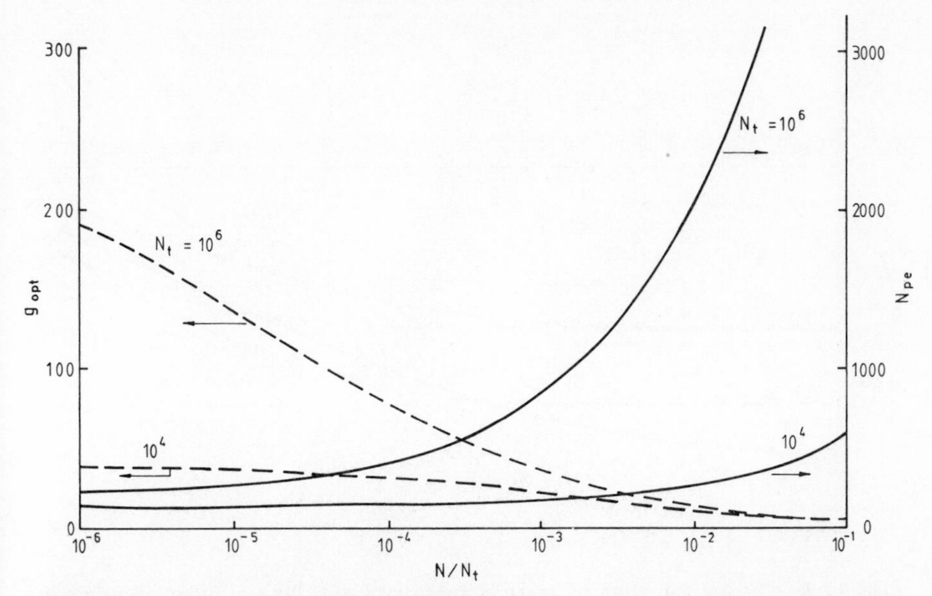

Figure 6.4. A plot of the optimum gain (g_{opt}) of an avalanche photodetector and the parameter N_{pe}, the number of photoelectrons per bit, versus N/N_t, the ratio of the dark current, intersymbol interference and nonzero extinction ratio noise contributions, to the thermal-noise component. A perfect fibre system would thus have $N/N_t = 0$, there being no degradation from the transmitter or transmission medium. The curves are plotted for $N_t = 10^6$ and 10^4. Note for a low-noise amplifier ($N_t = 10^4$) the optimum gain remains very low

avalanche gain, so long as the noise terms contributing to N are insignificant. But if N increases for any reason (increase in photodiode leakage current or in extinction ratio, for example), the receiver sensitivity is rapidly degraded.

At the other extreme, if the receiver noise parameter Z or N_t can be reduced, the optimum gain is not large. On the one hand, the receiver performance is less sensitive to photodiode leakage current, intersymbol interference, etc.; on the other hand, there is only a small penalty in using a PIN photodiode with unity gain. Smith *et al.*[8] have discussed ways in which the receiver noise parameter Z can be reduced to around 1000 by using a high-performance microwave FET in the input stage of the amplifier and using hybrid circuit technology rather than discrete packaged components. Such an approach to receiver design makes a PIN photodiode detector quite competitive with an APD. The technical performance of the APD is then only slightly better, even for a silicon APD with very low excess noise. APDs for wavelengths beyond about 1.1 μm, where silicon is transparent, have not been made with such low excess noise, and the PIN/FET hybrid receiver is a strong competitor on technical performance. It also offers economic and operational benefits, as it does not need the high bias voltage (200 to 400 V) demanded

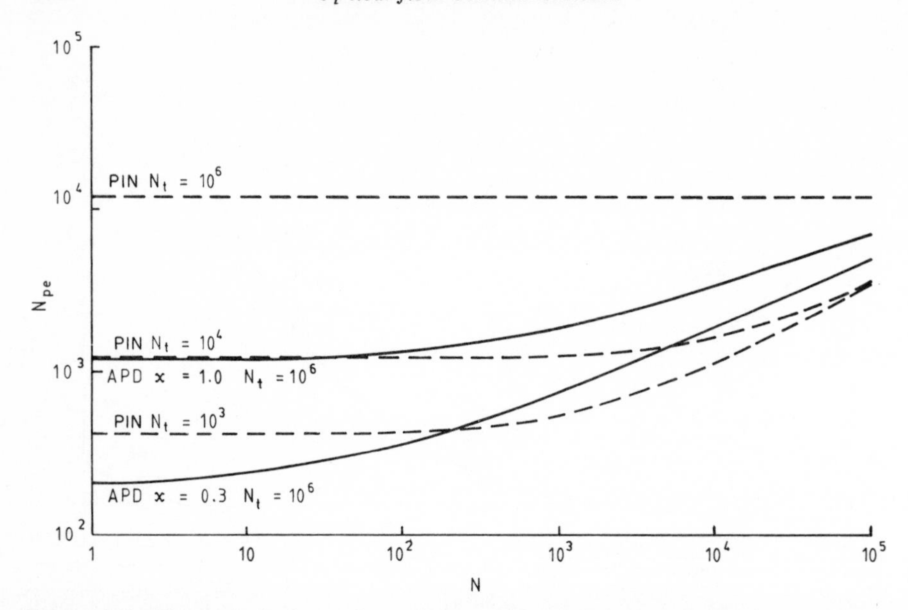

Figure 6.5. Theoretical plots of receiver sensitivity N_{pe} (the number of primary photoelectrons in the bit interval) for PIN and avalanche photodiodes as functions of the total number of equivalent 'noise primary electrons' N. The curves are calculated for amplifier noise figures characterized by $N_t = 10^6$, 10^4 and 10^3. Note that for very low-noise amplifiers, the PIN diode is at worst nearly equivalent to the avalanche photodiode, and under conditions of poor extinction ratio or high intersymbol interference it can be superior

by APDs, nor is there any demand to stabilize the gain against temperature fluctuations, for example.

In Figure 6.5, the receiver sensitivity N_{pe} is plotted as a function of N for both PIN and APD receivers with various values of N_t. On the left side, where N is insignificant, the receiver sensitivity is determined by the noise parameter N_t, which can be effectively countered by avalanche gain when using a low-noise APD. With a PIN diode detector the only way of making the receiver more sensitive is by reducing N_t: we see that a value of 10^4 produces comparable performance to an APD with $x = 1$ and $N_t = 10^6$. On the right of Figure 6.5, the sensitivity of the receiver is determined by N, and all receivers become equal as N becomes comparable with N_t.

The effect of intersymbol interference or bandwidth limitation in the transmission medium is vividly illustrated by Equation 6.9. Using it, we have plotted in Figure 6.6 a curve of the power penalty the receiver incurs to hold the error rate constant as a function of γ, the fraction of energy in the bit interval, and we see that the curve of penalty in dB received power versus γ is very steeply rising. The implication of this is clear; it is not generally sensible to operate an optical-fibre

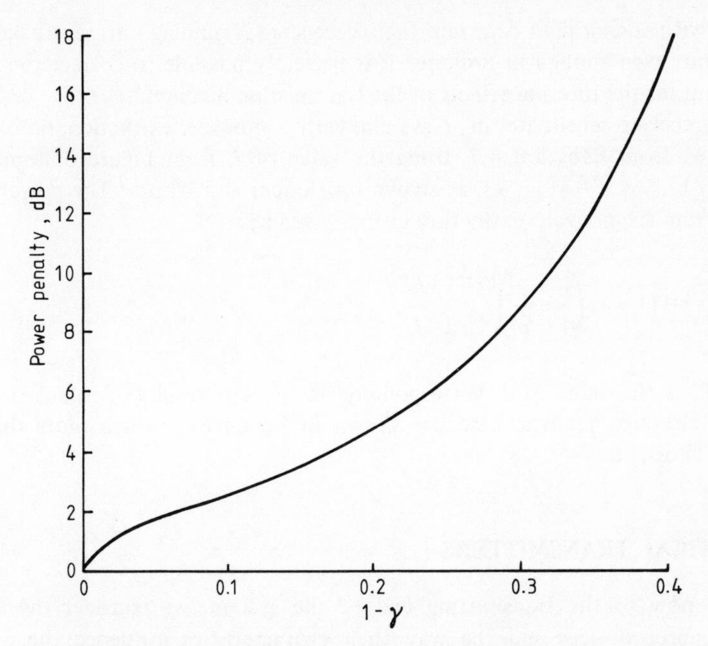

Figure 6.6. Plot of the total power penalty at the optical receiver versus the fraction of energy received in the bit interval, showing how rapidly the fibre system is degraded by the shot-noise interference from bandwidth limitation in the transmission medium[5]

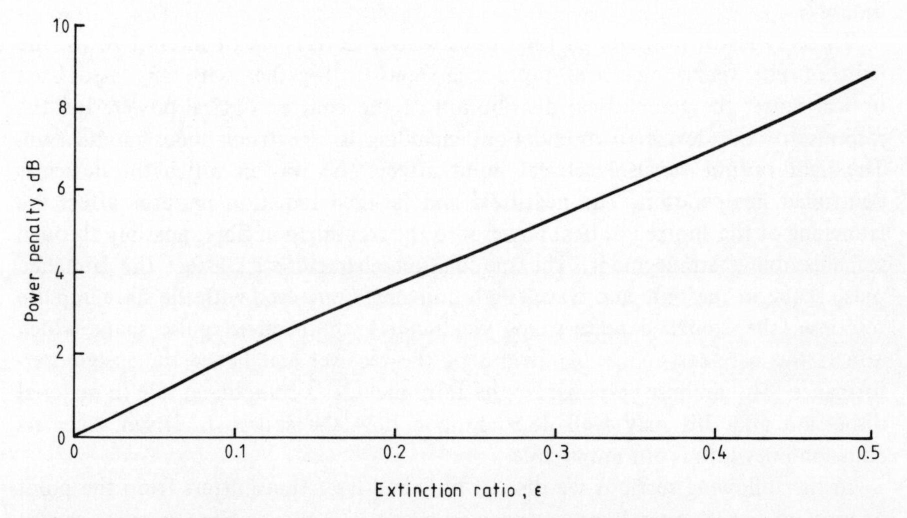

Figure 6.7. The receiver power penalty as a function of extinction ratio (defined as the ratio of average power on all zero pulses to all one pulses)[7]

system with such a high data rate that the system is running into severe bandwidth limitation, even though in principle it is perfectly possible to correct the received waveform for the filtering effects of the transmission medium.

The receiver sensitivity in a system with a nonzero extinction ratio may be calculated from Equation 6.7, using the value of L from Figure 6.3 appropriate not to γ but to $\gamma' = \gamma(1 - \epsilon)$, as shown by Hooper and White.[7] The power penalty arising from the nonzero extinction ratio is given by:

$$y(\epsilon) = (1 + \epsilon)\left(\frac{1}{1 - \epsilon}\right)^{\left(\frac{2+x}{1+x}\right)} \frac{L'}{L} \tag{6.10}$$

where L' is the value of L corresponding to γ'. The results of calculations for a special, although realistic case are shown in Figure 6.7, which plots the power penalty in dB, etc.

6.3 OPTICAL TRANSMITTERS

Turning now to the transmitting end of the system, we consider the available optical-source devices and the way their characteristics influence the design of the system. We consider light-emitting diodes and semiconductor injection lasers of various types as being the most suitable electrical-to-optical transducers, as they are directly modulated and fairly efficient, particularly the laser.[9–18] Other types of laser will not be considered here, although some interesting work has been carried out on Nd:YAG lasers directly pumped and modulated with light-emitting diodes.

Three types of property of the optical-source devices are of interest to us: the light output versus electrical input characteristic together with any associated optical noise; the geometrical distribution of the emitted optical power; and the response of the device to modulation, including its spectrum under modulation. The light output versus electrical input affects the way in which the device is controlled in operation. The near-field and far-field radiation patterns affect the launching of the emitted optical power into the transmission fibre, possibly through some coupling arrangement. The modulation characteristics affect the launched pulse shape in the time and wavelength domains. Convolved with the fibre impulse response, the launched pulse shape yields $h_p(t)$, the received pulse shape which affects the necessary noise bandwidth of the receiver and hence the system performance. The impulse response of the fibre includes a component due to material dispersion and this may well depend upon how the source is driven, since its emission linewidth is not immutable.

In the following sections we discuss LED and laser transmitters from the point of view of device control and system performance. Device-to-fibre coupling as part of overall transmitter design is briefly treated.

6.3.1 LED transmitters

LEDs are simple devices to use by comparison to lasers since their curves of light output versus drive current tend to be substantially linear and to change little with time. Because there is no sharp threshold point, it is not necessary to control the peak drive current with such care since the device is not likely to be destroyed by a slight overload current, while one can compensate for temperature effects by the use of simple temperature sensors and open-loop control circuits. In the system environment, the device is less attractive than the laser since the linewidth is greater and the radiance less, so that compared with the laser, material disperson is of greater concern and the launched power is usually less. Nevertheless, the low cost and simplicity of operation of the devices are great attractions and for many purposes they can be perfectly adequate, meeting the requirements of the system designer where the laser would only give him additional operating margins that are not needed. Since the emission characteristics of the devices are very insensitive to the operating conditions, there are fewer design considerations and options open to the designer. In general, it is necessary to evaluate the launched power into the fibre by calculating the product of the radiance of the source with the acceptance angle and area of the fibre, and then estimating the efficiency with which one can be coupled to the other. The linewidth of the source is not significantly variable, so that the material dispersion is readily calculated. This can be used with the mode dispersion and pulse width to assess the receiver sensitivity, and the drive pulse length may be shortened a little to reduce the receiver power penalty.

The LED response time is nonzero and if the device is to be used in a high bit rate system, this must be taken into account. The rise-time of typical high-radiance devices is in the region of 2 to 4 ns so that this can become a limiting factor. If the user has the freedom to manufacture custom-made devices, then it is possible to trade radiance for speed of response, and for special applications this may be worth while.

The conversion of electrical power into optical power is of great interest to the designer of an efficient system. Within the LED itself, the electrical power is converted into optical power with very high efficiency. Much of this optical power is subsequently lost from the device or is absorbed to produce thermal energy. By making the device of smaller area, smaller than the cross-sectional area of the fibre, the radiance for a given electrical current is increased. Then the use of a high numerical aperture lens system can magnify the LED emitting area to fill the fibre end and the effective collection aperture of the device that is coupled to the fibre will be increased. In this way, the launched power for a given electrical drive current into the device can be increased and devices with integral fibre tails and lenses have been made utilizing this principle.

Three types of LED have attraction for fibre systems. The simple 'Burrus' type device is generally made with an emitting area similar to that of the fibre core with

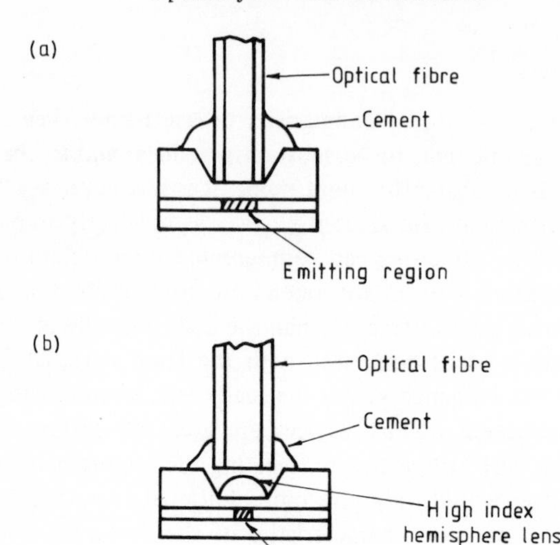

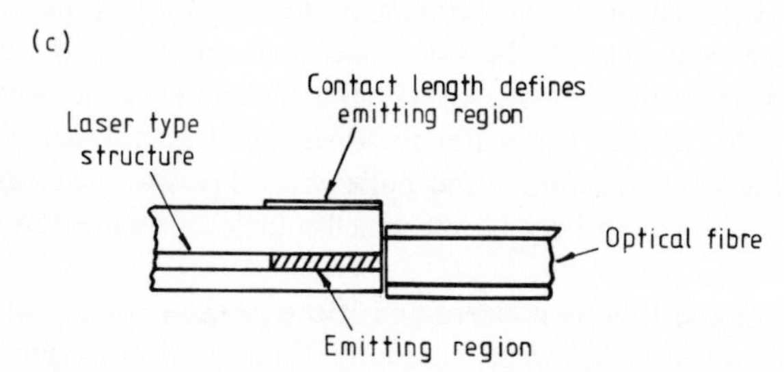

Figure 6.8. Types of LEDs used for optical fibre systems: (a) The surface-emitting Burrus type with a fibre cemented to the emitting surface. (b) The same type but with an integral high-index lens to improve the conversion of electrical power to launched optical powder. (c) The edge-emitting type (ELED)

the fibre end mounted or imaged close to it, as in Figure 6.8a. This design can then be modified to a smaller-area device with higher effective numerical aperture collecting optics along the lines shown in Figure 6.8b as discussed above. Finally, an altogether different structure which is more closely akin to the lasers used for fibre systems is the edge-emitting LED shown in Figure 6.8c. This relies on emission along the junction coupled into the fibre just as the emission from a laser device would be. The emission is incoherent although some super-radiant gain-narrowing of the line may occur. In the super-radiant state, greater radiance and narrow line-width are obtained simultaneously with a device that avoids the extreme non-linearities of the laser and thus still allows the use of very simple control circuits.

There has been much debate about the relative merits of the edge-emitting and 'Burrus' type LEDs in the context of fibre systems, but no clear conclusion appears at present. An attraction of the ELED is that it can make use of the same or a very similar package to that of the laser when used with a fibre system and since it involves a great deal of fabrication technology in common with the laser, this may lead to its acceptance for production systems.

Edge-emitting LEDs with a relatively thick 'guiding layer' next to the active layer and of very slightly lower refractive index have been developed to optimize coupling into a fibre. In these devices, super-radiance is suppressed since light travelling in the guiding layer is only weakly coupled to that in the gain region, so they are very linear. Their output is little affected by absorbing regions formed in the active layer as the device ages, in contrast to devices which are super-radiant to a significant extent.

6.3.2 The laser transmitter

Most commercially available semiconductor lasers of the type that are considered for use in fibre-optic systems leave much to be desired when viewed as sources. Generally we would be discussing in this context double-heterostructure CW lasers fabricated from the GaAs/GaAlAs system, although increasingly in the future we will probably be considering lasers in the InP/GaInAsP system, the former emitting in the wavelength range of 800 to 900 nm and the latter in the range 1000 to 1600 nm. We will first examine the reasons why such lasers present problems to the system designer.

In Figure 6.9 we show the optical power output versus drive current characteristics for a typical laser, indicating the effect of temperature. Generally, the system

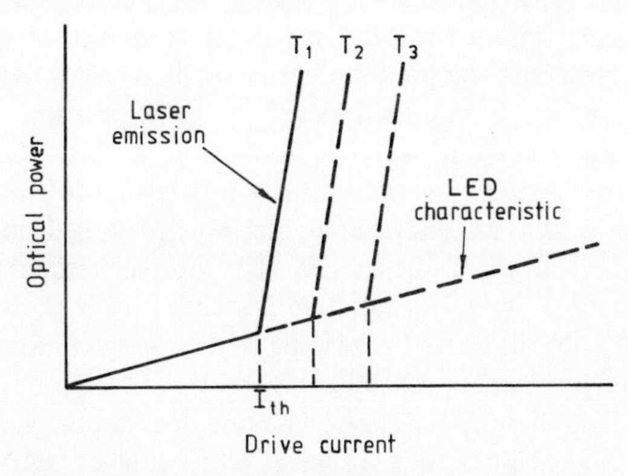

Figure 6.9. Schematic diagram of a laser output power versus drive current for different temperatures (T_i), showing the definition of threshold at the onset of laser action

designer will wish his transmitter to emit a constant peak power into the fibre during its operating life and regardless of operating temperature. Evidently, some control circuit is required to correct for the effects of temperature. The effects of ageing on the device are generally similar to the effects of temperature, and in general both will occur. We note also that the light-output curve in the lasing mode when operating as a laser is not always linear and can exhibit distinct kinks, particularly as the device ages. These kinks are sometimes associated with the onset of higher-order modes in the device. Our control circuits must also take these effects into account if they occur.

If we examine the optical (spectral) linewidth of the device, we will again find a complex situation. Below threshold for laser oscillation, the device behaves as an LED, emitting spontaneously across a broad spectrum (40 to 100 nm full-width half maximum (FWHM)). As threshold is approached, super-radiant gain starts to narrow this line and when laser action commences, the linewidth changes considerably into a series of discrete lines (modes) which spread over less range. If the device is biased above threshold, in some cases the oscillation of the laser will restrict itself to a single longitudinal mode of the laser cavity, giving essentially a pure single frequency, the ideal situation. However, if the device is pulsed with a random bit stream, every time it turns on, a number of modes or frequencies build up and most only decay very slowly, if at all, during the length of the drive pulse. Thus the emission linewidth of the device can depend upon how it is biased and driven and on the bit rate of the system, the effect being most serious in a high bit rate system when it is of great significance.

This broadening of the laser spectrum under pulsed or transient drive conditions is closely related to another modulation characteristic of the laser, transient oscillation or ringing. If the drive current is raised rapidly from below the threshold current to a value above threshold, a transient oscillation in the light output is frequently observed. This oscillation in light output, corresponding to an oscillating photon density within the laser cavity, is accompanied by an oscillating carrier inversion, preceding the photon density oscillation in phase. Whenever the carrier density is above the threshold value, the gain spectrum for the cavity is above threshold over a range of wavelengths and the spectrum is broadened. The decay time constant for this transient oscillation is related to the effective carrier lifetime, the photon lifetime, and the spontaneous photon density in the lasing mode. Lasers with relatively short carrier and photon lifetimes and large spontaneous photon density in the lasing mode have been made showing no transient oscillation and negligible transient broadening of the spectrum. Such devices appear very suitable for use up to about 200 Mbit/s.

Finally, we should note that the laser oscillation does not commence at the same time as the drive current pulse, but is subject to a switch-on delay period (Figure 6.10) which itself depends upon a number of operating parameters. This effect can lead to embarrassing patterning effects in the system if care is not taken to control

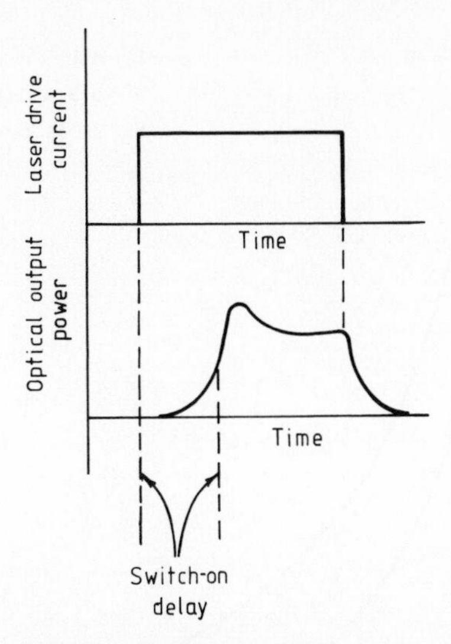

Figure 6.10. Schematic diagram of the laser switch-on delay

it. The switch-on delay of the laser is given by the expression

$$t_d = \tau \ln\left[1 + \frac{I_t - I_b}{I_p - (I_t - I_b)}\right] \tag{6.11}$$

where t_d is the delay, τ the carrier lifetime, I_t the threshold current, I_b the bias current and I_p the peak pulse current. For the GaAlAs devices, τ is typically in the range 2.5 to 6 ns. Taking a value of τ of 3.5 ns as being representative and thinking in terms of 100 Mbit/s system (10 ns per bit time), then we see that the delay in terms of the bit period means that the device cannot be run without substantial bias current unless I_t is very low (see Figure 6.11).

Many lasers today operate with threshold currents of 100 to 200 mA and pulse-drive currents of 10 to 100 mA to produce 1 mW output power per facet. Under such conditions, it is frequently undesirable to drive the device with a pulse current from zero, but rather it is preferable to apply a d.c. bias close to threshold (possibly above but usually just below) and then apply a separate pulse current. The advantages of this approach are lower pulse-drive currents, and lower switch-on delay, but the penalty is paid of higher power consumption, higher operating temperature and larger extinction ratio since the device emits significant power into the fibre during the 'off' period.

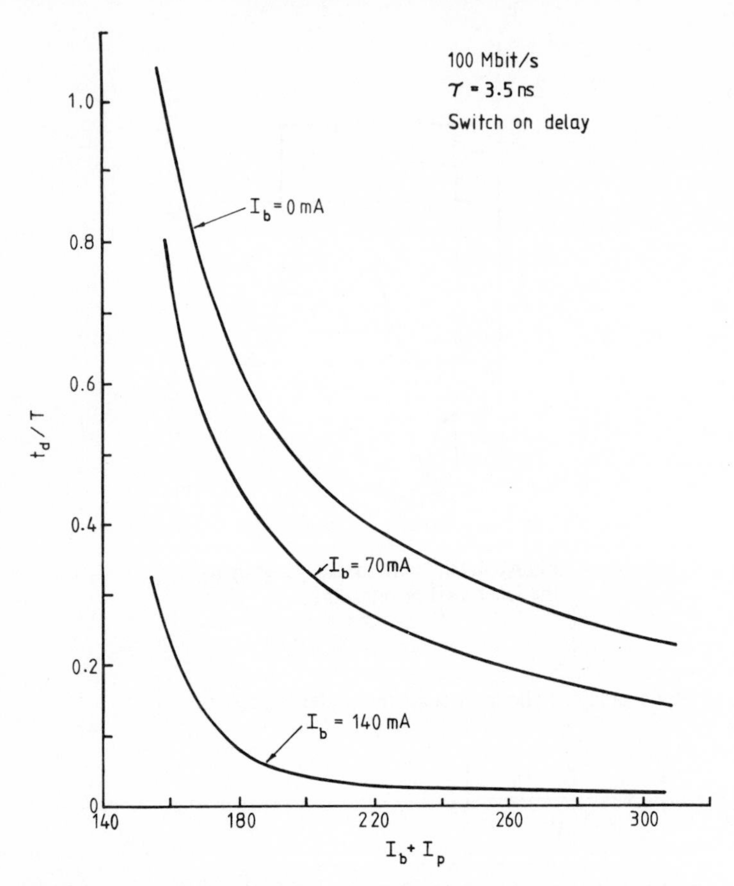

Figure 6.11. Plot of calculated switch-on delay normalized to the bit interval for 100 Mbit/s transmission rate using parameters typical of GaAlAs devices

6.3.3 Optical power control

Several strategies are available of varying complexity and sophistication to provide automatic control of a laser. In all cases some monitor is required for the output of the device. Physically this may take several forms, some of which are shown schematically in Figure 6.12, but all of which involve a detector monitoring the optical power of the device. A simple mean-power feedback control circuit is shown in Figure 6.12a. The data signals and the detector signals are each integrated and compared in an operational amplifier which is used to servo-control the d.c. bias applied to the laser. Constant current drive pulses are applied to the device of such an amplitude to ensure that there will always be sufficient drive range even if the slope efficiency of the device decays with age. It is necessary to integrate both data

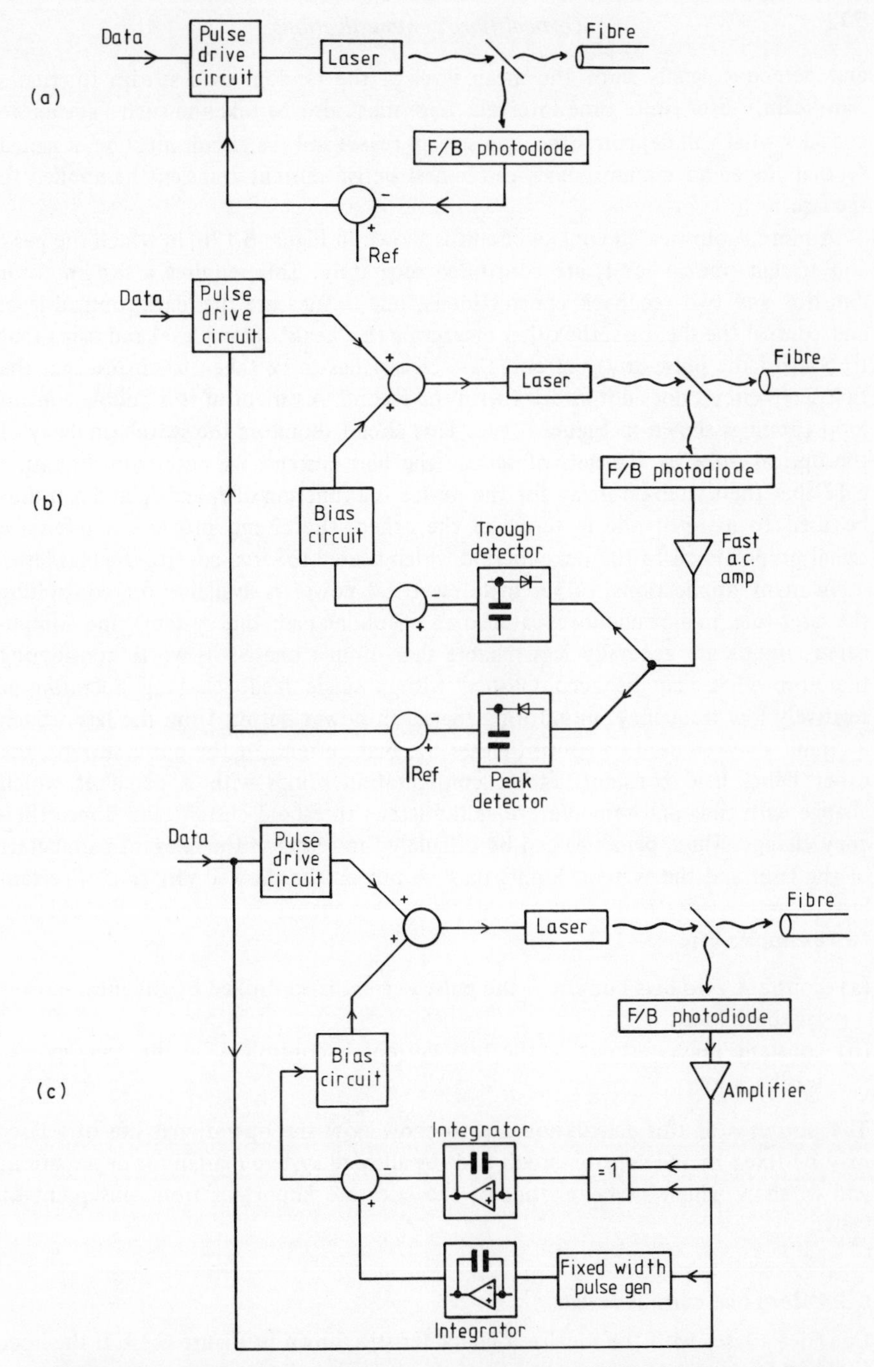

Figure 6.12. Three types of laser control circuits: (a) Mean-power feedback circuit. (b) Peak and pedestal control circuit. (c) Switch-on delay feedback circuit.[17] (Reproduced by permission of VDE-Verlag GmbH, Berlin)

and detector signals since the mean level of the random data stream fluctuates significantly over finite time intervals. Care must also be taken in such a system to consider what will happen if the data stream ceases and the circuit must be designed so that under no circumstances can a destructive current transient be applied to the laser.

A more sophisticated control circuit is shown in Figure 6.12b, in which the peak and trough optical levels are controlled separately. This requires a fast monitor detector and two feedback control loops, one to measure the 'dark' optical level and control the d.c. bias, the other measuring the 'peak' optical level and using that to control the pulse-drive current. Care clearly has to be taken to ensure that the first servo-circuit does not interact with the second. A variant of this double control loop circuit is shown in Figure 6.12c. This circuit monitors the switch-on delay of the optical pulse as a means of setting the bias current. We note from Equation 6.11 that the switch-on delay for the device is a function of I_t and I_b and can thus be used to monitor one in terms of the other. The circuit provides a reference signal proportional to the delay period which is used to servo-control the bias level.

In many applications, rather little electrical power is available for controlling the laser (e.g. in dependent repeaters on a telephone main-line system), and complicated circuits are generally less reliable than simple ones. It is worth considering therefore what may be accomplished with a single feedback loop operating at relatively low frequency, monitoring the mean power output from the laser. Such a circuit may be used to control either the bias current, or the pulse current, the other being held constant. Either configuration brings with it penalties which change with time and temperature as the laser's threshold current and slope efficiency change. These penalties can be calculated in terms of the relevant parameters of the laser and the system. Limits may be put on the allowed variation of certain laser characteristics before an acceptable system margin is exceeded. We will give two examples here:

(a) constant zero-bias current – the pulse current is controlled by the mean-power feedback loop;
(b) constant pulse current – the bias current is controlled by the mean-power feedback loop.

The purposes of this calculation are to show how the operational life of a laser may be fixed by the control circuit and the allowed system margin for degradation, and to show which characteristics of the laser are important from this point of view.

6.3.4 Zero-bias current design

Consider a laser with the idealized characteristic shown in Figure 6.13. If the laser is pulsed from zero bias, the switch-on delay is given by

$$t_d = \tau \ln(1 + \alpha I_t/I_p) = \ln(1 + \zeta) \tag{6.12}$$

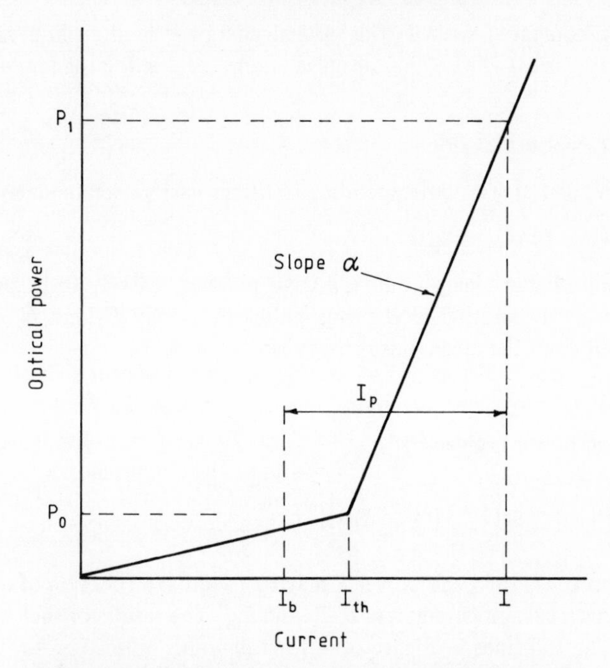

Figure 6.13. Schematic of the laser output versus current, defining the parameters used in the analysis of laser performance with a single feedback loop

where τ is the effective carrier lifetime and α is the mean slope efficiency above threshold. If the carrier inversion is not zero in the laser cavity at the onset of the current pulse but has some normalized value x_{p} due to pre-pumping, the switch-on delay is reduced:

$$t'_{\mathrm{d}} = \tau \ln (1 + \zeta x_{\mathrm{p}}) \tag{6.13}$$

The value of x_{p} depends on the decay of the carrier inversion while the laser is off between pulses. At 140 Mbit/s the decay is virtually complete after one unit interval, so we need only consider the decay between neighbouring ONE pulses. We will consider the general case of reduced-width current pulses of duration T_{c} in a unit interval T. The special case of full-width pulses follows simply:

let

$$\tau_{\mathrm{c}} = T - T_{\mathrm{c}}$$

then

$$x_{\mathrm{p}} = 1 - \exp(-\tau_{\mathrm{c}}/\tau) \tag{6.14}$$

Because the current pulse and hence the output power P are held constant on a timescale long compared with T, the optical energy E in the pulse varies as the switch-on delay varies. The minimum pulse energy occurs for the longer switch-on delay:

$$E_1 = P[T_c - \tau \ln(1 + \zeta)] \qquad (6.15)$$

The maximum pulse energy, corresponding to the shorter switch-on delay is:

$$E_2 = P[T_c - \tau \ln(1 + \zeta x_p)] \qquad (6.16)$$

Assuming equiprobable binary coding, a ONE pulse is as likely to be preceded by a ZERO pulse as by another ONE; thus pulses of energy E_1 and E_2 occur with equal probability and the mean pulse energy is:

$$E_m = \tfrac{1}{2}(E_1 + E_2)$$

Hence the mean power is given by:

$$P_m = P\left[T_c - \frac{\tau}{2} \ln(1 + \zeta)(1 + \zeta x_p)\right] / 2T \qquad (6.17)$$

From Equation 6.17, the peak power P may be found as a function of αI_t in terms of τ and the given system parameters T, T_c and P_m. The results of such calculations are shown as the full lines in Figure 6.14 for full-width pulses ($x_p = \tau_c = 0$) and in Figure 6.15 for half-width pulses. We have taken $T = 6.25$ ns appropriate to 160 Mbit/s systems, and $\tau = 3$ ns, typical of current lasers.

Although the mean power $P_m = E_m/2T$ is held constant by the feedback control, the system must cater for pulse energies as low as E_1. This 'patterning penalty' may be expressed as a required system margin $10 \log(E_1/E_m)$, which is a function of αI_t and is shown on Figures 6.14 and 6.15 as the full curves. The parameter αI_t varies with time and temperature, but not monotonically, so the horizontal axis on Figures 6.14 and 6.15 is not to be interpreted simply as a time-scale. Rather, αI_t may vary within a range of values. The permitted range depends on the maximum power output that can be obtained from the laser without unacceptable degradation, and on the system margin allowed for the patterning penalty. Table 6.1 lists the maximum values of αI_t assuming a maximum power of 6 mW for full-width pulses and 9 mW for half-width pulses, and a system margin of 3 dB.

We have allowed a higher maximum power output for half-width pulses on the assumption that the degradation rate depends on the mean power output more than on the peak. Once the maximum power output is reached, any further increase in αI_t results in the mean power decreasing. A penalty is thus incurred, but provided that this penalty added to the patterning penalty does not exceed the allotted system margin, the laser may be considered to be still within its permitted range of operation. The extension to the range of αI_t is shown as $(\alpha I_t)_{ext}$ in Table 6.1; it is very useful in most cases. Similar tables may be drawn up for other maximum output powers and system margins by inspection of Figures 6.14 and 6.15.

Table 6.1 Maximum values of I_t

Full-width pulses			
P_m (mW)	1.5	2.5	3.5
$(\alpha I_t)_{max}$	13.5 (a)	6.2 (b)	–
$(\alpha I_t)_{ext}$	13.5 (a)	17.0 (c)	12.0 (c)
Half-width pulses			
P_m (mW)	1.0	1.5	2.0
$(\alpha I_t)_{max}$	8.6 (b)	4.5 (b)	1.35 (b)
$(\alpha I_t)_{ext}$	15.0 (c)	14.0 (c)	12.0 (c)

Limitations: (a) patterning; (b) peak power; (c) combination

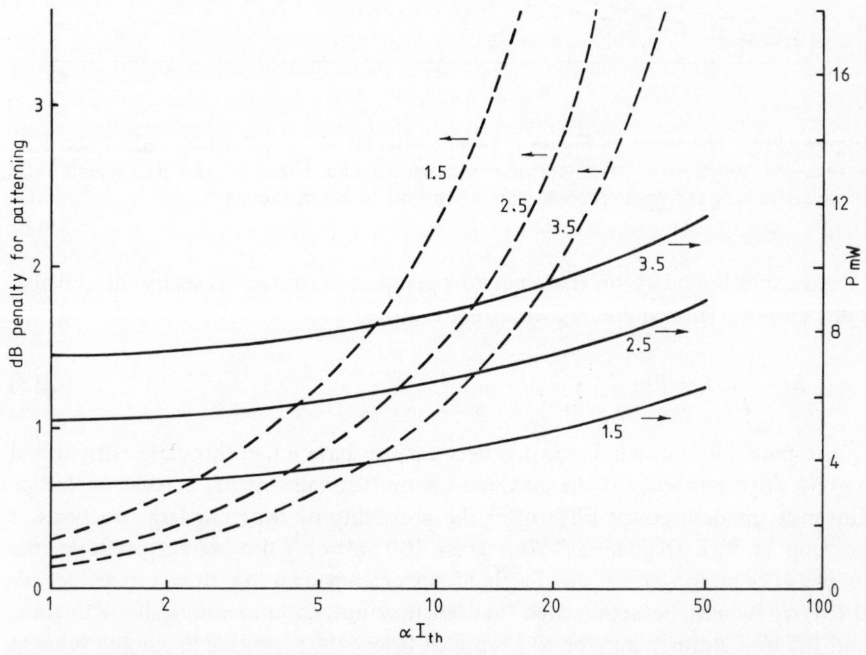

Figure 6.14. Plot of calculated penalty due to patterning (broken lines) as a function of the parameter αI_{th} the full-width transmit pulses, holding mean power constant and with zero-bias current. Also shown (solid lines) is the peak power P. The different curves are for the various values of the mean power in mW. As expected, the patterning with a device of low threshold current and low slope is much less serious than with a device of high threshold current and high slope, indicating the desirability of lower threshold current

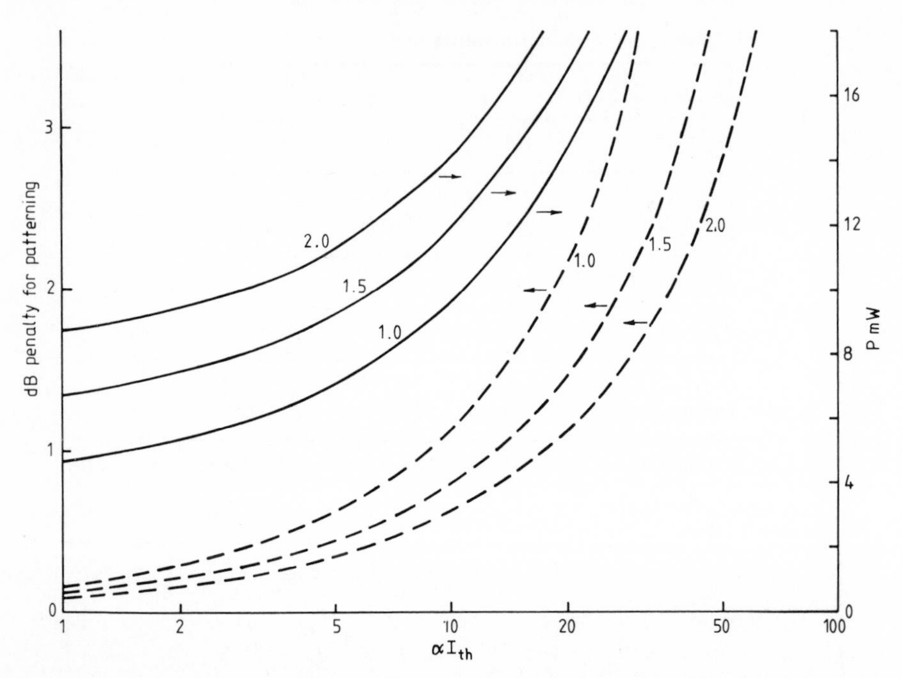

Figure 6.15. The same as Figure 6.14 but calculated for half-width pulses, showing a reduced sensitivity to patterning because of the effect of the half-width 'off' period allowing for some recovery independent of bit sequence

A further boundary on the range of operation of the laser is set by the available pulse current. The required pulse current is given by:

$$I_p = \frac{1}{\alpha}(\alpha I_t + P) \tag{6.18}$$

If the pulse current is limited, it is necessary to have a low threshold current and a good slope efficiency if the maximum permitted value of αI_t is to be exploited. However, modern power FETs offer the possibility of switching large fractions of an amp at high frequencies. With even 100 mA of pulse current available, the emphasis is no longer on low threshold current but on a low slope efficiency, say 0.1 W/A. It must be appreciated that αI_t may not vary monotonically with time, and the best strategy may be to keep well below the permitted maximum value to allow the greatest range within which αI_t can vary.

6.3.5 Constant pulse current design

The optical power–current characteristic is also shown in Figure 6.13 which explains the notation. The bias current I_b is adjusted by the feedback circuit to give a constant mean power output. We will assume that, so long as the bias is below threshold, the power output on the zero level P_0 remains nearly constant. With this control configuration, so long as I_b is less than I_t, there is again a patterning penalty as in the zero-bias current case, but the switch-on delays involved are different:

$$t_d = \tau \ln(\alpha I_p/\Delta P) = \tau \ln \zeta \tag{6.19}$$

where

$$\Delta P = P_1 - P_0$$

$$t_d' = \tau \ln(\zeta x_p + e^{-T_c/\tau}) \tag{6.20}$$

The minimum and maximum pulse energies are:

$$E_1 = \Delta P[T_c - \tau \ln \zeta] \tag{6.21}$$

$$E_2 = \Delta P[T_c - \tau \ln(\zeta x_p + e^{-T_c/\tau})] \tag{6.22}$$

The mean power is thus:

$$P_m = P_0 + \frac{\Delta P}{2T}\left[T_c - \frac{\tau}{2} \ln \zeta(\zeta x_p + e^{-T_c/\tau})\right] \tag{6.23}$$

This equation can be solved for ΔP as a function of αI_p in terms of P_m, P_c, T, T_c and τ. The results of such calculations are shown as the broken lines in Figure 6.16 for full-width pulses and in Figure 6.17 for half-width pulses, using the same parameters as before. The patterning penalty is shown as the full curves. This penalty becomes zero for some value of αI_p at which the bias current reaches threshold and the switch-on delay becomes zero:

$$(\alpha I_p)_c = \frac{2T}{T_c}(P_m - P_0) \tag{6.24}$$

If αI_p decreases below this value, a different penalty is incurred because the extinction ratio rises. The extinction ratio is given by:

$$\epsilon = \frac{TP_m - \alpha I_p T_c/2}{TP_m + \alpha I_p T_c/2} \tag{6.25}$$

The penalty for a nonzero extinction ratio has been calculated by Personick[4] and by Hooper and White[7] as a function of fibre bandwidth. The resulting penalty in dB is shown on Figures 6.16 and 6.17 as the sharply rising lines.

The parameter αI_p decreases during the life of a laser (I_p is assumed constant here, of course); it may go up and down with temperature fluctuations, but its long-term change is downwards. Thus the abscissae of Figures 6.16 and 6.17 may be interpreted as (nonlinear) time-scales. The parameter αI_p may decrease over a range which may be read from Figures 6.16 and 6.17, if the maximum output power and

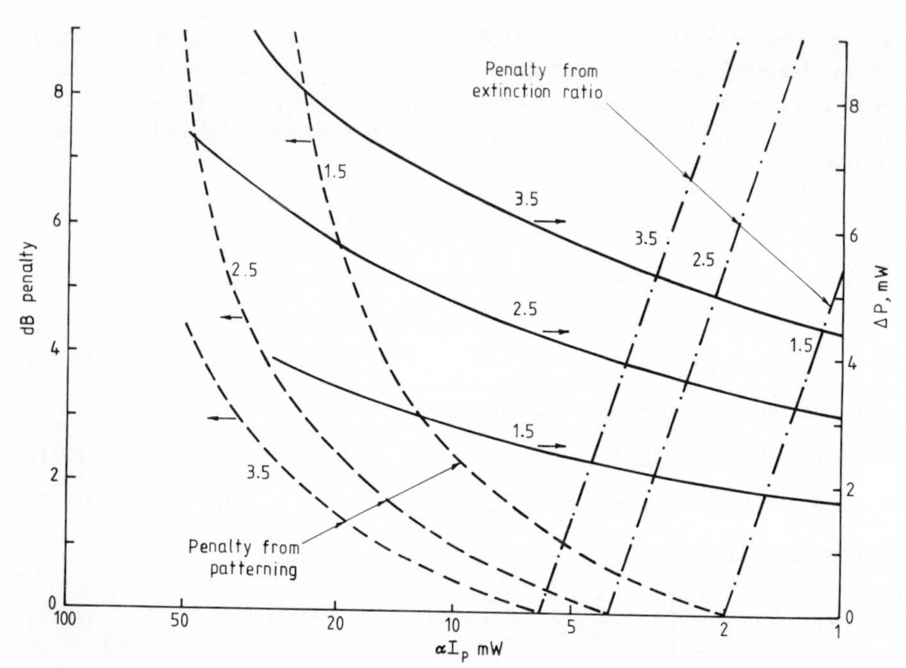

Figure 6.16. Plot of power penalty versus αI_p for a constant bias current transmitter design, assuming full-width transmit pulses. The curves are again plotted for various mean power values (labelled). The patterning penalty is zero when $I_b = I_{th}$; below this it increases as I_b decreases (shown dashed) and above this there is a sharp increase in power penalty at the receiver because of the finite extinction ratio at the source (shown chain-dotted). The solid curve shows the peak power required from the transmitter to achieve the given mean power as a function of αI_p

Table 6.2 Ranges for αI_p

Full-width pulses

P_m	$(\alpha I_p)_{min}$	$(\alpha I_p)_{max}$	$(\alpha I_p)_{ext}$	Range	Ext range
1.5	1.36	12.2 (a)	12.2 (a)	9.0	9.0
2.5	2.76	24.5 (a, b)	24.5 (a, b)	8.9	8.9
3.5	4.10	6.0 (b)	24.0 (c)	1.5	5.9

Half-width pulses

P_m	$(\alpha I_p)_{min}$	$(\alpha I_p)_{max}$	$(\alpha I_p)_{ext}$	Range	Ext range
1.0	1.36	22 (a, b)	22 (a, b)	16.2	16.2
1.5	2.80	18 (b)		6.4	
2.0	4.20	14 (b)		3.3	

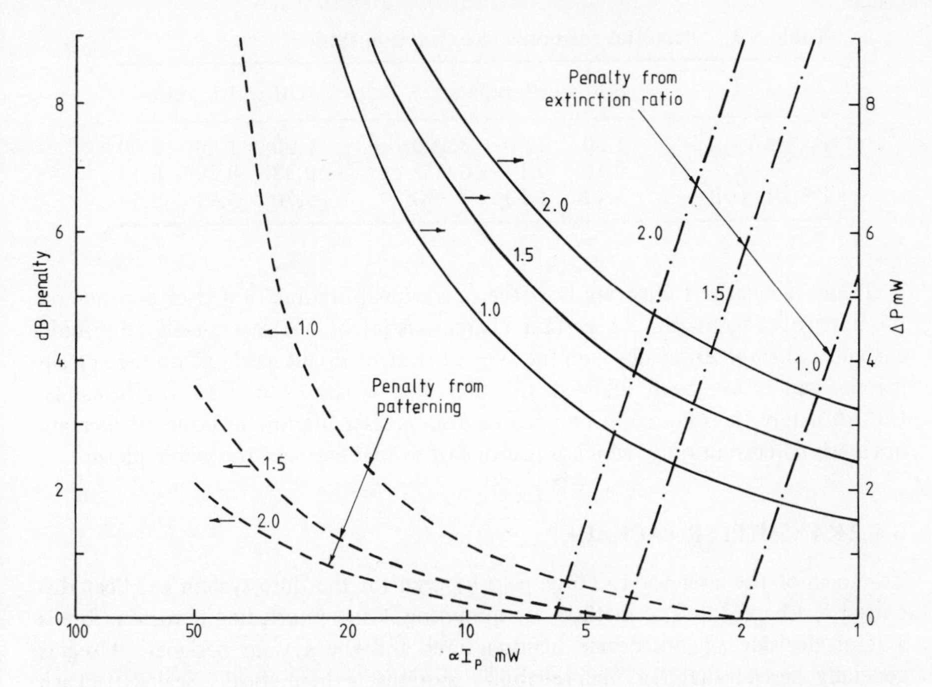

Figure 6.17. The same plot as Figure 6.16 but for half-width transmit pulses. As before, the patterning penalty is reduced because of the recovery time afforded by the half-width 'off' period, regardless of bit sequence

system margin are stated. In Table 6.2, such ranges are given, assuming the same limits as before. As before, the range can be extended by allowing the mean power to decrease when the peak power is at its maximum, provided the total penalty remains within the system margin.

A further boundary on the operational life of a laser is imposed by the threshold current, as the required bias is close to the threshold current in most cases. The difference between I_t and I_b is given by

$$I_t - I_b = (\alpha I_p - \Delta P) \tag{6.26}$$

So long as the d.c. bias can be supplied, there is no particular advantage in using a laser with a low threshold. More important is a high slope efficiency so that the upper limit of αI_p can be realized for practical pulse currents. It should be noted that at no time is the extinction ratio zero in this configuration (unless by chance the bias current should go down to zero). At the point when αI_p reaches the critical value and the switch-on delay goes to zero, the extinction ratio and the corresponding penalty are as detailed in Table 6.3. It can be seen that although the low values of P_m give the widest range of αI_p, they cause a larger extinction ratio penalty.

Table 6.3 Penalties for nonzero extinction ratio

	Full-width pulses			Half-width pulses		
P_m (mW)	1.50	2.50	3.500	1.00	1.50	2.00
ϵ	0.20	0.11	0.077	0.33	0.20	0.14
Penalty (dB)	3.42	1.93	1.350	5.70	3.42	2.47

These calculations illustrate how the operational lifetime of a laser depends on the rate of degradation of certain characteristics of the laser itself (threshold current and slope efficiency), on the type of control circuit used and on the system margin which has been allotted to cater for variations in laser performance. Unfortunately, it is common practice to discuss laser lifetime in terms of increase of threshold current only, which is just one of several facets of the whole picture.

6.4 TRANSMITTER PACKAGE

The form of the laser (or LED) chip to be used for the fibre system has been discussed in Chapter 2. The method for mounting it and interfacing it with a fibre is a joint decision of the device manufacturer and the system designer. It seems generally agreed that for high-reliability systems, a hermetically sealed package is desirable for the device. Such a device is characterized by electrical inputs and optical output so the package must cater for both. The initial devices have tended to use packages with large-area windows through which the optical output of the device passes. These packages were then attached to a suitable heat sink and lenses used to couple the emitted light into the fibre. A number of objections are apparent in this approach. The package itself, when integrated with lenses and fibre, tends to be rather cumbersome. Precision alignment of the fibre to the device has to be carried out when the device in its package is mounted or replaced. Monitoring of the optical output of the device adds further to these problems, since either a beam splitter has to be inserted in the output beam, further adding to the size, weight and alignment difficulties, or power is taken out of the package in the back direction by the addition of a second window, greatly complicating the sealing and assembly problems.

The solution to these problems that appears to be gaining favour is to mount the laser chip in a composite package with the monitor detector for its control circuits, some or all of the control and drive circuit, and to take the optical power out of the package by means of an optical fibre which itself passes through the hermetic seal. A transmit package of this type is shown schematically in Figure 6.18. Two further advantages accrue to this type of package. The whole assembly is of the dimensions of a large-scale integrated circuit and can thus be placed on a printed circuit board along with other components wherever this is mechanically convenient, while the compact nature of the drive circuit reduces stray capacitance

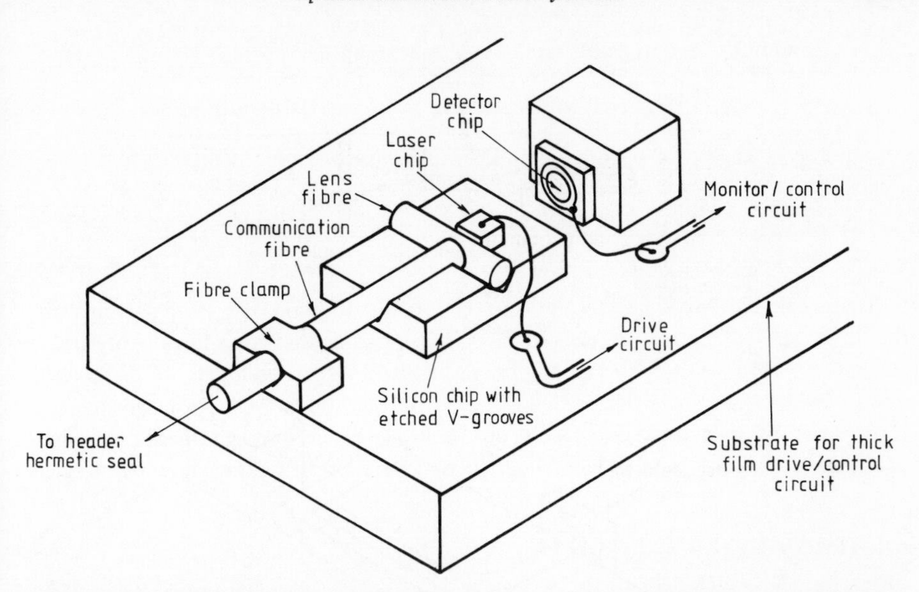

Figure 6.18. Schematic layout for a transmitter module, showing the laser chip mounted on a heat sink which locates the fibre, cylindrical lens (cross fibre) and chip in optical alignment, and is mounted on a thick-film integral drive circuit. The whole assembly would be contained within a hermatic package, providing for electrical inputs and optical output via a fibre tail

and eases the problems of driving a laser with high current pulses at high bit rates. From the manufacturers' point of view, the package also has the attraction that all the precision assembly and alignment work is done at the one place in the semiconductor plant which provides almost ideal conditions for that type of work, while the installation of the device requires little skill additional to that required for mounting LSIs, other than in the fitting of the fibre tail to some suitable bulkhead mounted fibre/fibre connector.

Techniques for aligning the fibre to the device chip in such a package are not yet universally established, although one approach certainly looks attractive. This relies on a silicon mount for the fibre, lens and device chip, all of which are located by grooves formed lithographically in the silicon. This technology allows very accurate grooves to be formed cheaply and leads to a structure of the type shown schematically in Figure 6.18. Having attached the device chip plus fibre and lens to this, the whole assembly is then mounted in the header package for the complete transmitter with a thick-film drive and control circuit, and electrical connections are made where necessary by bonding wires between the appropriate points, and to the header pins for external connections. Notice also that the same package is shown with the monitor detector included. This can, in principle, monitor the power emerging from the back face of the chip assuming that a 100% reflector has not

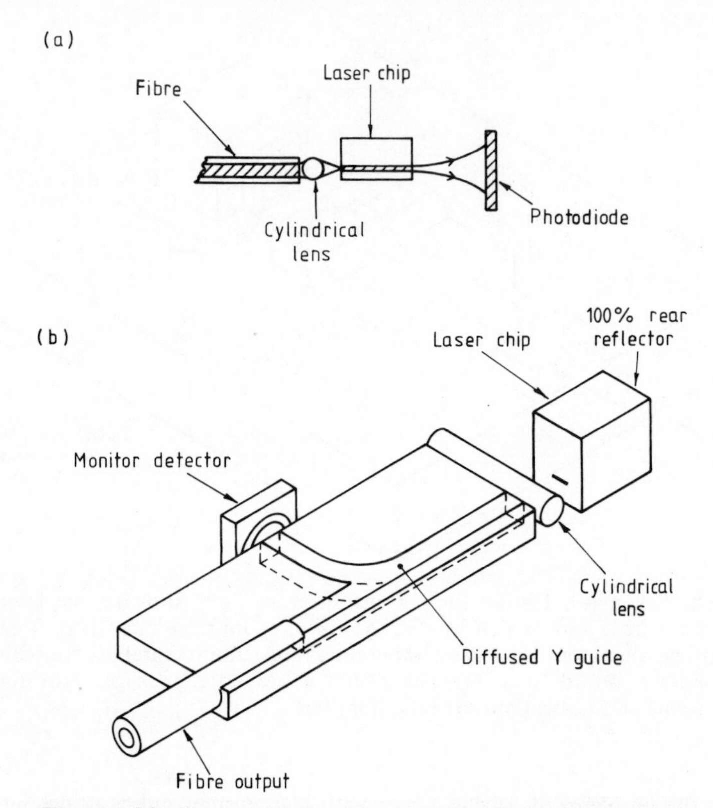

Figure 6.19. Two methods of monitoring the laser power output in a transmitter module: (a) using the power emitted from the back facet of the chip; and (b) using a specially fabricated beam splitter in the main output path. In principle, a method of the type (b) seems the correct way but few such techniques have been reported

been applied, or it can monitor the front-face power or the power launched into the fibre by suitable variations in mounting. The two most attractive configurations appear to be those shown in Figure 6.19.

6.5 TRANSMITTER DESIGN AND OPTIMIZATION

We have discussed some techniques for controlling a laser transmitter, but said very little about how one should choose the optimum operating conditions for it. Evidently, this cannot be done in isolation, since it is not meaningful to ask such questions outside the systems context. We will initially examine how to assess the optimum operating conditions for a transmitter in a system when new. In Figure 6.20 we show the block schematic of the laser transmitter with the various factors

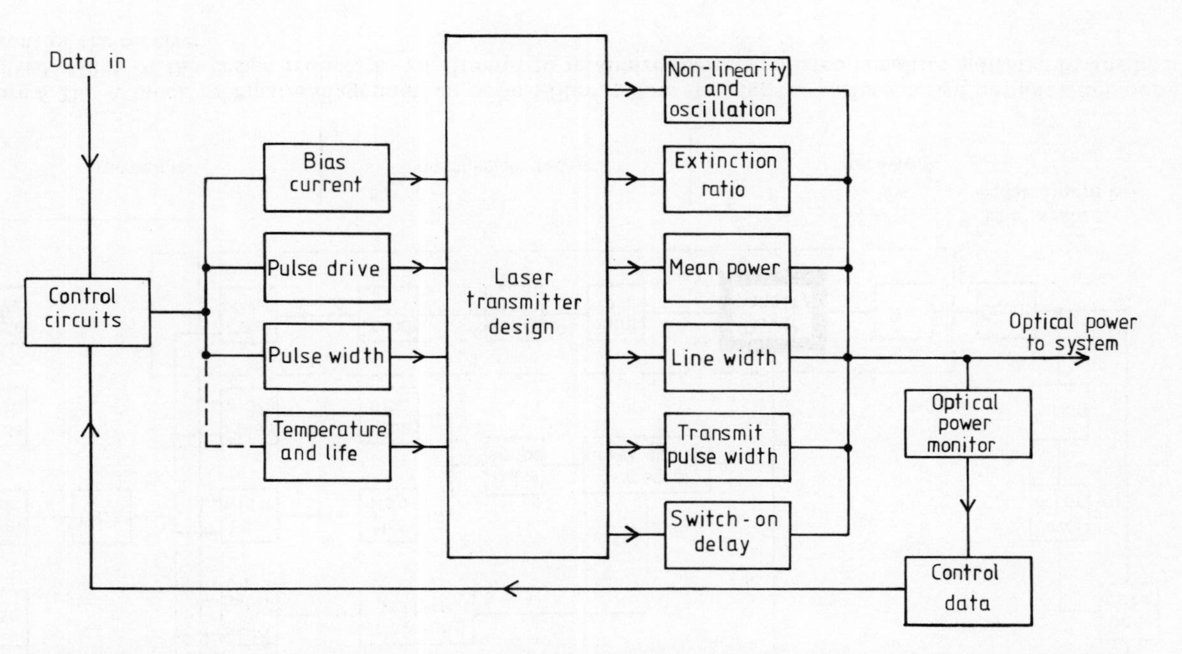

Figure 6.20. A block schematic diagram of a laser transmitter illustrating the various parameters involved in the control and performance specification

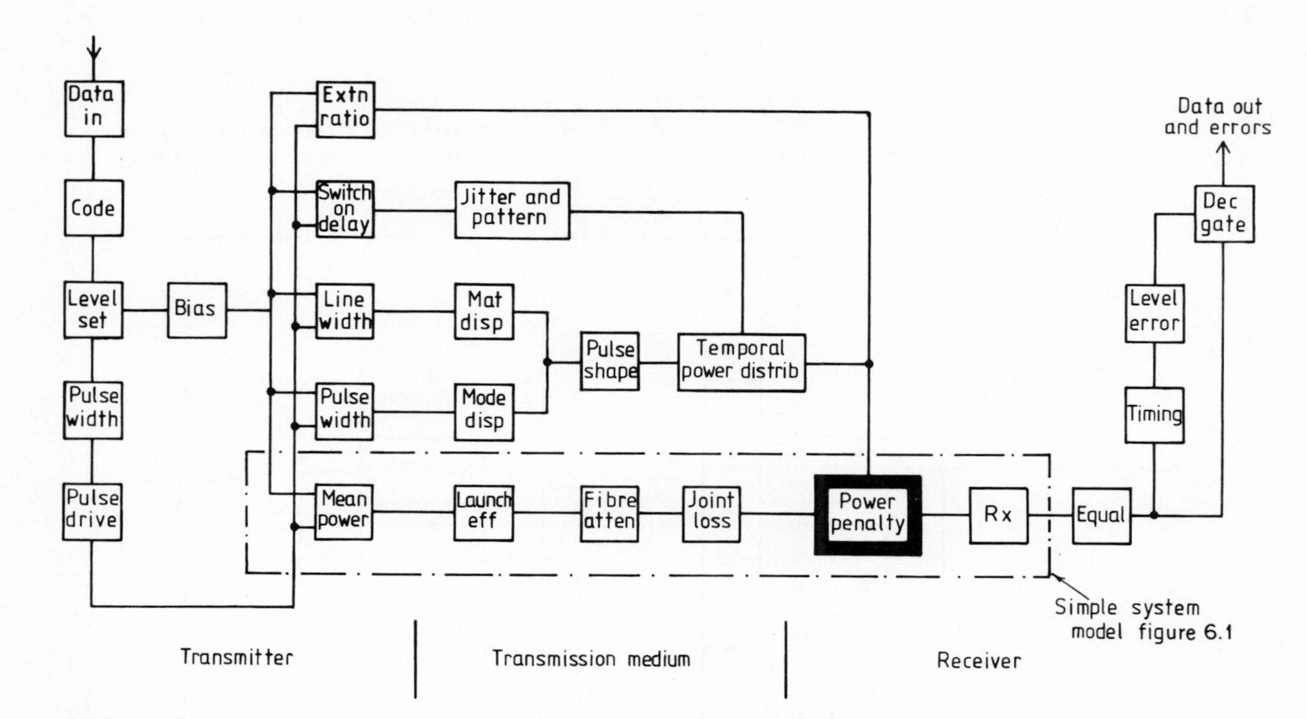

Figure 6.21. A block schematic diagram of an optical-fibre system showing the various design optimization loops involved. Much of the design reduces to an attempt to minimize the accumulated penalties indicated by the box in front of the receiver

that interact shown by connections. To discover how to optimize the various parameters in the transmitter we must connect it to our system, as shown in Figure 6.21, and decide on some optimization criterion. An obvious one would be to seek the minimum error rate for a given source, fibre and detector combination while staying within safe operating range for the components. This requirement reduces to the task of maximizing the power at the receiver as shown in Figure 6.21, as a function of the variables available to the designer, if we ignore the effects of jitter for which slightly different optimizations apply. This generally means minimizing the power penalty.

Some interesting design loops occur. Increasing the laser bias reduces the switch-on delay, giving a longer output pulse and hence greater mean power; it reduces the linewidth generally and it gives a less desirable extinction ratio and higher operating temperature and degradation rate for chip. Increasing the pulse-drive current would give a faster rise-time pulse and potentially a shorter drive pulse, higher peak power and better extinction ratio although it would also be accompanied by greater linewidth and hence material dispersion. A longer drive pulse leads to a greater mean output power, better extinction ratio, possibly narrower average linewidth and less material dispersion, but when convolved with total dispersion, a longer received pulse. We see that in general each input parameter produces both desirable and undesirable results and in practice these have to be balanced. Some simple first-order loops are shown in the next few examples and apply when a single mechanism dominates.

The first is shown in Figure 6.22 which illustrates an LED system operating over good graded-index fibre so that material dispersion dominates the pulse dispersion. Since the material's dispersion is linearly proportional to the source linewidth, filtering that line with a narrow-band filter will reduce the total pulse dispersion.

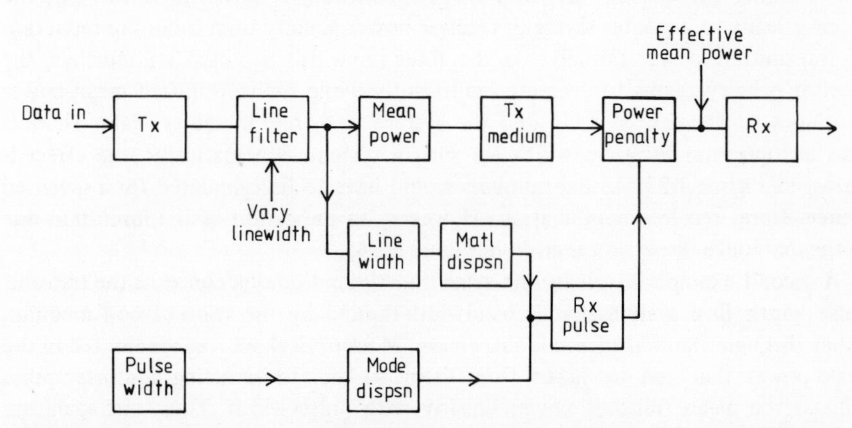

Figure 6.22. A simple optimization loop showing the effects of filtering the LED source linewidth, so that launch power decreases but so also does the receiver power penalty because of reduced material dispersion

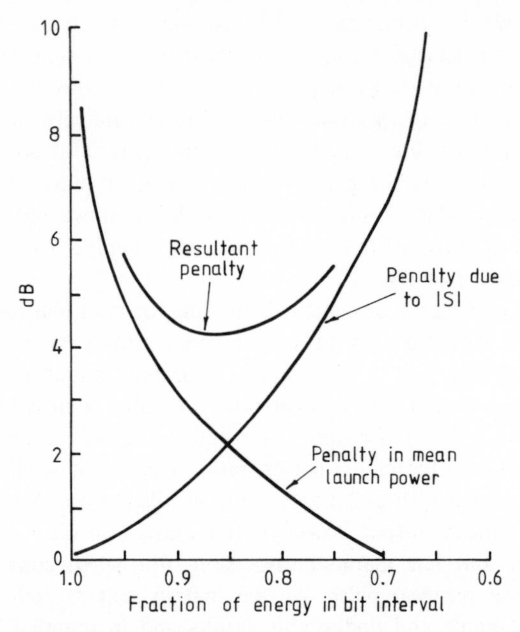

Figure 6.23. A breakdown of the form of loss penalties for the case of Figure 6.22 showing how an optimal situation is reached

If the system is significantly material-dispersion limited, with a receiver penalty due to intersymbol interference of a number of decibels, then it is usually found that filtering the source linewidth before launching it into the fibre or at the receiver leads to a greater saving of receiver power penalty than it does in reduction of transmitted power initially. As the filter linewidth is closed still further, the receiver penalty ceases to decrease significantly while the transmitted mean power continues to decrease steadily and the system error rate increases again. There is thus an optimum filter linewidth for such a system. Schematically, the effect is shown in Figure 6.23. Actual numbers would have to be calculated for a specified source–fibre–receiver combination. However, an indication of the limitation due to the material's dispersion is given in Figure 6.24.

A second example involving the same underlying thinking concerns the transmit pulse width in a system that is bandwidth-limited by the transmission medium, either through material or mode dispersion. Most optical sources are limited in the peak power that can be taken from them, so that transmitting a shorter pulse reduces the mean transmit power linearly with pulse width. Thus, one again has a situation very similar to that in the LED system discussed above. Shortening the transmit pulse will lead to shorter received pulse after it has been convolved with the line dispersion. This in turn leads to a reduced power penalty, which can more

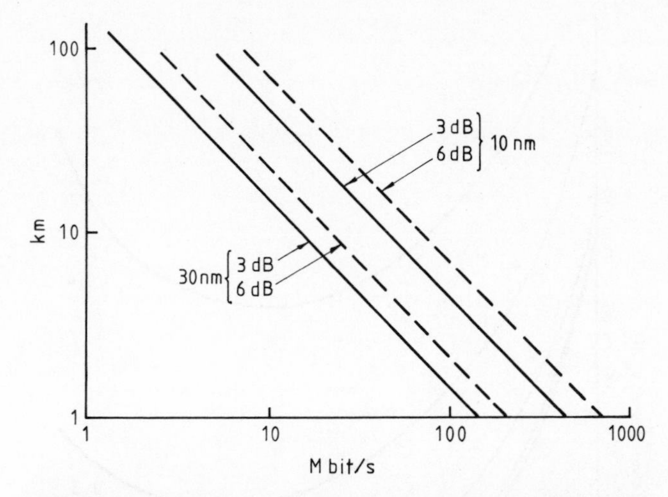

Figure 6.24. The material dispersion limits for LEDs operating at about 840 to 900 nm. The 6 dB and 3 dB power penalty lines are plotted for half-width transmit pulses assuming linewidths of 30 and 10 nm

than compensate for the decrease in transmitted power resulting from the shortened pulse. Error rate again goes through a minimum as the transmit pulse is reduced from a full-width pulse (non-return-to-zero, NRZ) to a pulse that is short compared to the bit interval (return-to-zero, RTZ). In Figure 6.25 we show the result of some calculations in which the effect of shortening the transmit pulse is shown, indicating that the optimum pulse width for a wide variety of line dispersions lies in the range 0.5 to 0.8 of the full-width pulse, with the error rate going through a broad minimum in that range. These results are plotted for a variety of transmission medium bandwidths, characterized in terms of the parameter α_F, the rms received pulse width that would be obtained when a delta-function pulse was launched into the fibre section, and the launched pulse width, α_L. For simple comparison, the value of α_F is related to the $1/e$ full width of the (assumed) Gaussian fibre response by the relation

$$\tau_e = 2\sqrt{2}\alpha_F T \qquad \text{(FW } 1/e \text{ optical pulse width, } T = \text{bit interval)}$$

while a value of $\alpha_L = 1$ corresponds to a full-width rectangular transmit pulse.

Both these cases are somewhat idealized, however. A moment's thought shows that in the latter case, for example, unless the laser is pulsed from zero, then the decrease in mean power accompanying the shortened drive pulse will also be accompanied by a worsened extinction ratio so that a further penalty from this must be added. A further example of a multiple effect is shown in Figure 6.26 which shows the effect of laser bias current on both extinction ratio and linewidth and their relative effects at the receiver. The curves are calculated for an 800 Mbit/s

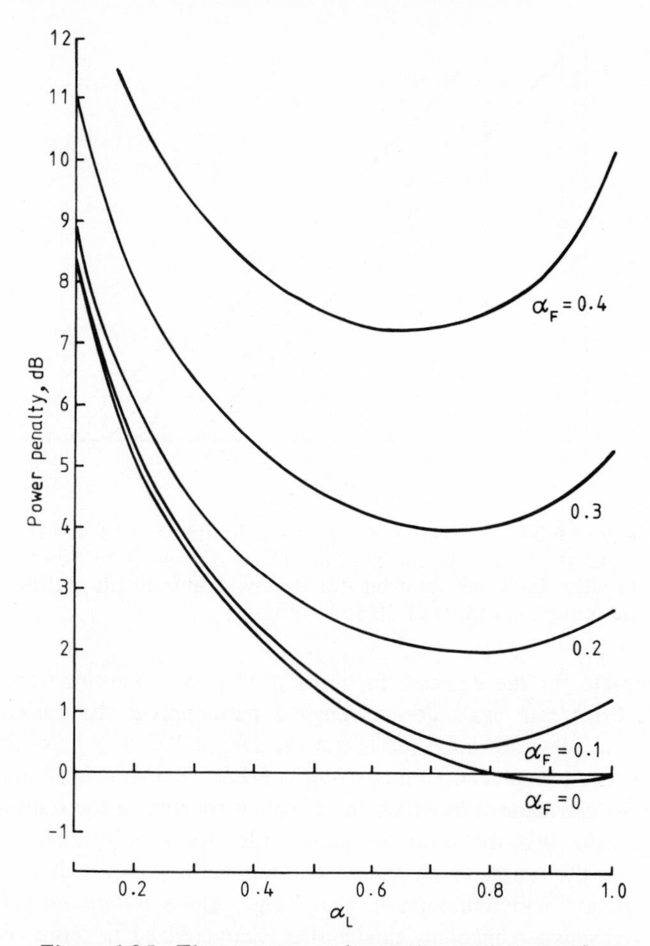

Figure 6.25. The power penalty at the receiver as a function of transmit pulse width for various degrees of bandwidth limitation in the transmission medium. The parameter α_L describes the transmit pulse width, a value of 1 corresponding to full-width rectangular launched pulse, 0.5 to half-width. The parameter α_F defines the fibre impulse response[5]

system operating over 7.3 km of fibre. While either parameter on its own shows a rapid variation in receiver power penalty with bias current, the combined effect in the total system is rather small in this case.

In summary, we may say that the model schematically set out in Figure 6.21 and discussed briefly from a theoretical point of view earlier, does provide a basis for optimizing the operation of a digital fibre system. However, in applying the model, very great care must be exercised and a thorough knowledge of the detailed operating characteristics of each of the devices employed is essential before they are 'inserted' into the system and their interactions examined.

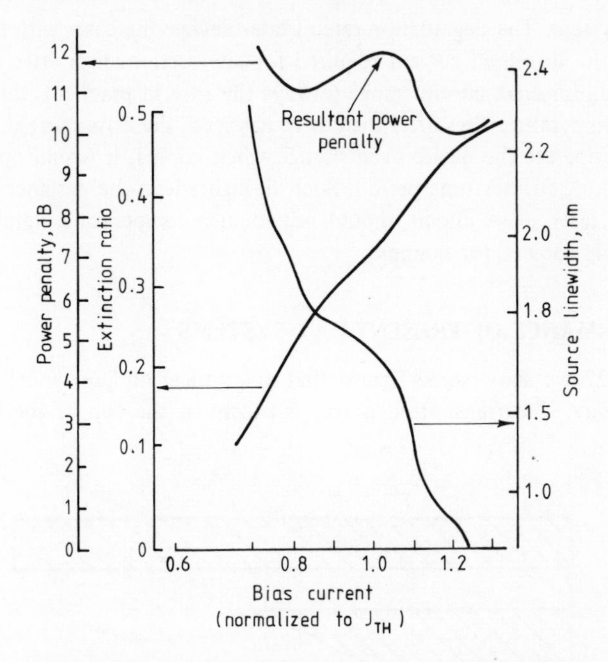

Figure 6.26. The results of a particular evaluation of the effect of laser bias current on linewidth and extinction ratio for an 800 Mbit/s transmission system over 7.3 km of fibre. Both are plotted against equivalent dB receiver power penalty scales so that the total resultant power penalty can be seen as a function of bias current. (Reproduced by permission of the IEEE)

6.6 DEVICE LIFETIME

In the above discussions we have ignored lifetime considerations in the design process. In reality these must be taken into account in two ways. The transmitter circuits must be designed to accommodate the range of device operation that will be experienced during its operating life. To a first approximation, this is a matter of straight design, allowing adequate operating margins in bias and drive current, together with receiver sensitivity to accommodate effects. However, there may also be more positive actions open to the designer. For example, for degradation that is proportional to the operating life of the device, one may choose to use RTZ operation, with deliberately short drive pulse so that the operating period may be reduced to 0.25 or less of the total system time. This also suggests the use of zero-bias current and with lower threshold current lasers that are now becoming available (down to 10 mA or less), or larger pulse currents, this may well become an attractive option. We should also note that the effects of temperature are doubly severe in a system. The degradation rate of laser devices increases with temperature and so does the threshold current required for laser action. In a drive circuit with a finite maximum drive current range (always the case in practice), this will mean that if the temperature rises after the device has aged, the drive current may not be sufficient to operate the device even though when cooled, it would operate effectively for a long further time period. Such thoughts lead the designer to examine whether the laser drive circuit should not include temperature stabilization, by thermo-electric coolers, for example.

6.7 PERFORMANCE OF PRESENT-DAY SYSTEMS

In Figure 6.27 we show some figures that are typical of laser and LED systems operating today. The transmitted power is shown at the top of the figure while

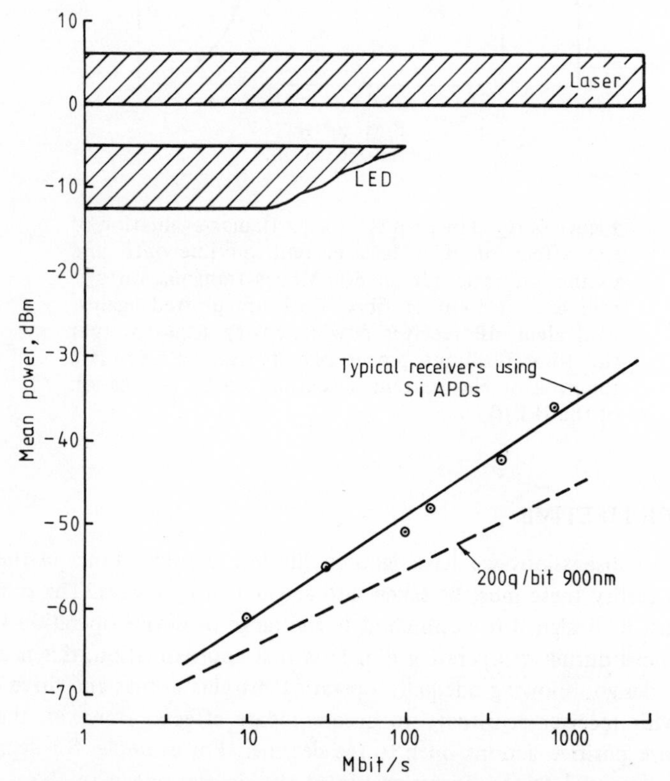

Figure 6.27. A simple illustration of the system power budget for typical fibre systems. The upper curves show the typical launch power from laser and LED sources while the lower curves show typical receiver sensitivities for 10^{-9} error rate. Also shown is the theoretical curve for 200 quanta per bit

the receiver power found necessary for 10^{-9} error rate is given at the bottom, together with a line for 200 quanta/bit received power calculated for 900 nm wavelength. The receiver points shown represent optical receivers in numerous experimental systems around the world, all of which use avalanche photodiodes.

In Figure 6.28 we show some of the repeater separations achieved in experimental systems versus bit rate, and using the data of Figure 6.27 for laser sources, we plot on the same figure the expected repeater separations for fibres of

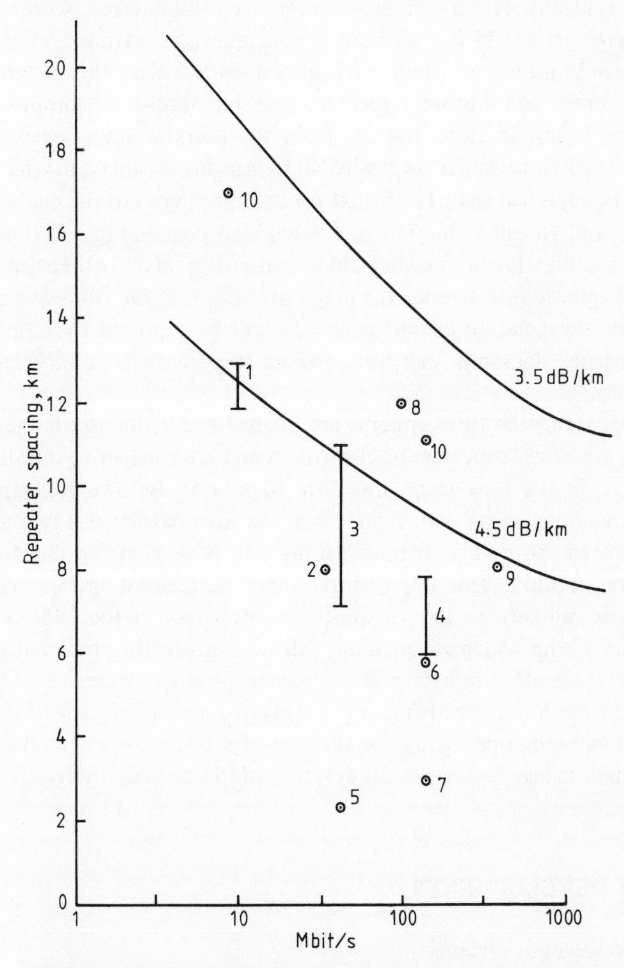

Figure 6.28. A summary of some first-generation fibre system performances, illustrating the long repeater spacings already achieved. Also noteworthy is the GEC 8 Mbit/s system[10] which uses an LED at 900 nm source but fibre cable with a loss significantly under 3 dB/km

various mean losses (including joints) after allowing a 5 dB operating margin. We see that most of the current results fall on the line for 4.5 dB/km fibre loss, a figure which is broadly representative of present fibre cable[19,20] although much better results will undoubtedly be achieved in the future. The reader can easily repeat the calculation to obtain a feel for what performance is likely to become possible as the fibre and cable losses fall still further.

Figure 6.28 also demonstrates one of the major attractions of using optical-fibre systems when we note that the present-day coaxial cable systems, the fibres' competitors, typically require repeaters every few kilometres. Within cities, the ability to travel 10 to 15 km without a repeater is even more attractive since switching-centre buildings are frequently spaced by less than this distance, so that remotely powered and housed repeaters can be abolished completely. Other attractions are found in fibre systems from the users' point of view. The small size and weight of fibre cables, typically 40 kg/km for a cable carrying 8000 two-way conversations, is less than 1% of that for an equivalent coaxial cable. Stemming from this size and weight reduction goes easier transport and installation and much more efficient utilization of existing duct space. For many other users, freedom from electromagnetic interference is a major attraction of the fibre system; and just as electromagnetic radiation cannot penetrate and be captured by a fibre cable, so the light it carries does not leak out, making it potentially a very secure transmission medium.

In the short term, the fibre systems are likely to save the user money primarily because they use fewer repeaters or electronics and are cheaper to install. However, it is clear that, in the long term, the fibre is potentially a very cheap wideband transmission medium in its own right, since the materials cost is typically 50p to £1 for a kilometre, so that its manufacturing cost is very susceptible to reduction by scale of production. This is probably where its greatest impact will be seen, although this is unlikely to happen until the latter half of the 1980s. The availability of a very cheap wideband medium can be confidently predicted to open up new markets for both business and consumer products based upon the ready transfer of information in electronic (optical) form, and making use of the immense possibilities now being opened up for the very cheap storage, retrieval and processing of such data using electro-optical systems and large-scale integration of digital circuits.

6.8 FUTURE DEVELOPMENTS

6.8.1 Long-wavelength systems

The systems described in the foregoing section have all been based upon operation in the 800 to 900 nm wavelength band and use devices based upon GaAs and Si materials systems. Of increasing interest for future systems is the possibility of operation at a longer wavelength, in the range 1100 to 1400 nm. There are two

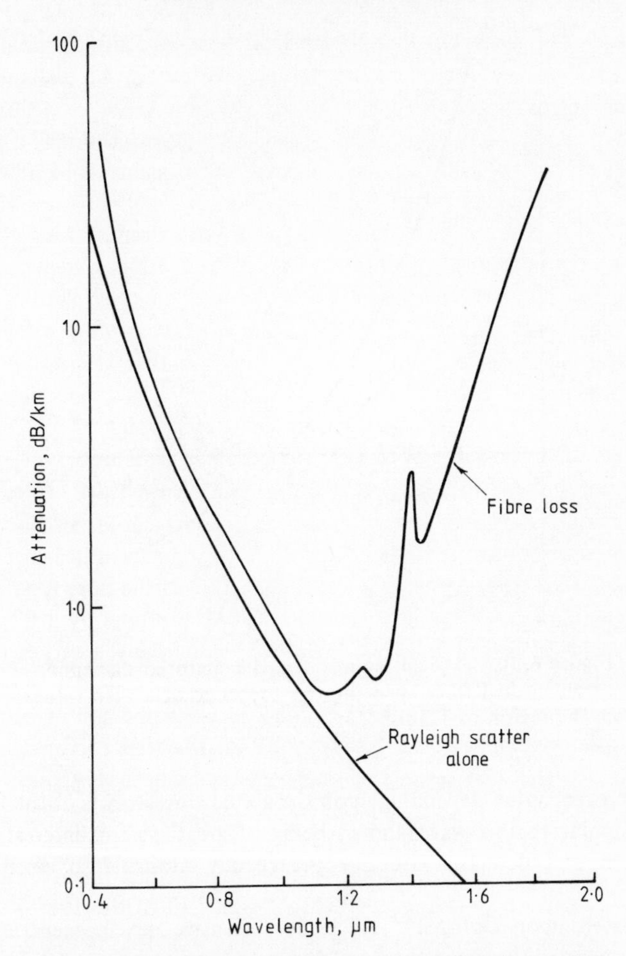

Figure 6.29. A recent illustration of the fibre losses that can be achieved by the removal of absorption and imperfections, showing a minimum loss of about 0.5 dB/km at 1.2 to 1.3 μm[21]

major reasons for this. The first reason concerns the loss of optical fibres. Figure 6.29 shows the loss of a silica-based fibre which is essentially free from absorptive contamination,[21] so that the short-wavelength loss is controlled largely by Rayleigh scattering from the glass constituents, while the longer-wavelength loss (beyond 1500 to 1600 nm) is controlled by absorption from the glass materials. The fibre shows a loss minimum in the region of 0.5 dB/km at about 1300 nm with a fairly broad transmission band centred around that wavelength and below 1 dB/km. In Figure 6.30 we show the second reason for choosing to operate in this waveband, namely that the first-order material dispersion[22] goes through zero at about

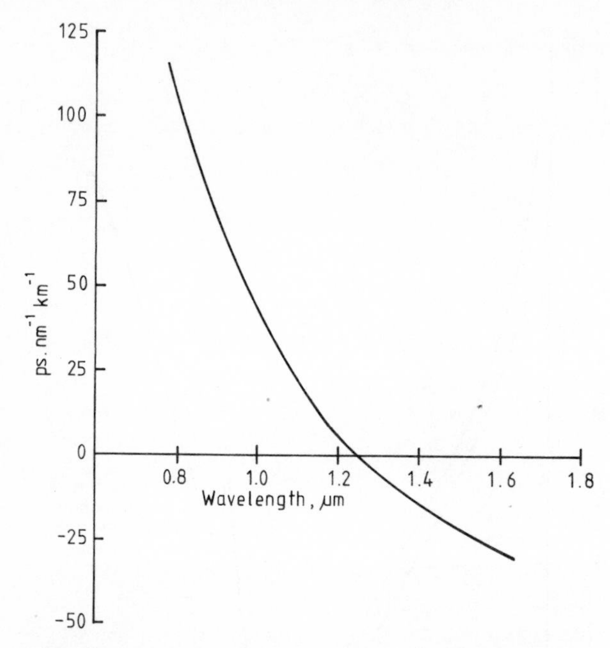

Figure 6.30. Typical figures for the material dispersion of an optical fibre showing the presence of a zero value in the region of 1.3 μm[22]

1300 nm, the exact value depending upon the actual constituents of the fibre used. Both factors point to this waveband as being of great systems interest. Unfortunately, sources and detectors are not yet readily available to construct such systems.

Sources based upon GaInAsP[23] are under intensive development and can be fabricated by varying the constituents of the active layer, grown on InP substrates, to operate at wavelengths anywhere in the band of interest while still matching the lattice parameter of the substrate. Detectors are also being developed[24] but as yet proven devices have not been reported. The prospects for such systems when all the components are available are most enticing. Assuming that similar launch powers are possible to those already achieved and that these are coupled with similar receiver sensitivities, then repeater spacings of between four and eight times those currently experienced should become possible subject to the fibre bandwidth not imposing a substantial power penalty through intersymbol interference. In practice, this requirement will probably demand monomode fibres for high-performance laser-based systems, although LED sources with graded-index fibres could make possible some very simple systems of performance that would look impressive by comparison with today's GaAlAs laser-based systems. Some early demonstrations are now being reported and will no doubt be followed by many more in the near future.

6.8.2 Some other systems possibilities

We have seen in the foregoing sections how fibre systems offer the planner a new flexibility, in terms of longer unrepeatered spans, smaller and more compact cables, much cheaper equipment and cable in the longer term, and freedom from interference and earth-loop problems. In addition to these features, some other interesting possibilities also arise. If we examine the loss versus wavelength curve for an optical fibre and note that many semiconductor laser sources have pulsed linewidths of 1 to 2 nm approximately, we see that the possibility exists of transmitting through a single fibre a number of separate wavelengths, each carrying a separate information channel quite independently of the others. The principle can be carried further so that two-way transmission can be achieved on a single fibre. The number of channels that can be carried depends upon several factors which are listed in Table 6.4 and must be evaluated for any particular case.

Several methods of separating wavelengths at the detector end or combining them at the transmitter end have been described.[25-27] In Figure 6.31 we show two. Figure 6.31a shows a miniature prism spectrometer separating a number of discrete wavelength channels, each to its own detector. The channel spacing in wavelength must obviously take into account the dispersive properties of the separator, the linewidth of source, its variation with temperature, and the amount of cross-talk that can be tolerated, together with the availability of suitable sources. Figure 6.31b shows a simpler device with immediate application to splitting two wavelengths, either for duplex or single-direction transmission. It uses a multilayer dielectric reflector as the wavelength-separating element sandwiched between two Selfoc lenses which serve to collimate the radiation emitted from the fibre. Clearly this design could be elaborated to split the signals into more than two separate wavelength channels. Wavelength multiplexing makes possible a number of separate channels over a single fibre. Typically we will probably be concerned with not more than 10 such channels. How might they be used?

One attractive use of such an approach is to accommodate traffic growth along a transmission system. The fibre cable would be installed and initially operated

Table 6.4 Factors affecting the number of channels that can be carried in a single fibre

(a)	Wavelength band in fibre having acceptably low loss
(b)	Linewidth of sources used
(c)	Line shape of sources used
(d)	Monomode or multimode fibre
(e)	Dispersive properties of multiplexer
(f)	Acceptable cross-talk – near end or far end
(g)	Acceptable insertion loss
(h)	Acceptable reflection coefficient
(i)	Acceptable cost of component
(j)	Availability of suitable components

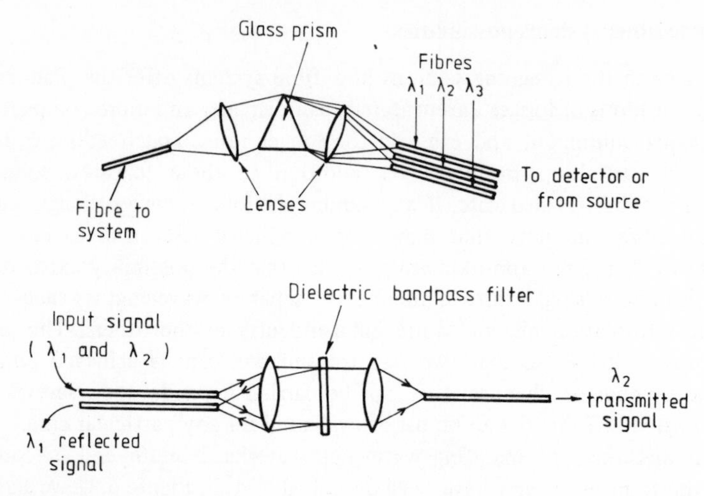

Figure 6.31. Some techniques for wavelength multiplexing several channels on to a single fibre: (a) using a prism as a wavelength dispersive element; and (b) using a multilayer dielectric filter as a wavelength-selective beam splitter

with a single channel (repeater) at each cable termination. As traffic built up, wavelength-dispersive elements would be interposed between the cable ends and the repeater, and additional repeaters added as needed. A second, quite different, application is to consider the provision of advanced wideband services to a customer. In Figure 6.32 we have sketched one such technical possibility, and show the use of separate wavelengths on the single fibre to provide discrete services, high-quality video in each direction consuming two wavelengths and high-quality, multiplexed audio and data channels taking up another two wavelengths, one in each direction. The reader can easily construct for himself a dozen other possibilities

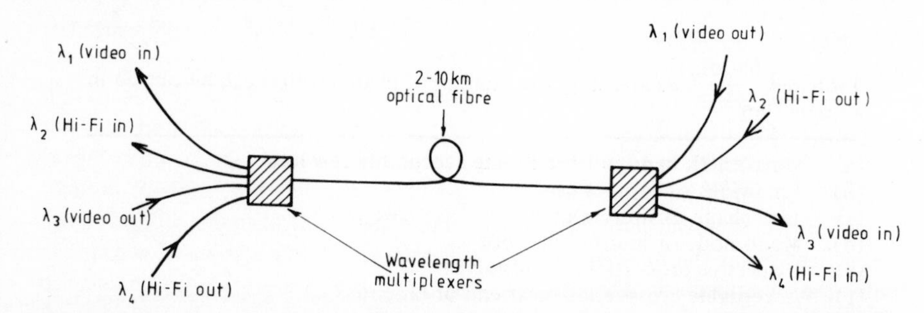

Figure 6.32. A schematic wavelength-multiplexed transmission system, suggesting the use of four wavelengths to provide separate analog video and 2 Mbit/s data channels in each direction on a single fibre

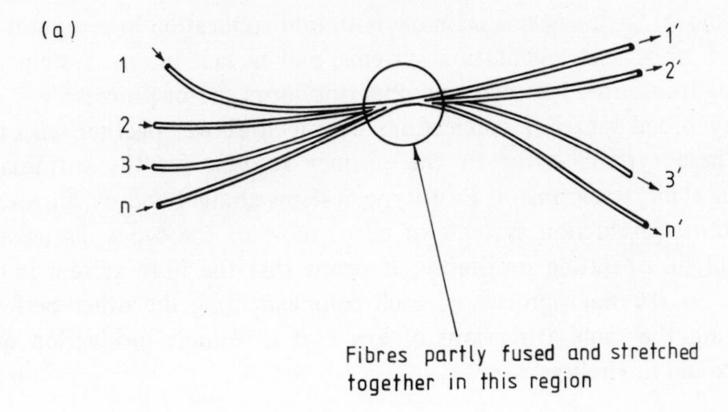

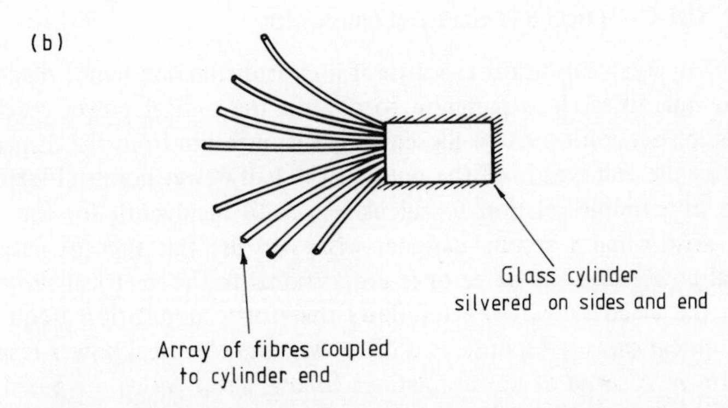

Figure 6.33. A schematic illustration of two types of star couplers: (a) the fused-fibre directional star type; and (b) the reflective mode scrambler type. Both provide a method of sharing optical energy on a single incoming fibre between a number of outgoing fibres, to allow multi-terminal links to be used over a shared fibre network

and only careful analysis of a complete system will show which combination is best for his situation.

A further design freedom the fibre system offers is the use of star or tee couplers in a wideband data highway configuration.[28] Two examples of stars are shown schematically in Figure 6.33, one being a directional star in which power from all the input ports is spread among all the output ports, and one bidirectional[29] in which power entering by any input fibre is spread among all the others. The former seems the most attractive for many highway applications. Tee couplers have been described for use with fibre-bundle systems but as yet the designs for single-fibre use seem to leave much to be desired. However, using the directional star coupler, it is already possible to assemble some attractive and compact data highway systems with all the advantages of the other fibre transmission systems described above. Such systems seem likely to find application in computer systems, aircraft and ships, instrumentation systems, and in fact in any system in which data is being transferred between a number of sensors and/or processors.

The very broad range of applications described above together with the great freedom these systems offer to the planner account for the enthusiasm now surrounding their development. Prototype systems abound and within a very short space of time, production systems of all or most of the types discussed in this chapter will be operating profitably. It seems that the fibre system is the ideal accomplice to the microprocessor, each complementing the other perfectly and both offering the same attractions of low cost in volume production with very great power and flexibility.

6.9 APPENDIX: Optical and electrical bandwidth

We note that great confusion can arise if a clear distinction is not made between these two quantities. It is common to observe the optical power received from a transmission section on an oscilloscope and to measure from the display a quantity such as the full width of the pulse at the half-power points (FWHP). This is then used in a simple relation to calculate a 3-dB bandwidth for the fibre. The problems arise when a system designer wishes to use this type of data, because the optical power into the detector is proportional to the electrical *current* leaving it, not to the electrical *power*. It follows that for a modulation frequency such that the optical power response is 3 dB down, the electrical power response will be 6 dB down. A series of useful relations follow, all of which are based upon the assumption that the transmission medium response is Gaussian.

$$f(t) = \frac{1}{\sqrt{2\pi}\delta} \exp(-t^2/2\delta^2) = \text{assumed optical power response}$$

$$g(w) = \frac{1}{\sqrt{2\pi}} \exp(-w^2\delta^2/2) = \text{assumed optical modulation response}$$

$$\tau_e = 2\sqrt{2}\delta \text{ seconds} = [\text{FW } 1/e \text{ optical power}]$$

$$W(3 \text{ dB optical}) = 0.8326\sqrt{2}/\delta \text{ rad/sec}$$

$$f(3 \text{ dB optical}) = 0.53/\tau_e \text{ Hz}$$

$$f(3 \text{ dB electrical}) = 0.375/\tau_e \text{ Hz}$$

For non-Gaussian pulses of the general form $h(t)$ the rms pulse width is defined as:

$$\delta_p = \int_{-\infty}^{\infty} t^2 h(t)\, dt - \left[\int_{-\infty}^{\infty} t h(t)\, dt\right]^2$$

where $h(t)$ has already been normalized so that

$$\int_{-\infty}^{\infty} h(t)\,\mathrm{d}t = 1$$

In general it is found that the impulse response of long fibre links is usually given to a good approximation by a Gaussian pulse shape and a simple measure of τ_e is adequate to describe the system performance for all first-order design purposes using the above relations.

REFERENCES

1. W. M. Hubbard, 'Utilisation of optical frequency carriers for low and moderate bandwidth channels', *Bell Syst. Tech. J.*, **52**, 731–765, 1973.
2. R. Dogliotti, A. Guardincerri and A. Luvison, 'Baseband equalisation in fibre optic digital transmission', *Opt. Quant. Electron.*, **8**, 343–353, 1976.
3. J. E. Midwinter, 'A study of intersymbol interference and transmission medium instability for an optical fibre system', *Opt. Quant. Electron.*, **9**, 299–304, 1977.
4. S. D. Personick, 'Receiver design for digital fiber optic communication systems', Parts I and II, *Bell Syst. Tech. J.*, **52**, 843–886, 1973.
5. D. R. Smith and I. Garrett, 'A simplified approach to digital optical receiver design', *Opt. Quant. Electron.*, **10**, 211–221, 1978.
6. A. B. Carlson, *Communication Systems* (2nd edn), McGraw-Hill, 1975.
7. R. C. Hooper and B. R. White, 'Digital optical receiver design for non-zero extinction ratio using a simplified approach', *Opt. Quant. Electron.*, **10**, 279–282, 1978.
8. D. R. Smith, R. C. Hooper and I. Garrett, 'Receivers for optical communications: a comparison of avalanche photodiodes and PIN–FET hybrids', *Opt. Quant. Electron.*, **10**, 293–300, 1978.
9. Y. Seki, 'Light extraction efficiency of the LED with guide layers', *Jap. J. Appl. Phys.*, **15**, 327–338, 1976.
10. A. G. Steventon and M. R. Matthews, 'Spectral and transient response of low threshold proton isolated GaAlAs lasers', *Electron. Lett.*, 1978 (to be published).
11. C. A. Burrus, 'Radiance of small-area high-current density electroluminescent diodes', *Proc. IEEE*, **60**, 231–232, 1972.
12. D. Gloge, 'LED design for fibre systems', *Electron. Lett.*, **13**, 399–400, 1977.
13. D. Marcuse, 'LED fundamentals: comparison of front and edge emitting diodes', *IEEE J. Quant. Electron.*, **QE-13**, 819–827, 1977.
14. Y. Horikoshi and G. Iwane, 'Efficient coupling between chalcogenide glass radiator bonded LEDs with optical guide structure and optical fibres', *Jap. J. Appl. Phys.*, **16**, 531–532, 1976.
15. K. Nawata, S. Machida and T. Ito, 'An 800 Mbit/s optical transmission experiment using single mode fibre', *IEEE J. Quant. Electron.*, **QE-14**, 98–103, 1978.
16. K. Konnerth and C. Lanya, 'Delay between current pulse and light emission of a GaAs injection laser', *Appl. Phys. Lett.*, **4**, 120, 1964.

17. S. R. Salter, D. R. Smith, B. R. White and R. P. Webb, 'Laser automatic level control for optical communications systems', *Third European Conference on Optical Communications*, Munich, September 1977, VDE-Verlag GmbH, Berlin, 1977.
18. J. E. Midwinter, 'Optical communications, today and tomorrow', *Electronics and Power*, **24**, 442–447, 1978.
19. J. E. Midwinter and J. R. Stern, 'Propagation studies of 40 km of graded index fibre installed in an operational duct route', *IEEE Trans. Comm.*, **COM-26**, 1015–1020, 1978.
20. R. W. Berry, D. J. Brace and I. A. Ravenscroft, 'Optical fibre system trials at 8 Mbit/s and 140 Mbit/s', *IEEE Trans. Comm.*, **COM-26**, 1020–1027, 1978.
21. M. Horiguchi, 'Spectral losses of low OH content optical fibres', *Electron. Lett.*, **12**, 310–311, 1976.
22. D. N. Payne and W. A. Gambling, 'Zero material dispersion in optical fibres', *Electron. Lett.*, **11**, 176–178, 1975.
23. J. J. Hsieh and C. C. Shen, 'Room temperature CW operation of buried stripe double heterostructure GaInAsP/InP diode lasers', *Appl. Phys. Lett.*, **30**, 429–431, 1977.
24. K. J. Bachman and J. L. Shay, 'An InGaAs detector for the 1.0 to 1.7 μm wavelength range', *Appl. Phys. Lett.*, **32**, 446–448, 1978.
25. W. J. Tomlinson and G. D. Aumiller, 'Optical multiplexer for multimode fibre transmission systems', *Appl. Phys. Lett.*, **31**, 169–171, 1977.
26. H. Ishio and T. Miki, 'A preliminary experiment on wavelength division multiplexing transmission systems', paper presented to IOOC meeting, Tokyo, July 1977.
27. S. Sugimoto, K. Minemura, K. Kobayashi, M. Seki, M. Shikada, A. Ueki, T. Yanese and T. Miki, 'High speed digital signal transmission experiments by optical wavelength division multiplexing', paper presented to IOOC meeting, Tokyo, July 1977.
28. E. G. Rawson and R. M. Metcalfe, 'Fibernet – multimode optical fibres for local computer networks', *IEEE Trans. Comm.*, **COM-26**, 983–990, 1978.
29. B. S. Kawasaki and K. O. Hill, 'Low loss access coupler for multimode optical fibre distribution networks', *Appl. Opt.*, **16**, 1794, 1977.

Subject Index